Computing in Scilab

Scilab is free and open-source for numerical computation. It gives a powerful computing environment for engineering and scientific applications including numerical analysis, data analysis, algorithm development, and application development. It was released in 1994 and since then it has been continuously used across the globe with around 100,000 monthly downloads. Scilab has grabbed a lot of attention in recent years because of its free availability and MATLAB-like features.

While focusing on the curriculum requirements of undergraduate students in physics and electronics, *Computing in Scilab* caters to a wide audience including undergraduate students in chemistry, computer science and mathematics too. In this book, advanced physical problems have been solved by making Scilab programs and wherever necessary, sufficient background theory of these problems is also given. The book is written in an easy-to-understand language and in the style of classroom teaching. Illustrations, diagrams, and graphs will definitely make learning more interesting and easier. This book has an extensive coverage and it primarily focuses on applications of Scilab in improving the problem-solving skills of the reader.

Key Features

- Coverage as per UGC's CBCS–LOCF syllabus
- Physical interpretation of complex calculations and results using mathematical and computational tools for enhanced comprehension
- Use of descriptive approach to explain the concepts, including step-by-step derivations, diagrams, graphs, and examples
- Application-based examples for better understanding of concepts
- Practice exercises at the end of each chapter for sharpening coding skills

Chetana Jain is Professor of Physics at Hansraj College, University of Delhi. She obtained her graduation and post-graduation degree in Physics from Hansraj College, University of Delhi. She has obtained her doctorate degree from the Department of Physics and Astrophysics, University of Delhi. Her main area of research is astrophysics. She has authored several research papers in various national and international journals of repute. She has written a well-appreciated undergraduate book, *Principles of Electromagnetic Theory*. You may reach the author at: drchetanajain11@gmail.com.

Computing in Scilab

Chetana Jain

CAMBRIDGE
UNIVERSITY PRESS

CAMBRIDGE
UNIVERSITY PRESS

University Printing House, Cambridge CB2 8BS, United Kingdom

One Liberty Plaza, 20th Floor, New York, NY 10006, USA

477 Williamstown Road, Port Melbourne, VIC 3207, Australia

314 to 321, 3rd Floor, Plot No.3, Splendor Forum, Jasola District Centre, New Delhi 110025, India

103 Penang Road, #05–06/07, Visioncrest Commercial, Singapore 238467

Cambridge University Press is part of the University of Cambridge.

It furthers the University's mission by disseminating knowledge in the pursuit of education, learning and research at the highest international levels of excellence.

www.cambridge.org
Information on this title: www.cambridge.org/9781009214193

First published 2022

Printed in India by Nutech Print Services, New Delhi 110020

A catalogue record for this publication is available from the British Library

ISBN 978-1-009-21419-3 Paperback

Contents

Figures

Tables

Matrices and Vector Spaces

After going through this chapter, the reader should be able to

◊ Create different types of vectors and matrices in Scilab.

◊ Perform arithmetic operations on vectors and matrices.

◊ Solve linear and vector algebra in Scilab

◊ Construct and solve mathematical expressions that involve vectors, matrices and a system of linear equations.

◊ Solve advanced problems in physics using matrices.

◊ Construct differential and Laplace operators in Scilab.

◊ Determine the wave function of stationary states using the Hermitian differential operator.

1.1 Introduction

Matrices are often aptly described as key to solve everything in the scientific world. This chapter expounds the usefulness of vectors and matrices that occur in many kinds of problems across the disciplines. They are used to study innumerable physical phenomena such as, motion of rigid bodies, eigen states of a quantum mechanical system, electrical networks and coordinate system conversion.

In Scilab, matrix computation forms the basis of all calculations. This chapter recapitulates the basic Scilab rules that have to be followed for creating and editing matrices. It also summarizes the arithmetic operations that can be performed on matrices. *Section 1.2* gives an overview on different ways of generating a matrix and its elements. Some special types of matrices such as row/column vector, diagonal matrix, identity matrix and triangular matrices have been introduced in *Section 1.3*. Matrix operations such as row/column operation, conjugation, scalar/vector multiplication and division have been explained in *Section 1.4*. The laws of vector algebra have been outlined in *Section 1.5*. Some interesting examples of applications and use of matrices in physical sciences have been discussed in *Section 1.6*.

1.2 Creation of a Matrix

Matrices are rectangular arrangements of '*m*' rows and '*n*' columns; an arrangement of *m* rows and *n* columns is called an (*m* × *n*) matrix. If it contains only one row or only one column, then it is called a vector. There are several ways of defining vectors and matrices in Scilab. Some of them have been explained as follows.

1. The elements of a matrix are defined by writing them inside a square bracket, such that the elements of a row are separated by a comma or a white space. The elements of consecutive rows are separated by a semi-colon.

2. The elements of a matrix can be of several types and have been listed in *Table 1.1*. As can be seen in this table,

   a. The elements can be real numbers.

   b. The elements can be complex numbers. The complex number consists of a real part and/or an imaginary part.

   c. The elements can be rational numbers, which are defined using the 'rlist' command of Scilab.

   d. The elements can be random numbers, which are generated using the random number generator, 'rand' command of Scilab. As can be seen in the table, it is also possible to format the number of significant digits in the random numbers.

   e. The elements can be character strings.

   f. The elements can be polynomials. In Scilab, '%z' is used as a variable for defining polynomials. As explained in the table, it is also possible to represent a polynomial by other variables using the 'poly(0,"variable")' command.

   g. All the elements can be made equal to zero.

   h. All the elements can be made equal to ones.

   i. Elements of the matrix can follow a certain progression rule.

Table 1.1 Creation of Matrices

S. No. (Method)	Scilab Command	Output Matrix
2 (a)	A = [1,2;3,4] Or A = [1 2;3 4]	A = 1. 2. 3. 4.
2 (b)	A = [%i 2 ; 3 4*%i]	A = i 2. 3. 4.i
2 (c)	Num = [1 1 ; 1 1]; Den = [1 2 ; 3 4]; A = rlist (Num,Den,[])	A = $\dfrac{1}{1}$ $\dfrac{1}{2}$ $\dfrac{1}{3}$ $\dfrac{1}{4}$
2 (d)	A = rand(2,2)	A = 0.6856896 0.8415518 0.1531217 0.4062025
	format(5); A = rand(2,2)	A = 0.68 0.84 0.15 0.40
2 (e)	A = ["This" "is" ; "SciLab" "Program"]	A = !This is ! ! ! !SciLab Program !
2 (f)	A = [3 5+3*%z ; 7+%z^2 2+4*%z^2]	A = 3 5 + 3z $7 + z^2$ $2 + 4z^2$
	p = poly(0,'p'); A = [3*p 5+3*p ; 7+p^2 2+4*p^2]	A = 3p 5 + 3p $7 + p^2$ $2 + 4p^2$
	p = poly(0,'p'); A = [7 3*p ; 1/p^3 1/(1+2*p)]	A = $\dfrac{7}{1}$ $\dfrac{3p}{1}$ $\dfrac{1}{p^3}$ $\dfrac{1}{1 + 2p}$

S. No. (Method)	Scilab Command	Output Matrix
2 (g)	A = zeros(2 , 2) (A matrix of size 2 × 2 and all the elements are equal to zero)	A = 0. 0. 0. 0.
2 (h)	A = ones(2 , 2) (A matrix of size 2 × 2 and all the elements are equal to one)	A = 1. 1. 1. 1.
2 (i)	A = [1:4]	A = 1. 2. 3. 4.
	A = [1:2:10]	A = 1. 3. 5. 7. 9.

3. Elements of a matrix can also be defined from the console by writing the following Scilab program.

```
rows_num = input("How many rows?")
columns_num = input("How many columns?")
for i = 1:rows_num;
    for j=1:columns_num;
      A(i,j)= input("Enter the elements");
    end
end
```

1.3 Nature of the Matrix

It is straightforward to define different types of matrices in Scilab. The commands for some of them have been listed in *Table 1.2*.

Table 1.2 Types of Matrices

Type of Matrix	Scilab Command	Output Matrix
Row matrix (1 × 3)	A = [5 3 8]	A = 5. 3. 8.

Column matrix (3 × 1)	A = [5; 3; 8]	A = 5.
		      3.
		      8.

Type of Matrix	Scilab Command	Output Matrix
Diagonal matrix (2 × 2)	A = diag ([1 2])	A = 1. 0. 0. 2.
	A = diag ([rand(),rand()])	A = 0.349361 0. 0. 0.387377
Identity matrix (2× 2)	A = eye (2,2)	A = 1. 0. 0. 1.

Triangular matrix
(Lower triangle)
	A = tril([1 2 3 ; 4 5 6 ; 7 8 9])	A = 1. 0. 0.
		     4. 5. 0.
		     7. 8. 9.

	A = tril ([1 2 3 ; 4 5 6 ; 7 8 9],[-1])	A = 0. 0. 0.
		     4. 0. 0.
		     7. 8. 0.

	A = tril ([1 2 3 ; 4 5 6 ; 7 8 9],[1])	A = 1. 2. 0.
		     4. 5. 6.
		     7. 8. 9.

Triangular matrix
(Upper triangle)
	A = triu ([1 2 3 ; 4 5 6 ; 7 8 9])	A = 1. 2. 3.
		     0. 5. 6.
		     0. 0. 9.

1.4 Matrix Operation

The elementary matrix operations in Scilab have been listed in *Table 1.3*.

Table 1.3   Matrix Operations

Matrix Operation	Scilab Command	Output Matrix

Transpose of a row matrix	A = [7 9 1]
	B = A.'
		B = 7.
		    9.
		    1.

Transpose of a column matrix	A = [3 ; 7 ; 2] B = A.'	B = 3. 7. 2.

Matrix Operation	Scilab Command	Output Matrix

Conjugate of a matrix	A = [1+2*%i 5*%i ; 11*%i 4-%i]
	
	B = conj (A)
		A = 1.+2.i 5.i
		   11.i 4. - i
		
		B = 1.-2.i - 5.i
		   - 11.i 4.+i

Conjugate transpose of a matrix	A = [1+2*%i 5*%i ; 11*%i 4-%i]
	
	B = A'
		A = 1.+2.i 5.i
		   11.i 4. - i
		B = 1. - 2.i - 11.i
		   - 5.i 4. + i

Interchange of first and second row	A=[1 2 3;4 5 6;7 8 9]
	
	A([1,2],:) = A([2,1],:)
		A = 1. 2. 3.
		   4. 5. 6.
		   7. 8. 9.
		A = 4. 5. 6.
		   1. 2. 3.
		   7. 8. 9.

Interchange of first and second column	A = [1 2 3; 4 5 6;7 8 9]
	
	A(:,[1,2]) = A(:,[2,1])
		A = 1. 2. 3.
		   4. 5. 6.
		   7. 8. 9.
		A = 2. 1. 3.
		   5. 4. 6.
		   8. 7. 9.

Square of matrix	A = [2 1; 4 3] B = A^2	B = 8. 5. 20. 13.
Square of matrix	A = [2 1;4 3] B = A**2	B = 8. 5. 20. 13.
Square of elements of the matrix	A = [2 1;4 3] B = A.^2	B = 4. 1. 16. 9.
Square root of elements of the matrix	A = [2 1;4 3] B = sqrt (A)	B = 1.414 1. 2. 1.732
Product of all the elements of the matrix	A = [2 1;4 3] B = prod (A)	B = 24

Matrix Operation	Scilab Command	Output Matrix

Sum of two matrices	A = [1 3 ; 5 2]
	
	B = [4 7 ; 6 4]
	
	C = A + B
		A = 1. 3.
		      5. 2.
		B = 4. 7.
		      6. 4.
		C = 5. 10.
		      11. 6.

Difference of two matrices	A = [1 3 ; 5 2]
	
	B = [4 7 ; 6 4]
	
	C = B - A
		A = 1. 3.
		      5. 2.
		B = 4. 7.
		      6. 4.
		C = 3. 4.
		      1. 2.

Product of two matrices	A = [1 3 ; 5 2]
	
	B = [4 7 ; 6 4]
	
	C = A * B
	
	D = B * A
		A = 1. 3.
		      5. 2.
		B = 4. 7.
		      6. 4.
		C = 22. 19.
		      32. 43.
		D = 39. 26.
		      26. 26.

Product of two matrices (element wise)	A = [1 3 ; 5 2]
	
	B = [4 7 ; 6 4]
	
	C = A.*B
	
	D = B.*A
		A = 1. 3.
		      5. 2.
		B = 4. 7.
		      6. 4.
		C = 4. 21.
		      30. 8.
		D = 4. 21.
		      30. 8.

Matrix division	A = [7 9 ; 4 6]
	
	B = [3 1 ; 5 9]
	
	C = B/A
		A = 7. 9.
		     4. 6.
		B = 3. 1.
		     5. 9.
		C = 2.333 - 3.333
		     - 1. 3.

Matrix Operation	Scilab Command	Output Matrix

Matrix division (element wise)	A = [7 9 ; 4 6]
	
	B = [3 1 ; 5 9]
	
	C = B./A
		A = 7. 9.
		     4. 6.
		B = 3. 1.
		     5. 9.
		C = 0.428 0.111
		     1.25 1.5

Trace of the matrix	A=[7 3 3 ; 4 7 6 ; 7 8 1]
	
	B = trace(A)
		A = 7. 3. 3.
		     4. 7. 6.
		     7. 8. 1.
		B = 15.

	A=[7 3 3;4 7 6;7 8 1]
	
	B = sum(diag(A))
	(Sum of diagonal elements)
		A = 7. 3. 3.
		     4. 7. 6.
		     7. 8. 1.
		B = 15.

Determinant of the matrix	A=[1 0 0;0 2 0;0 0 3]
	
	B = det(A)
		A = 1. 0. 0.
		     0. 2. 0.
		     0. 0. 3.
		B = 6.

	A=[1 0 0;0 2 0;0 0 3]
	
	B = spec(A)
		A = 1. 0. 0.
		     0. 2. 0.
		     0. 0. 3.
		B = 1.
		     2.
		     3.

Eigen value and Eigen vector of the matrix	[B,C] = spec(A)
	
	(Matrix B contains Eigen vectors)
	
	(Matrix C contains Eigen values along the diagonal)
		A = 1. 0. 0.
		     0. 2. 0.
		     0. 0. 3.
		B = 1 0. 0.
		     0. 1. 0.
		     0. 0. 1.
		C = 1. 0. 0.
		     0. 2. 0.
		     0. 0. 3.

Matrix Operation	Scilab Command	Output Matrix

Matrix inverse	A = [2 1 ; 4 3]
	
	B = inv(A)
		A = 2. 1.
		       4. 3.
		
		B = 1.5 - 0.5
		       - 2. 1.

1.5 Vector Algebra

This section describes the use of Scilab for applications in vector algebra. The most commonly used laws of vector algebra have been listed in *Table 1.4*.

Table 1.4   Vector Algebra

Vector Algebra	Scilab Command	Output
Sum of elements of a vector	A = [2 1 3]; B = sum (A)	B = 6
Product of elements of a vector	A = [2 3 4]; B = prod (A)	B = 24
Maximum value in a vector	A = [2 3 1]; B = max (A)	B = 3
Minimum value in a vector	A = [2 3 1]; B = min (A)	B = 1
Number of elements	A = [2 1 4]; B = length (A)	B = 3
Value of second element of a vector	A = [2 3 1]; B = A (2)	B = 3
Magnitude of a vector	A = [2 1 4]; B = norm (A)	B = 4.5825757
Unit vector	A = [2 1 4]; B = A/norm (A)	B = 0.4364358 0.2182179 0.8728716

Vector Algebra	Scilab Command	Output
Projection of a vector B over vector A	A = [2 1 4]; B = [5 4 7]; C = A*B'/norm (A)	C = 9.1651514
Dot product of two vectors	A = [2 1 4]; B = [5 4 7]; C = sum (A.*B)	C = 42
Cross product of two vectors	A = [2; 1; 4]; B = [5; 4; 7]; C = cross (A,B)	C = - 9. 6. 3.

1.6 Applications

1.6.1 Coordinate conversion (Cartesian to cylindrical coordinate system)

The Cartesian coordinates are the simplest and, as far as calculations are concerned, the most sought after system of coordinates. However, there are several physical phenomena and systems that involve rotational symmetry about an axis. For example, flow of current through a long straight wire and flow of water through a long straight pipe having circular cross-section. In these problems, the use of the Cartesian coordinate system becomes tedious and it is preferable to perform the calculations in the cylindrical coordinate system. As will be explained in this section, matrices are easy and useful for converting the coordinates of a point from one coordinate system to another coordinate system.

Suppose the coordinates of a point (P) in Cartesian system are denoted by (x, y, z). The corresponding cylindrical coordinates are denoted by (r, θ, z). Here,

- 'r' is the perpendicular distance of point P from the z-axis.

- 'θ' is the azimuthal angle. It is the angle between the x-axis and the line joining the origin with the projection of point P on the x - y plane.

- 'z' is the perpendicular distance of point P from the x - y plane

The Cartesian and cylindrical coordinates of point P are related by *Eqn. 1.1–1.6.*

- Cartesian → Cylindrical

$$r = \left(x^2 + y^2\right)^{1/2} \tag{1.1}$$

$$\theta = \tan^{-1}\left(\frac{y}{x}\right) \tag{1.2}$$

$$z = z \tag{1.3}$$

- Cylindrical → Cartesian

$$x = r\cos\theta \tag{1.4}$$

$$y = r\sin\theta \tag{1.5}$$

$$z = z \tag{1.6}$$

The following user-defined Scilab function converts Cartesian coordinates (x, y, z) of a point to its cylindrical coordinates (r, θ, z). The conversion rules given in this function are in accordance with *Eqn. 1.1–1.3*. The input variable for this function is a vector whose components are Cartesian coordinates of a given vector.

```
function [coordinates] = cartesian_to_cylindrical(A)
x = A(1);
y = A(2);
z = A(3);
r = (x.^2 + y.^2)^0.5;
theta = atand(y/x);                      //angle in degree
if theta < 0 then
    theta = theta + 180
end
coordinates = [r theta z]
endfunction
```

As a simple application, suppose a position vector is,

$$\vec{A} = \hat{i} + 4\hat{j} + 4\hat{k}$$

The coordinates of the point can be converted to a cylindrical system using the following Scilab program.

```
A = [1 4 4];
cartesian_to_cylindrical(A)
```

The result will be equal to

$$(r, \theta, z) = (4.123, 75.96, 4)$$

1.6.2 Coordinate conversion (Cartesian to spherical coordinate system)

The spherical coordinate system is useful while studying systems having symmetry about a point. It can also be used, for example, in situations involving calculation of potential energy due to a point charge or a uniformly charged sphere and problems involving flow of heat inside a sphere

Suppose coordinates of a point (P) in Cartesian system are denoted by (x, y, z). The corresponding spherical coordinates are denoted by (r, θ, φ). Here,

- 'r' is the radial distance of the point from origin.

- 'θ' is the polar angle. It is measured in the clockwise direction from the z-axis

- 'φ' is the azimuthal angle. It is measured in the anticlockwise direction from the x-axis.

The Cartesian and spherical coordinates of the point P are related by *Eqn. 1.7–1.12*.

- Cartesian $\rightarrow$ Spherical

$$r = \left(x^2 + y^2 + z^2 \right)^{1/2} \tag{1.7}$$

$$\theta = \tan^{-1} \left(\frac{\sqrt{x^2 + y^2}}{z} \right) \tag{1.8}$$

$$\varphi = \tan^{-1} \left(\frac{y}{x} \right) \tag{1.9}$$

- Spherical $\rightarrow$ Cartesian

$$x = r \sin \theta \cos \varphi \tag{1.10}$$

$$y = r \sin \theta \sin \varphi \tag{1.11}$$

$$z = r \cos \theta \tag{1.12}$$

The following user-defined Scilab function converts the Cartesian coordinates (x, y, z) of a point to spherical coordinates (r, θ, φ). The conversion rules given in this function are in accordance with *Eqn. 1.7–1.9*. The input variable is a vector whose components are Cartesian coordinates of a given vector.

```
function [coordinates] = cartesian_to_spherical(A)
x = A(1);
y = A(2);
z = A(3);
r = (x.^2 + y.^2 + z.^2)^0.5;
theta = atand(((x.^2 + y.^2)^0.5)/(z));
phi = atand(y/x);
coordinates = [r theta phi]
endfunction
```

Therefore, if the position vector is

$$\vec{A} = \hat{i} + 4\hat{j} + 4\hat{k}$$

then, the coordinates of the point can be converted to the spherical system using the following Scilab program.

```
A = [1 4 4];
cartesian_to_spherical(A)
```

The result will be equal to

$$(r, \theta, \varphi) = (5.744, 45.868, 75.963)$$

1.6.3 Orthogonal vectors

Two vectors are orthogonal if they are perpendicular to each other, i.e., their scalar product is zero. Suppose the two vectors are

$$\vec{A} = 2\hat{i} + p\hat{j} + 3\hat{k}$$

$$\vec{B} = 4\hat{i} - \hat{j} - \hat{k}$$

The following Scilab program determines the value of the variable 'p' that makes the vectors $\vec{A}$ and $\vec{B}$ perpendicular to each other. In this program:

- The 'poly' command is used to form a polynomial in variable 'p'.
- The 'roots' command is used to determine the solution of the polynomial formed after taking the scalar product of the two vectors.

```
c = poly(0,'p')
A = [2, c, 3];
B = [4, -1, -1];
roots(A*B')
```

The value of the constant 'p' will be equal to 5.

1.6.4 Centre of mass of a system

Consider a system of four particles having mass m_1, m_2, m_3 and m_4. The position coordinates of these particles are $(x_1, y_1, z_1), (x_2, y_2, z_2), (x_3, y_3, z_3)$, and (x_4, y_4, z_4) respectively. The coordinates of the centre of mass of the system are given by *Eqn. 1.13–1.15*.

$$X_{COM} = \frac{\sum_{i=1}^{4} x_i m_i}{\sum_{i=1}^{4} m_i} \tag{1.13}$$

$$Y_{COM} = \frac{\sum_{i=1}^{4} y_i m_i}{\sum_{i=1}^{4} m_i} \tag{1.14}$$

$$Z_{COM} = \frac{\sum_{i=1}^{4} z_i m_i}{\sum_{i=1}^{4} m_i} \tag{1.15}$$

The following Scilab program calculates the coordinates of the centre of mass of the given system of particles using *Eqn. 1.13–1.15*.

```
x = [2 3 4 5];        //x-coordinate of all the particles
y = [-1 3 2 4];       //y-coordinate of all the particles
z = [2 4 -4 5];       //z-coordinate of all the particles
m = [10 15 20 25];    //mass of all the particles
x_com = sum(x.*m)/sum(m)
y_com = sum(y.*m)/sum(m)
z_com = sum(z.*m)/sum(m)
```

The answer will be equal to

(3.857, 2.5, 1.786)

1.6.5 Electrical circuits (Mesh analysis)

Mesh analysis is used to solve for the current in electrical circuits. A mesh is a loop in the circuit that does not contain any other loop. Each mesh is assigned a current. This method makes use of Kirchhoff's voltage law to determine a set of solvable equations (one equation from each mesh), which are the sums of the voltage drops in the complete mesh. For example, consider the circuit given in *Figure 1.1*. From Kirchhoff's voltage law, the mesh equations are

$$i_1 + i_1 + 2(i_1 - i_2) = 5 \tag{1.16}$$

$$2i_2 + 2i_2 + 2(i_2 - i_1) = 10 \tag{1.17}$$

Eqns. 1.16–1.17 imply that

$$4i_1 - 2i_2 = 5 \tag{1.18}$$

$$-2i_1 + 6i_2 = 10 \tag{1.19}$$

In the matrix form, *Eqns. 1.18–1.19* can be written as

$$A \begin{bmatrix} i_1 \\ i_2 \end{bmatrix} = B$$

Here, A and B are coefficient matrices such that

$$\begin{bmatrix} 4 & -2 \\ -2 & 6 \end{bmatrix} \begin{bmatrix} i_1 \\ i_2 \end{bmatrix} = \begin{bmatrix} 5 \\ 10 \end{bmatrix}$$

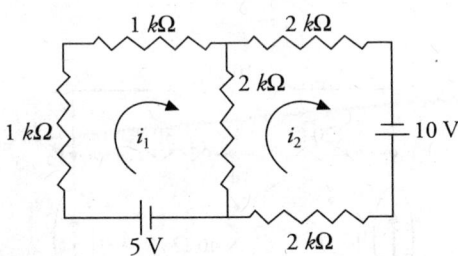

Figure 1.1   Mesh circuit analysis using matrices

The following Scilab program uses the matrix method for solving the two linear equations (*Eqns. 1.18–1.19*) and gives the values of i_1 and i_2. In this program, two methods have been shown:

- In the first method, the built-in Scilab function 'linsolve' has been used. This function solves linear equations of the form $AX + B = 0$. The result is stored in the form of vector X.

- In the second method, a single line code 'A\B' has been used.

```
A = [4 -2; -2 6];
B = [5 10]';
C = -B
Answer = linsolve(A,C)
OR
Answer = A\B
```

In both the methods, the answer will be equal to 2.5 and 2.5.

1.6.6 Electrical circuits (Nodal analysis)

Nodal analysis is used for determining the voltage at 'nodes' where branches of a circuit meet. Kirchhoff's current law is used for obtaining a set of solvable equations, such that the net sum of current incident on any node is zero.

Consider the electrical circuit given in *Figure 1.2*. From Kirchhoff's current law,

At Node 1:

$$\frac{V_1 - V_2}{20} + \frac{V_1 - V_3}{10} - 0.5 = 0 \tag{1.20}$$

This implies

$$3V_1 - V_2 - 2V_3 = 10 \tag{1.21}$$

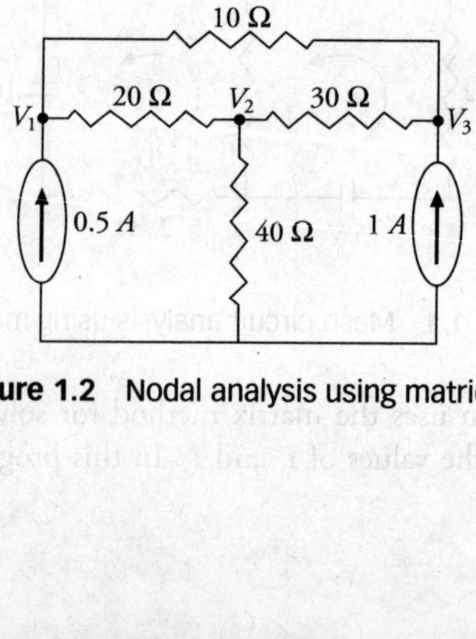

Figure 1.2 Nodal analysis using matrices

At Node 2:

$$\frac{V_2 - V_1}{20} + \frac{V_2 - V_3}{30} + \frac{V_2}{40} = 0 \qquad (1.22)$$

This implies

$$-6V_1 + 13V_2 - 4V_3 = 0 \qquad (1.23)$$

At Node 3:

$$\frac{V_3 - V_1}{10} + \frac{V_3 - V_2}{30} - 1 = 0 \qquad (1.24)$$

This implies

$$-3V_1 - V_2 + 4V_3 = 30 \qquad (1.25)$$

In the matrix form, *Eqn. 1.21, 1.23* and *1.25* can be written as

$$A \begin{bmatrix} V_1 \\ V_2 \\ V_3 \end{bmatrix} = B$$

Here, A and B are coefficient matrices such that

$$\begin{bmatrix} 3 & -1 & -2 \\ -6 & 13 & -4 \\ -3 & -1 & 4 \end{bmatrix} \begin{bmatrix} V_1 \\ V_2 \\ V_3 \end{bmatrix} = \begin{bmatrix} 10 \\ 0 \\ 30 \end{bmatrix}$$

The following Scilab program solves these linear equations and gives the values of node voltages.

```
A = [3 -1 -2 ; -6 13 -4 ; -3 -1 4];
C = [10 ; 0 ; 30];
B = A\C
```

The answer will be equal to $V_1 = 76.66$, $V_2 = 60$ and $V_3 = 80$.

1.6.7   Force on a test charge

Consider *Figure 1.3* wherein two charges of magnitude 1 nC and 2 nC exert an electric force on a test charge having magnitude 3 nC. The coordinates of the location of these charges are shown in the diagram. Force exerted by the two charges on the test charge is given by *Eqn. 1.26*. The following Scilab program calculates this force

$$F = \frac{Q_t}{4\pi\varepsilon_o} \sum_{i=1}^{2} \frac{Q_i\left(\vec{r_t} - \vec{r_i}\right)}{\left|\vec{r_t} - \vec{r_i}\right|^3} \qquad\qquad (1.26)$$

```
//Load *.sci file which contains the function for dis-
tance calculation
exec('vectors.sci',-1);
A = [1 3 0];              //Coordinates of first charge
B = [2 -1 -1];           //Coordinates of second charge
C_test = [4 0 9];         //Coordinates of test charge
charge_1 = 1d-9;          //Magnitude of first charge
charge_2 = 2d-9;          //Magnitude of second charge
charge_test = 3d-9;       //Magnitude of test charge
AC = C_test - A;                              // r_t - r_1
BC = C_test - B;                              // r_t - r_2

distance_AC = distance_between_points(A,C_test);
distance_AC = distance_AC^3;
distance_BC = distance_between_points(B,C_test);
distance_BC = distance_BC^3;
force = (charge_test*36*%pi/(4*%pi*1d-
9))*((charge_1*(AC)/distance_AC)+((charge_2*(BC)/dis-
tance_BC)))
```

The answer will be equal to $F = 0.1826\widehat{a}_x - 0.032\widehat{a}_y + 0.7485\widehat{a}_z$ nN

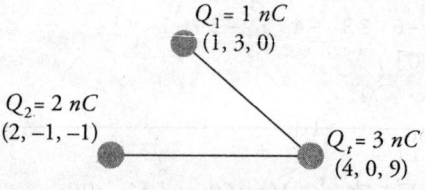

Figure 1.3   Electric force on a test charge

1.6.8 Principal axes of moment of inertia

This is a very compelling application involving the concept of diagonalization of square matrices. In a rotating rigid body, there are three orthogonal principal axes of moment of inertia. However, their direction may not be obvious from the geometry and the axis of rotation. This section describes a trivial method to determine the direction of the principal axes.

Consider a rigid body rotating with fixed angular velocity (ω) about an unknown principal axis. The angular momentum of the body is given by *Eqn. 1.27*.

$$L = I\omega \tag{1.27}$$

Here, 'I' is the moment of inertia tensor of rank 2. The total angular momentum can be written as

$$L_i = \sum_j I_{ij}\omega_j \tag{1.28}$$

In *Eqn. 1.28*, the moment of inertia tensor is given by

$$I_{ij} = \sum_\alpha m_\alpha \left[\delta_{ij} \left\{ \sum_k X_{\alpha,k}^2 \right\} - X_{\alpha,i} X_{\alpha,j} \right] \tag{1.29}$$

The component form of the moment of inertia tensor (of *Eqn. 1.29*) can be deduced in the following manner (*Eqn. 1.30–1.35*).

$$I_{xx} = \sum_i m_i \left(y_i^2 + z_i^2 \right) \tag{1.30}$$

$$I_{yy} = \sum_i m_i \left(x_i^2 + z_i^2 \right) \tag{1.31}$$

$$I_{zz} = \sum_i m_i \left(x_i^2 + y_i^2 \right) \tag{1.32}$$

$$I_{xy} = I_{yx} = -\sum_i m_i x_i y_i \tag{1.33}$$

$$I_{xz} = I_{zx} = -\sum_i m_i x_i z_i \tag{1.34}$$

$$I_{yz} = I_{zy} = -\sum_i m_i y_i z_i \tag{1.35}$$

If the body is rotating about the principal axis of rotation, then the total angular momentum will be in the same direction as the angular velocity, i.e.,

$$L = I\omega = \lambda\omega \qquad\qquad (1.36)$$

Therefore, from *Eqn. 1.36*, it is important to determine the moment of inertia tensor for the rotating rigid body and then determine the eigen value and its corresponding eigen vector. The eigen values of the diagonalized matrix (λ_i) are called the principal moments. In general, if a rigid body is symmetrical about the origin, then that axis is the principal axis. In addition, there are two more principal axes that are perpendicular to the symmetry axis.

The following Scilab program determines the direction of the principal axes of a dumbbell (Refer to *Figure 1.4*). It is assumed that the dumbbell is kept in the x - y plane and the coordinates of the two masses have been mentioned.

The moment of inertia tensor for this configuration will be equal to

$$I = m \begin{bmatrix} 2a^2 & 2a^2 & 0 \\ 2a^2 & 2a^2 & 0 \\ 0 & 0 & 4a^2 \end{bmatrix} = 2ma^2 \begin{bmatrix} 1 & 1 & 0 \\ 1 & 1 & 0 \\ 0 & 0 & 2 \end{bmatrix}$$

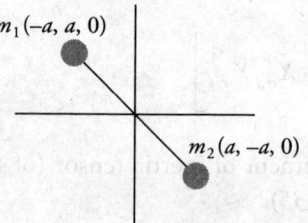

$m_1(-a, a, 0)$

$m_2(a, -a, 0)$

Figure 1.4   Diagram of a rotating dumbbell

```
A = [1 1 0 ; 1 1 0 ; 0 0 2];
[eigen_vector,eigen_value] = spec(A);
```

The result of this short Scilab program is as follows.

- The eigen value matrix will be equal to

$$\begin{pmatrix} 0 & 0 & 0 \\ 0 & 2 & 0 \\ 0 & 0 & 2 \end{pmatrix}$$

- This implies that the eigen values are equal to $(0, 4ma^2, 4ma^2)$.
- The eigen vector matrix will be equal to

$$\begin{pmatrix} -0.7071068 & 0.7071068 & 0 \\ 0.7071068 & 0.7071068 & 0 \\ 0 & 0 & 1 \end{pmatrix}$$

- The eigen vector corresponding to the first eigen value can be written as

$$\begin{pmatrix} -1 \\ 1 \\ 0 \end{pmatrix}$$

This implies that the direction of principal axis is along $\widehat{e}_y - \widehat{e}_x$. This is in fact the symmetry axis of the dumbbell about the origin.

- Eigen vector corresponding to eigen value, $4ma^2$ can be written as

$$\begin{pmatrix} 1 \\ 1 \\ 0 \end{pmatrix}$$

This implies that the direction of the principal axis is along $\widehat{e}_x + \widehat{e}_y$. This is perpendicular to the previous principal axis.

- There is no information about the third principal axis and the z-component can have any value.

1.6.9 Matrix representation of differential operator

Matrix representation of the differential operator $\left(D = \dfrac{d}{dx} \right)$ can be determined using the numerical technique of the finite difference approximation of derivatives. The basic idea behind this approximation is as follows. Suppose x is a coordinate vector having N equally spaced values and assume that the difference between the consecutive values of this vector is h. The first derivative of a function $f(x)$ at point x can be determined by three different forms of the finite difference approximation rule,

- Forward difference (*Eqn. 1.37*)

$$f'(x) \approx \lim_{h \to 0} \left\{ \frac{f(x+h) - f(x)}{h} \right\} \tag{1.37}$$

- Backward difference (*Eqn. 1.38*)

$$f'(x) \approx \lim_{h \to 0} \left\{ \frac{f(x) - f(x-h)}{h} \right\} \tag{1.38}$$

- Central difference (*Eqn. 1.39*)

$$f'(x) \approx \lim_{h \to 0} \left\{ \frac{f(x+h) - f(x-h)}{2h} \right\} \qquad (1.39)$$

The application of this method to create the matrix representation of differential operator is described here for the function $f(x) = \cos x$ followed by the Scilab program.

The algorithm of this program is as follows.

- Specify the range of x-variable and length (N) of the x-vector.

- Determine the step size (interval between consecutive values of x).

- Define the function $f(x)$.

- Calculate the value of $f(x)$ at every value of the x-variable.

- This program creates the matrix representation of the differential operator.

- For the first value of 'x', $f(x-h)$ is not defined; therefore, the forward difference approximation has been used, such that,

$$D(1,1) = -\frac{1}{h}$$

$$D(1,2) = \frac{1}{h}$$

- For the last value of 'x', $f(x+h)$ is not defined; therefore the backward difference approximation has been used.

Therefore,

$$D(N,N) = \frac{1}{h}$$

$$D(N,N-1) = -\frac{1}{h}$$

- For the rest of the 'x' values, the central difference approximation has been used. For this case, a diagonal matrix is constructed whose elements are placed above and below the central diagonal, such that,

  - The elements above the central diagonal are $\frac{1}{2h}$.

  - The elements below the central diagonal are $-\frac{1}{2h}$.

- The matrix operator thus generated operates on the function $f(x)$.

- *Figure 1.5* shows the result.

```
minimum_x = 0;
maximum_x = 2*%pi;
N = 500;
x = linspace(minimum_x,maximum_x,N)';
h = x(2) - x(1);

function alpha = func(x)
alpha = cos(x);
endfunction
D = (diag(ones((N-1),1),1)-diag(ones((N-1),1),-1))/(2*h);
D(1,1) = -1/h;
D(1,2) = 1/h;
D(2,1) = -1/(2*h);
D(N-1,N) = 1/(2*h);
D(N,N-1) = -1/h;
D(N,N) = 1/h;

plot2d(x,func(x))
plot2d(x,D*func(x))
```

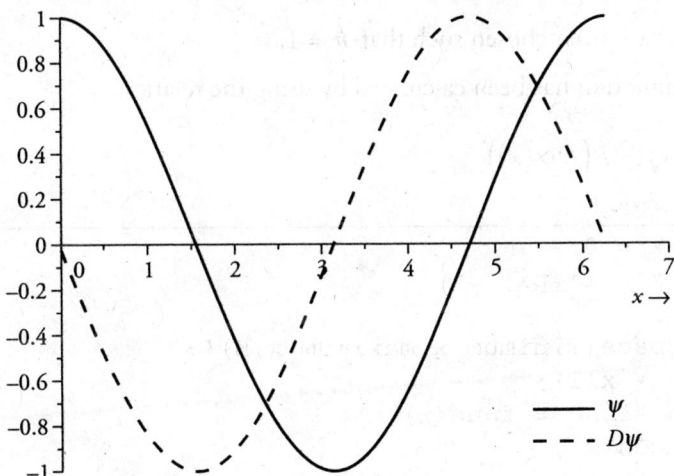

Figure 1.5 Application of the matrix representation of the differential operator

1.6.10 Position–Momentum commutation

The commutation of position and momentum of a point particle is an interesting application of the differential operator. In one dimension, this commutation relation can be expressed as shown in *Eqn. 1.40*.

$$[p_x, x] = p_x x - x p_x \tag{1.40}$$

In *Eqn. 1.40*,

- The momentum operator in the x-direction is p_x. It is equal to $-i\hbar \dfrac{\partial}{\partial x}$.
- The position operator in the x-direction is x.

Therefore, for a wave function $\psi(x)$, the commutation relation can be written as (*Eqn. 1.41*),

$$[p_x, x]\psi(x) = (p_x x - x p_x)\psi(x) = -i\hbar\frac{\partial}{\partial x}\big(x\psi(x)\big) + x i\hbar\frac{\partial}{\partial x}\psi(x) = -i\hbar\psi(x) \tag{1.41}$$

This commutation relation has been proved in the following Scilab program with the help of a cosine function. In this program:

- The x-range has been taken to be $[0, 2\pi]$.
- The x-range has been divided into 500 intervals of equal width.
- The differential operator (D) has been generated in a similar manner as discussed in the previous section.
- The position operator (P) is a diagonal matrix whose diagonal elements are equal to the value of x for the particular interval.
- The units have been chosen such that $\hbar = 1$.
- The commutation has been calculated by using the relation

$$D\big(P\psi(x)\big) - P\big(D\psi(x)\big)$$

```
minimum_x = 0;
maximum_x = 2*%pi;
N = 500;
x = linspace(minimum_x,maximum_x,N)';
h = x(2) - x(1);
function alpha = func(x)
alpha = cos(x);
endfunction


D = (diag(ones((N-1),1),1)-diag(ones((N-1),1),-1))/(2*h);
D(1,1) = -1/h;
D(1,2) = 1/h;
D(2,1) = -1/(2*h);
D(N,N-1) = -1/h;
D(N-1,N) = 1/(2*h);
D(N,N) = 1/h;
```

```
D = -%i*D;
P = diag(ones(N,1));
commutation = (D*(diag(x'*P)*func(x)))-
(diag(x'*P)*(D*func(x)));

subplot(211)
plot2d(x,func(x))
subplot(212)
plot2d(x,commutation/(-%i))
```

As can be seen in *Figure 1.6*, the original wave function is similar to the result of the commutation.

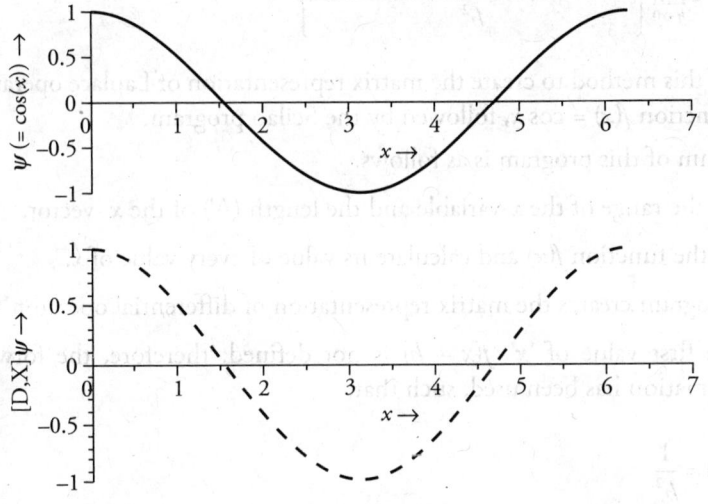

Figure 1.6 Graph showing the position–momentum commutation

1.6.11 Matrix representation of the Laplace operator

Matrix representation of the Laplace operator $\left(\dfrac{d^2}{dx^2}\right)$ can be determined using the numerical technique of the finite difference approximation of second order derivatives. The basic idea behind this approximation rule is as follows.

Suppose x is a coordinate vector having N equally spaced values. The difference between the consecutive values of this vector is equal to h. The second derivative of a function $f(x)$ at point x can be determined using the three different forms of the finite difference approximation rule,

- Forward difference (*Eqn. 1.42*)

$$f''(x) \approx \lim_{h \to 0} \left\{ \frac{f(x+2h) - 2f(x+h) + f(x)}{h^2} \right\}$$ (1.42)

- Backward difference (*Eqn. 1.43*)

$$f''(x) \approx \lim_{h \to 0} \left\{ \frac{f(x) - 2f(x-h) + f'(x-2h)}{h^2} \right\}$$ (1.43)

- Central difference (*Eqn. 1.44*)

$$f''(x) \approx \lim_{h \to 0} \left\{ \frac{f(x-h) - 2f(x) + f(x+h)}{h^2} \right\}$$ (1.44)

Application of this method to create the matrix representation of Laplace operator is described here for the function $f(x) = \cos x$, followed by the Scilab program.

The algorithm of this program is as follows.

- Specify the range of the x-variable and the length (N) of the x-vector.

- Define the function $f(x)$ and calculate its value of every value of x.

- This program creates the matrix representation of differential operator 'L'.

- For the first value of 'x', $f(x - h)$ is not defined; therefore, the forward difference approximation has been used, such that,

$$L(1,1) = \frac{1}{h^2}$$

$$L(1,2) = -\frac{2}{h^2}$$

$$L(1,3) = \frac{1}{h^2}$$

- For the last value of 'x', $f(x + h)$ is not defined; therefore, the backward difference approximation has been used, such that,

$$L(N, N-2) = \frac{1}{h^2}$$

$$L(N, N-1) = -\frac{2}{h^2}$$

$$L(N,N) = \frac{1}{h^2}$$

- For the rest of the 'x' values, the central difference approximation is used. For this case, a diagonal matrix is constructed whose elements are placed such that,
  - The elements along the central diagonal are $-\frac{2}{h^2}$.
  - The elements above the central diagonal are $\frac{1}{h^2}$.
  - The elements below the central diagonal are $\frac{1}{h^2}$.
- The matrix operator acts on $f(x)$.
- Plot the original function and the resultant of the Laplace operation. *Figure 1.7* shows the result.

```
minimum_x = 0;
maximum_x = 2*%pi;
N = 600;
x = linspace(minimum_x,maximum_x,N)';
h = x(2) - x(1);

function alpha = func(x)
alpha = cos(x);
endfunction

Lap = (-2*diag(ones(N,1),0) + diag(ones((N-1),1),1) +
diag(ones((N-1),1),-1))/(h^2);
Lap(1,1)   = 1/(h*h);
Lap(1,2)   = -2/(h*h);
Lap(1,3)   = 1/(h*h);
Lap(2,1)   = 1/(h*h);
Lap(N-1,N) = 1/(h*h);
Lap(N,N-2) = 1/(h*h);
Lap(N,N-1) = -2/(h*h);
Lap(N,N)   = 1/(h*h);

plot2d(x,func(x))
plot2d(x,Lap*func(x))
```

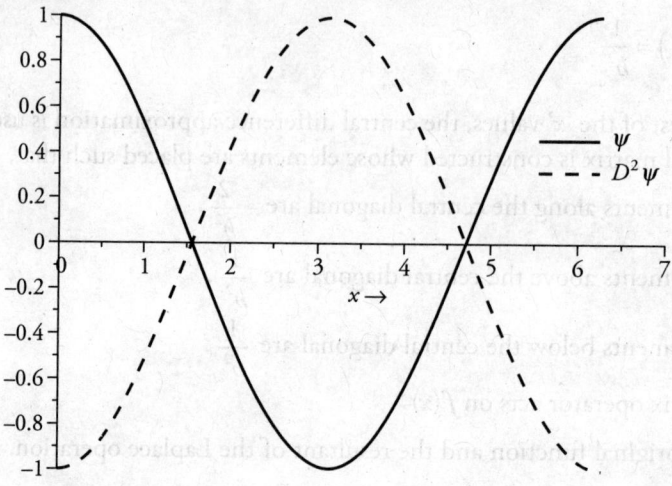

Figure 1.7   Application of the matrix representation of the Laplace operator

1.6.12 Wave function for stationary states

This is an interesting application of matrix representation of operators and has been used in this section to solve the particle in the one-dimensional box problem. Suppose a particle is confined to the region $0 < x < 5$. The potential profile is given in *Eqn. 1.45*.

$$V = \begin{cases} 0 & \text{for } 0 < x < 5 \\ \infty & \text{elsewhere} \end{cases} \tag{1.45}$$

As a result of infinite potential at the boundaries, the following conditions (given in *Eqn. 1.46* and *Eqn. 1.47*) should be satisfied so as to achieve the continuity of wave function and its first derivative.

$$\psi\left(\text{at } x = 0\right) = 0 \tag{1.46}$$

$$\psi\left(\text{at } x = 5\right) = 0 \tag{1.47}$$

The following Scilab program determines the wave function and calculates the energy eigen values for different stationary states of the particle. The wave functions of the first three energy states are shown in *Figure 1.8*. The algorithm of the program is as follows.

- The time independent (stationary) Schrödinger equation is given by *Eqn. 1.48*.

$$H\psi = E\psi \tag{1.48}$$

- The eigen vector of this energy eigen value equation is the wave function ψ.

- The energy eigen value is E.

- H is the Hamiltonian for the particle inside the box and is given by *Eqn. 1.49*

$$H = -\frac{\hbar^2}{2m}\frac{d^2}{dx^2}$$

(1.49)

- The range of x-axis is [0,5] and it is arbitrarily divided into 100 parts.

- The units have been chosen such that ($\hbar$) and mass of the particle is one.

- The Laplace operator is generated on similar lines as done in *Section 1.6.11*. However, in this case, some components of the Laplace operator have been purposely made equal to zero, so as to satisfy the boundary conditions

$$\psi\left(\text{at } x = 0\right) = \psi\left(\text{at } x = 5\right) = 0$$

- The Laplace operator is then multiplied by $-\dfrac{\hbar^2}{2m}$ to obtain the Hamiltonian matrix whose eigen vector and eigen value is determined by the usual Scilab commands, which have been explained before.

- The first three eigen states are plotted using the 'plot2d' command.

```
minimum_x = 0;
maximum_x = 5;
hbar = 1;
m = 1;
N = 100;
x = linspace(minimum_x,maximum_x,N)';
h = x(2) - x(1);
Lap = (-2*diag(ones(N,1),0) + diag(ones((N-1),1),1) +
diag(ones((N-1),1),-1))/(h^2);
Lap(1,1) = 0;
Lap(1,2) = 0;
Lap(2,1) = 0;
Lap(N-1,N) = 0;
Lap(N,N-1) = 0;
Lap(N,N) = 0;
Hamiltonian = -(hbar*hbar/(2*m))*Lap;
[eigen_vector,eigen_value] = spec(Hamiltonian);
plot2d(x,-eigen_vector(:,3));        //Ground state
plot2d(x,-eigen_vector(:,4));     //First excited state
plot2d(x,-eigen_vector(:,5));     //Second excited state
```

The energy eigen value is stored in the matrix *eigen_value* as its diagonal elements. These values can be determined in the following manner.

```
A = diag(eigen_value);
A(3)
A(4)
A(5)
```

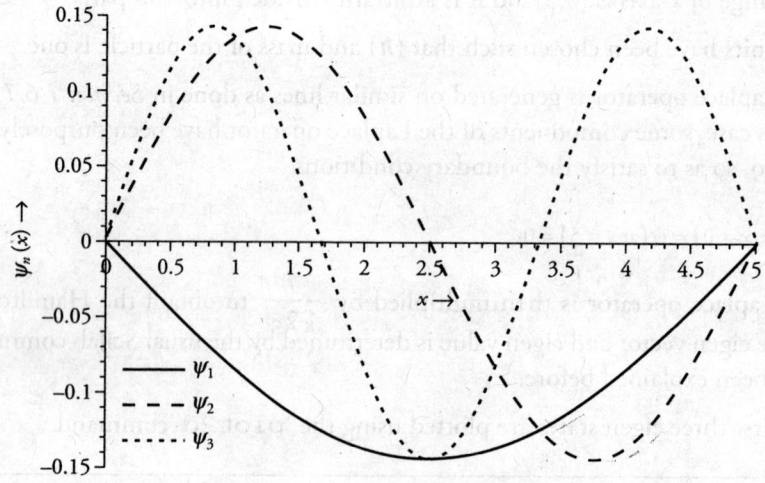

Figure 1.8   Graph for Section 1.6.12

The energies will be equal to the following:

- Ground state = 0.1973755
- First excited state = 0.7893034
- Second excited state = 1.7751875

The energy eigen values from the analytical result $\left(E_n = \dfrac{\hbar^2 n^2 \pi^2}{2ma^2} \right)$ are given by

- Ground state ($n = 1$) = 0.1973921
- First excited state ($n = 2$) = 0.7895684
- Second excited state ($n = 3$) = 1.7765288

1.7 Exercises

1.   Suppose two matrices A and B are defined in Scilab as

   A = [1 2 3];

   B = [4 5 6];

What will be A*B?

2. With reference to the previous question, what will be the outcome of the following operations?
    a) A. * B
    b) A * B'
    c) A' * B

3. For a matrix given by A = [1 2 3;4 5 6], what will be the output for the command, A(:,:)?

4. A matrix is given by
    A =

    8.84   9.32   3.61
    6.52   2.14   2.92
    3.07   3.12   5.66

    What will be the output from the following Scilab command?
    B = A(1,:)

5. With reference to the previous question, what will be the output from the following Scilab command?
    B = int(A(1,:))

6. Give the output result for operations on rows (R_n) and columns (C_n) of matrix A, given in *Table 1.5*.

Table 1.5 Data for *Exercise 6*

$$A = \begin{bmatrix} 1 & 2 & 3 \\ 4 & 5 & 6 \\ 7 & 8 & 9 \end{bmatrix}$$

Operation	Scilab Command
$R_1 \rightarrow 2R_2$	A(1,:) = 2*A(1,:)
$R_1 \rightarrow R_1 + R_2$	A(1,:) = A(1,:) + A(2,:)
$C_2 \rightarrow 3C_2$	A(:,2) = 3*A(:,2)
$C_1 \rightarrow C_1 - 0.5C_2$	A(:,1) = A(:,1) - 0.5*A(:,2)

7. Write a Scilab program for determining the last element of a vector A.

8. A is a vector given by A = [1 2 3]. What will be the output from the following Scilab command?
    A(:,1)= A(:,1)-2

9. A matrix is given by A = [1 2 3 ; 3 4 5 ; 5 6 7]. What will be the output from the Scilab command: A(:,1:2)?

10. A matrix is given by

$$A = \begin{bmatrix} 1 & 1 & 1 & 1 \\ 1 & 1 & 1 & 1 \\ 1 & 1 & 1 & 1 \\ 1 & 1 & 1 & 1 \\ 1 & 1 & 1 & 1 \end{bmatrix}$$

Write a single line Scilab command to change this matrix to

$$A = \begin{bmatrix} 1 & 2 & 2 & 2 \\ 1 & 2 & 2 & 2 \\ 1 & 1 & 1 & 1 \\ 1 & 1 & 1 & 1 \\ 1 & 1 & 1 & 1 \end{bmatrix}$$

11. Write a single line Scilab command to generate the following matrix.

$$A = \begin{bmatrix} 1 & 1 & 1 \\ 0 & 0 & 0 \\ 1 & 1 & 1 \end{bmatrix}$$

12. Write a Scilab function to determine the distance between two vectors. Illustrate it with the help of an example, where the two vectors are

$$A = \begin{bmatrix} 1 & 2 & 3 \end{bmatrix}$$
$$B = \begin{bmatrix} 4 & 5 & 6 \end{bmatrix}$$

13. Write a Scilab function for determining the sum of cube of all the elements of a vector. Illustrate it with the help of an example, where the vector is given by

$$A = \begin{bmatrix} 1 & 2 & 3 \end{bmatrix}$$

14. Write a Scilab function for determining the scalar product of two vectors.

15. Write a Scilab function for determining the cross product of two vectors. Use it to determine the cross product of

$$\hat{i} + 2\hat{j} + 3\hat{k}$$

$$4\hat{i} + 5\hat{j} + 6\hat{k}$$

16. Write a Scilab function for determining the angle between two vectors. Use it to find the angle between

$$\vec{A} = 3\hat{i} + 4\hat{j} + 2\hat{k}$$

$$\vec{B} = 2\hat{i} + 2\hat{j} - 3\hat{k}$$

17. Write a Scilab program for determining the volume of a parallelepiped.

18. Write a Scilab program to generate the vector $[a\ 2a\ 3a]$. Here, 'a' is a variable.

19. Write a Scilab program to generate the following polynomial vector.

$$\left[1 + a, 2 + 4a + 2a^2, 3 + 9a + 9a^2 + 3a^3\right]$$

Here, 'a' is a variable.

20. Write a Scilab program to determine the following indefinite integral.

$$\int dx$$

Determine the value of this integral at $x = 2$.

21. Repeat the previous question for evaluating the following integral.

$$\int x\,dx$$

22. Write a Scilab program to define a polynomial having root equal to 3.

23. Write a Scilab program to show that the determinant of a matrix and the determinant of transpose of that matrix are equal.

24. Write a Scilab program to show that a Hermitian matrix is equal to its complex conjugate-transpose.

25. Write a Scilab program to show that the following matrix is orthogonal.
    A = [sind(30) cosd(30) ; -cosd(30) sind(30)]

26. Write a Scilab program to show that the following matrix is unitary.
    A = [%i 0 ; 0 %i]

27. Write a Scilab program to show that the eigen vectors corresponding to two distinct eigen values of a Hermitian matrix are orthogonal.

28. Using Scilab program, express the vector $\vec{A} = xy\widehat{a_x} + z\widehat{a_y} + y\widehat{a_z}$ at the point (2, -3,4)in cylindrical coordinates.

29. Write a user-defined Scilab function to transform cylindrical coordinates of a point to Cartesian coordinates.

30. Write a user-defined Scilab function to transform spherical coordinates of a point to Cartesian coordinates.

31. Using Scilab program, solve the circuit given in *Figure 1.9* for the current flowing in every mesh.

32. Consider a uniform solid cube of mass 'M' with each side equal to 'a'. The cube is rotating about one of its corners. Write the moment of inertia tensor and determine the principal axes by writing a Scilab program.

33. Repeat the previous exercise. Assume that the cube is rotating about its centre.

34. Write a Scilab program to determine the effect of a differential operator (D) when it acts on the function $f(x) = \sin x$. Show the result with the help of a graph.

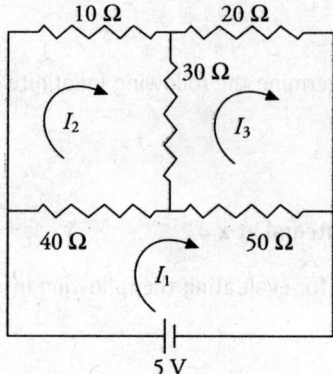

Figure 1.9   Circuit diagram for *Exercise 31*

35. Write a Scilab program to determine the effect of differential operator (D) when it acts on the function $f(x) = x^3$. Show the result with the help of a graph.

36. The position–momentum commutation relation is given by

$$\left[p_x, x\right] = p_x x - x p_x$$

The momentum operator along the x-direction is given by $p_x = \dfrac{\partial}{\partial x}$. Write a Scilab program to graphically show that

$$\left[p_x, x\right]\left(x^3 - 2x^2 - 2x + 5\right) = \left(x^3 - 2x^2 - 2x + 5\right)$$

37. Write a Scilab program for the following operator.

$$A = \frac{\partial}{\partial x}\left[\left(1 - x^2\right)\frac{\partial}{\partial x}\right]$$

38. Use the program in question 37 to show that in the x-range [0, 1],

$$Ax = -2x$$

39. Write a Scilab program to determine the effect of the Laplace operator (L) on the function $f(x) = \sin x$. Show the result with the help of a graph.

40. Write a Scilab program to determine the effect of the Laplace operator (L) on the function $f(x) = x^3$. Show the result with the help of a graph.

41. Write a Scilab program to determine the wave function and energy eigen values of harmonic oscillator by making use of a Hermitian differential operator. Assume that
    - Mass of electron = $0.511 \times 10^6 \, eV/c^2$
    - $\hbar c = 1973 \, eV \, \mathring{A}$
    - Potential $= V = \dfrac{1}{2} kx^2 \, (k = 1)$

42. Write a Scilab program to determine the wave function and energy eigen values of the hydrogen atom by making use of a Hermitian differential operator. Assume that
    - Mass of electron = $0.511 \times 10^6 eV/c^2$
    - Charge of electron = $e = 3.795 \, (eV\mathring{A})^{1/2}$
    - $\hbar c = 1973 \, eV \, \mathring{A}$
    - Potential $= V = \dfrac{e^2}{x}$

Plotting and Graphics Design

2

After going through this chapter, the reader should be able to

◊ Make self-explanatory graphs in Scilab.

◊ Format the coordinate axes of a graphics window in Scilab by customizing the appearance of the font (size, color, and typeface), placement of the tick marks and the scale of coordinate axes.

◊ Format the line style of a graph by controlling the thickness, style and colour of the curve.

◊ Format the data markers by choosing their style, size, colour and thickness.

◊ Format the title of the graph.

◊ Format the placement of legends on the graphs.

◊ Write user-defined Scilab functions for formatting the coordinate axes, line style, data markers and legend.

◊ Use plotting skills to generate meaningful graphs depicting physical phenomena and electric circuits.

◊ Use plotting skills to make level surfaces and visualize the gradient of scalar fields.

2.1 Introduction

The science of physics generally deals with physical phenomena where one quantity (called an independent variable) is related to another quantity (called a dependent variable) through a mathematical equation. The graphical representation of these data is a convenient tool for deciphering this scientific information. It is of utmost importance that the experimental data should be plotted very carefully so that it is easy to appropriately visualize and interpret the relationship between the dependent and independent variables. For example, it is always advisable to

- Choose the units of the coordinate axes in an appropriate manner.
- Choose the coordinate axes so that the entire data are accommodated.
- Choose logarithmic scales if the range of variables is large.
- Interpolate the data to generate a smooth curve traversing through the data points.
- Mark the data points with markers and error bars wherever available.
- Label the graph properly and write a concise title that summarizes the graph.
- Describe each part of the graph with the help of suitably placed legends.

This chapter introduces the reader to various plotting commands invariably used in this book for developing meaningful graphs. The importance of this chapter lies in the fact that it gives an overview on writing small user-defined functions for generating self-explanatory graphs, instead of writing long codes.

The graphical representation of data can be formatted by three methods in Scilab

- By using the figure and axes properties present in the main menu bar of the graphical window.
- By using Scilab instructions at the command line on the console.
- By writing user-defined functions and invoking them at the console.

The first method is trivial and is left for the reader to explore. In most of the following chapters, graphs and plots have been formatted using small functions that are executed in a script. The major focus of this chapter is to introduce the reader to this kind of formatting tool. However, for completeness, direct command line instructions have also been mentioned wherever possible.

The layout of this chapter is as follows. The Scilab commands 'plot' and 'plot2d,' have been used in this book for generating graphs. *Section 2.2* starts with highlighting the basic difference between these two commands and manipulating them so that they are on equal footing. This section also focuses on writing small functions for editing the coordinate axes. Thus, styling of font size, font colour, typeface, alignment and scale are described. *Section 2.3* discusses appearance of line styles. This includes formatting the thickness, colour and style of the curves. A data point marker is extremely useful to visually indicate the importance of each data point. The designing of data marker is discussed in *Section 2.4*. The formatting of the 'graph title' and placement of the 'graph legend' is discussed in *Sections 2.5* and *2.6*, respectively. The user-defined functions developed in these sections can be written in a *.sci* format and executed as explained in the chapter. The plotting skills developed in all these sections have been applied to various physical problems in *Section 2.7*. The reader is encouraged to reproduce all the graphs in the subsequent chapters by appropriately applying the formatting tools.

2.2 Formatting of the Coordinate Axes

Before starting with the formatting of the coordinate axes, it is important to recapitulate the basic plotting commands that have been used in this book. Suppose a data set is such that the dependent variable is equal to the independent variable (*Table 2.1*). This data set can be generated in Scilab by using the following commands.

```
x = -10:1:10;
y = -10:1:10;
```

Table 2.1   Sample Data

X	Y	X	Y
-10	-10	1	1
-9	-9	2	2
-8	-8	3	3
-7	-7	4	4
-6	-6	5	5
-5	-5	6	6
-4	-4	7	7
-3	-3	8	8
-2	-2	9	9
-1	-1	10	10
0	0		

In this book, Scilab commands 'plot' and 'plot2d' have been used for plotting. The difference between them under default Scilab settings is shown in *Figure 2.1*.

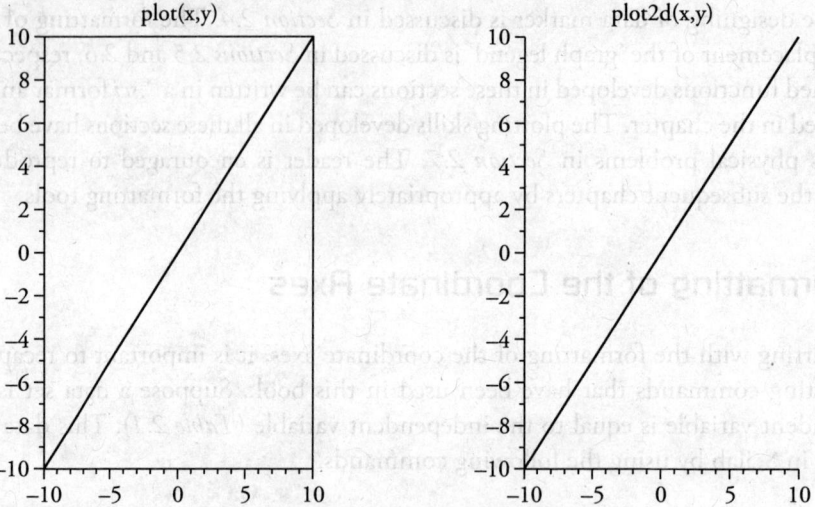

Figure 2.1 Plot of data points under default parameter settings

The Scilab program used for generating *Figure 2.1* is written as follows,

```
plot(x,y)
plot2d(x,y)
```

In *Figure 2.1*,

- The graph on the left has been generated using the 'plot' command. Under default settings, the graph is enclosed inside a box.

- The graph on the right has been drawn using the 'plot2d' command. Under the default settings, this graph is not enclosed inside a box.

This subtle difference between the commands 'plot' and 'plot2d' can be easily eliminated using the following Scilab program.

```
plot(x,y)
a = get("current_axes")
a.box = "off";

plot2d(x,y)
a = get("current_axes")
a.box = "on";
```

Figure 2.2 shows the effect of these commands on including/removing the border from the graph.

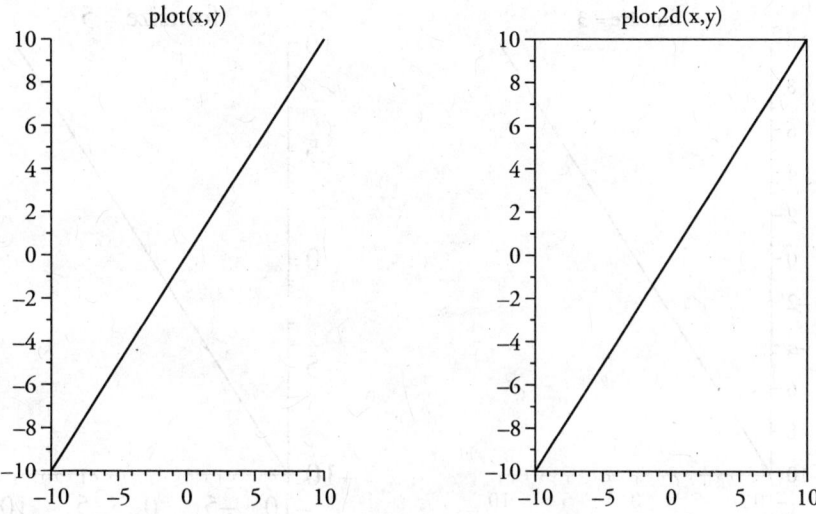

Figure 2.2   Plot of data points with modified settings

It is easy to observe that the graph generated with 'plot2d' is now enclosed inside a box, whereas the box outlining the graph generated from 'plot' command is not visible.

2.2.1 Font size

This section describes the formatting of the font size of the labels on the coordinate axes, by making use of a 'function'. In the following Scilab program, a user-defined 'function' is used to define the font size. It is then invoked after the plotting command. The size of the font is chosen through an integer whose value should be ≥ 0. This is shown in *Figure 2.3*.

```
function set_my_axes(fontsize)
a = get("current_axes")
a.font_size=fontsize;
endfunction

plot2d(x,y)
set_my_axes(3);              //Integer decides the font size
```

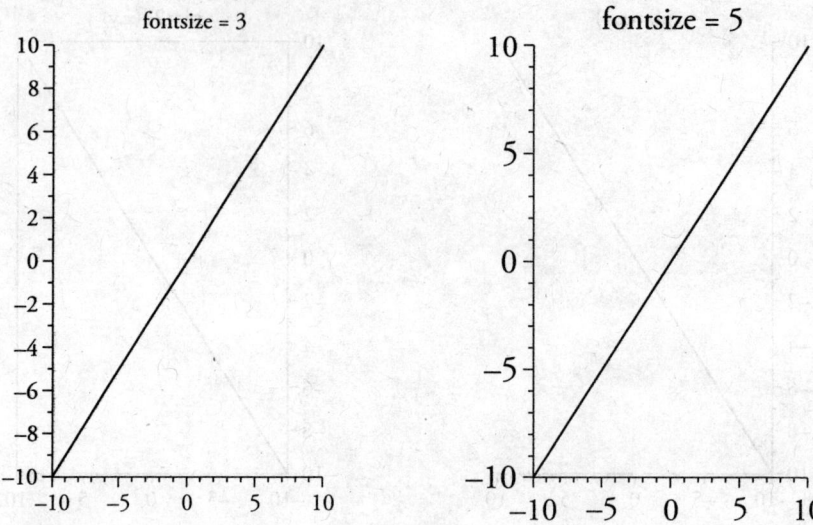

Figure 2.3   Difference in the font size

In *Figure 2.3*,

- The graph on the left has been generated using the font size '3' for the labels of the coordinate axes.

- The graph on the right has been drawn using the font size '5' for the labels of the coordinate axes.

- It is also possible to format the font size without writing a function. This is shown below for font size equal to five times the default value.

```
plot2d(x,y)
a = get("current_axes")
a.font_size = 5;
```

- The graph on the right can also be formatted directly from the command line. The program is written as follows. This method is now obsolete.

```
xset("font size",5)
plot2d(x,y) or plot(x,y)
```

2.2.2 Font colour

The default font colour for axes of a graphics window is black. This subsection describes how to format the colour of the font of labels on the coordinate axes, by making use of a 'function'.

In the following Scilab program, a 'function' is used to define the colour of the font along with the font size as discussed earlier. It is then invoked after the plotting command.

The colour of the font can be chosen through an integer whose value should be greater than 0. The resultant plot is shown in *Figure 2.4*. The numbers corresponding to some commonly used colours are given in *Table 2.2*.

```
function set_my_axes(fontsize, font_color)
a = get("current_axes")
a.font_size=fontsize;
a.labels_font_color=font_color;
endfunction

plot2d(x,y)
set_my_axes(3,2);
```

In *Figure 2.4*,

- Graph on the left has been generated using the font size '3'. The font colour is taken to be '2', which corresponds to colour 'blue'.

- Graph on the right has been drawn using the font size '4'. The font colour is taken to be '5', which corresponds to colour 'red'.

- Instead of writing a function, these graphs can also be generated through the following instructions. This program has been written for the graph on the right.

```
plot2d(x,y)
a = get("current_axes")
a.font_size = 4;
a.labels_font_color = 5;
```

- Graph on the right can also be generated by giving instructions directly through the command line. The program is written as follows. The 'xset' command is now obsolete.

```
xset("font size",4)
xset("foreground",5)
plot2d(x,y) or plot(x,y)
```

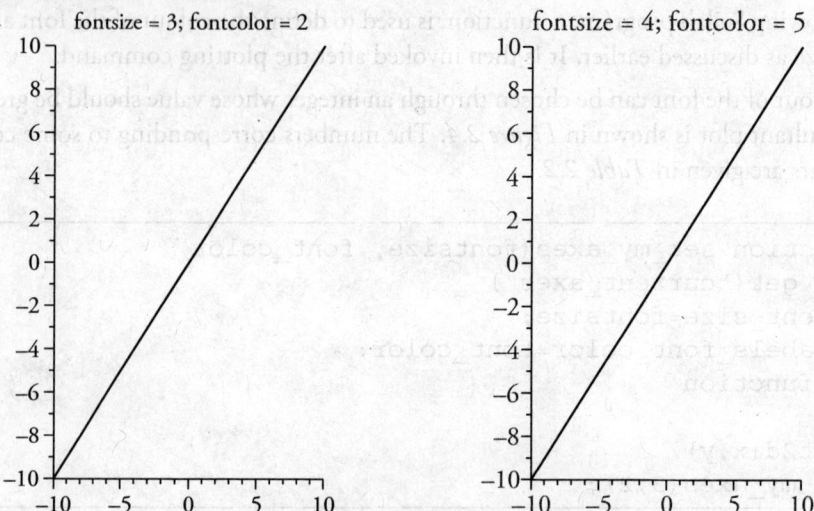

Figure 2.4   Difference in font size and font colour

Table 2.2   Number code for commonly used colours

Colour	Number Code
Black	1
Blue	2
Green	3
Cyan	4
Red	5
Magenta	6
Yellow	7
Pink	34
Brown	35
Violet	37
Orange	38

2.2.3 Typeface

Scilab provides a user friendly environment where one can easily manipulate the typeface. The following program shows how it is done; and the corresponding graph is shown in *Figure 2.5*. The numbers corresponding to some commonly used typeface are given in *Table 2.3*.

```
function set_my_axes(fontsize, fontstyle)
a = get("current_axes")
a.font_size = fontsize;
a.font_style=fontstyle;
endfunction

plot2d(x,y)
set_my_axes(3,0);
```

In *Figure 2.5*,

- Graph on the left has been generated using the font size '3'. The font style is taken to be '0', which corresponds to the typeface 'Monospaced (Courier)'.

- Graph on the right is also drawn using the font size '3'. The font style in this case is '5', which corresponds to the typeface 'Serif Bold Italic (Times)'.

- These graphs can also be generated without writing a function. For example, for graph on the right,

```
plot2d(x,y)
a = get("current_axes")
a.font_size = 3
a.font_style = 5;
```

- Graph on the right can also be produced directly through the command line. The program is written as follows. (The 'xset' command is now obsolete.)

```
xset("font",5,3);
plot2d(x,y) or plot(x,y)
```

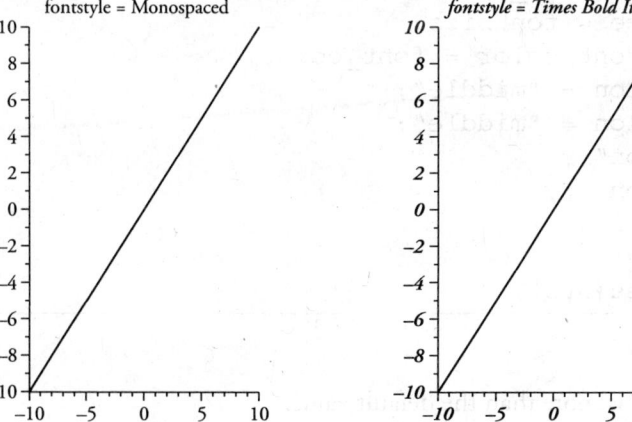

Figure 2.5 Use of different typeface

Table 2.3 Number code for commonly used typeface

Typeface	Number Code
Monospaced (Courier)	0
Symbol	1
Serif (Times)	2
Serif Italic (Times)	3
Serif Bold (Times)	4
Serif Bold Italic (Times)	5
Sans Serif (Helvetica)	6
Sans Serif Italic (Helvetica)	7
Sans Serif Bold (Helvetica)	8
Sans Serif Bold Italic (Helvetica)	9

2.2.4 Axis position

Sometimes it is desirable to plot the coordinate axes running through the middle of the graph. The following Scilab program shows the steps to align the coordinate axes. This is shown in *Figure 2.6*.

```
function set_my_axes(fontsize, font_color)
a = get("current_axes")
a.font_size = fontsize;
a.labels_font_color = font_color;
a.y_location = "middle";
a.x_location = "middle";
a.box = "on"
endfunction

plot2d(x,y)
set_my_axes(5,5);
```

In *Figure 2.6*,

- The font size is more than the default value.

- The font colour is different from the default black colour.

- The graph is enclosed inside a box.
- The coordinate axes run through the origin of the graph.
- Instead of the 'middle' option, one can use 'top' to plot the x-axis on top of the rectangle. Similarly, for the y axis, 'right' can be used to plot the y-axis on the right side.

2.2.5 Tick marks

This is another interesting feature regarding the organization of the coordinate axes, wherein the number of major and minor ticks on the axes can be controlled through a function as well as through a command line. The following Scilab program shows the use of 'function' to control the number of minor ticks in the graph, along with the font size and font colour. The corresponding graph is shown in *Figure 2.7*.

```
function set_my_axes(fontsize, font_color, sub_tics)
a = get("current_axes")
a.font_size = fontsize;
a.labels_font_color = font_color;
a.sub_tics = sub_tics;
endfunction

plot2d(x,y)
set_my_axes(3, 1, [0,0]);
plot2d(x,y)
set_my_axes(4, 2, [0,2]);
```

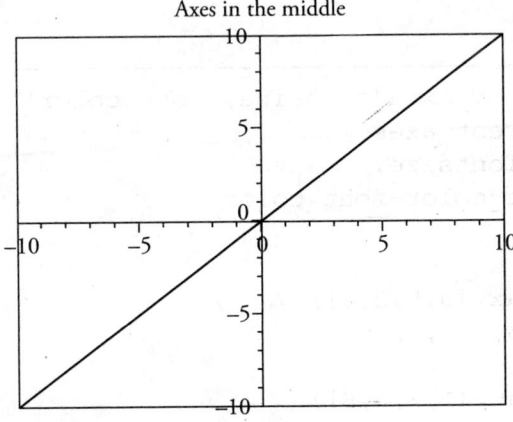

Figure 2.6   Alignment of coordinate axes

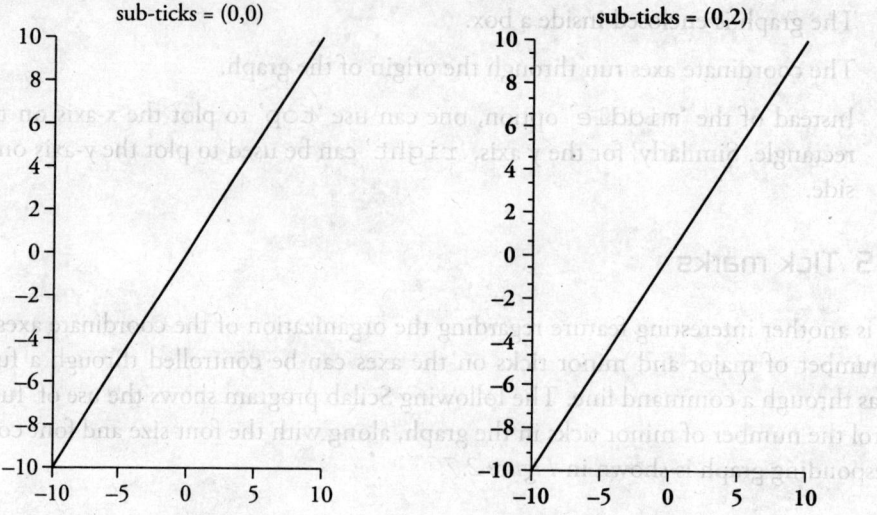

Figure 2.7 Different number of minor ticks

In *Figure 2.7*,

- Graph on the left has font size 3 and default font colour. The number of minor ticks on the x-axis as well as on the y-axis is zero.

- Graph on the right has font size 4 and font colour blue (number code 2). There are no minor ticks on the x-axis. The major ticks on the y-axis are separated by two minor ticks in between.

- The number of major and minor ticks can also be controlled through the command line as shown in the following Scilab program; the corresponding graph is shown in *Figure 2.8*.

```
function set_my_axes(fontsize, font_color)
a = get("current_axes")
a.font_size=fontsize;
a.labels_font_color=font_color;
endfunction

plot2d(x,y,nax=[0,5,2,3])
set_my_axes(3, 1);

plot2d(x,y,nax=[1,3,0,5])
set_my_axes(4, 2);
```

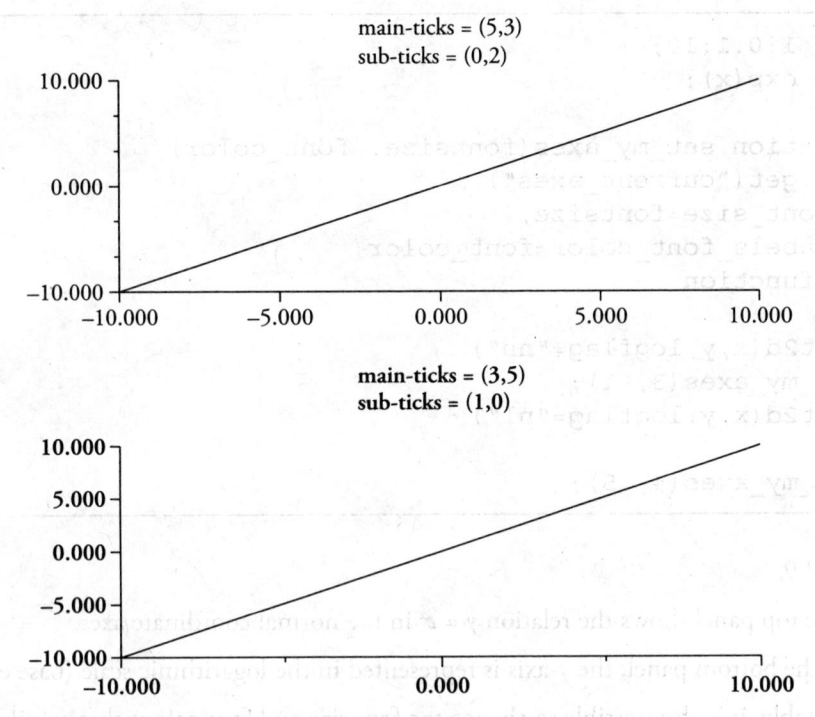

Figure 2.8   Control of number of major and minor ticks

In *Figure 2.8*,

- The font size and font colour is same as in *Figure 2.7*.

- For the top panel of the graph, 'nax = [0,5,2,3]'implies 0 minor and 5 major ticks on the x-axis; and 2 minor and 3 major ticks on the y-axis.

- For the bottom panel of the graph, 'nax = [1,3,0,5]' implies 1 minor and 3 major ticks on the x-axis; and 0 minor and 5 major ticks on the y-axis.

2.2.6 Logarithmic axes

It is often advantageous to format the coordinate axes in the logarithmic scale. The logarithmic scale is non-linear and is used to represent data having a large range of variables. The intensity of light, the loudness of sound and the magnitude scale of stellar brightness are some common examples where the logarithmic scale is more appropriate than the linear scale.

If both the coordinate axes are represented in the logarithmic scale, then it is called a log-log chart. If only one coordinate axes uses the logarithmic scale, then it is called a semi-log chart. The following Scilab program shows the representation of the axes in logarithmic scale. The corresponding graph is shown in *Figure 2.9*.

```
x = 1:0.1:10;
y = exp(x);

function set_my_axes(fontsize, font_color)
a = get("current_axes")
a.font_size=fontsize;
a.labels_font_color=font_color;
endfunction

plot2d(x,y,logflag="nn")
set_my_axes(3, 1);
plot2d(x,y,logflag="nl")

set_my_axes(4, 5);
```

In *Figure 2.9*,

- The top panel shows the relation $y = e^x$ in the normal coordinate axes.

- In the bottom panel, the y-axis is represented in the logarithmic scale (base e).

- Notably, it is also possible to change the font size and font colour through the function defined in the code.

2.2.7 Polar plot

In Scilab, 'polarplot' creates a plot of angle θ versus radius ρ. The angle θ is measured from the x-axis and is specified in radians. The radius vector ρ is specified in the data units. This is shown in the following example.

```
theta = 0:0.01:2*%pi;
rho = sin((3*theta));
polarplot(theta, rho);
```

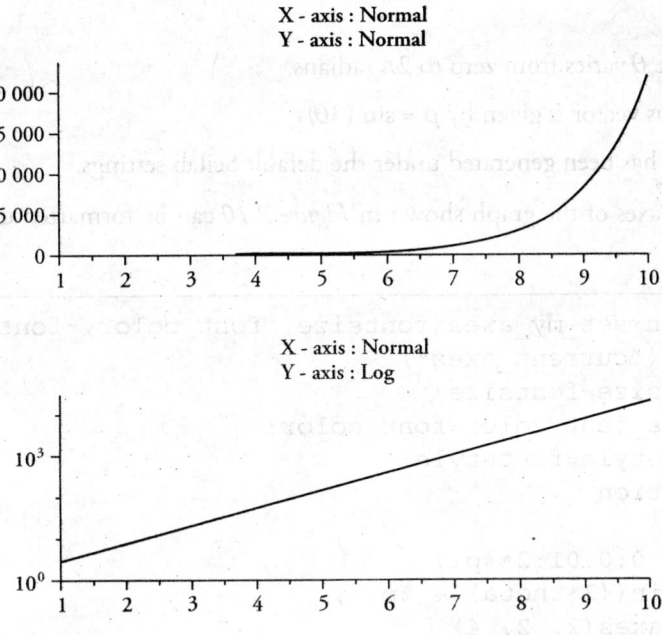

Figure 2.9   Data representation in logarithmic scale

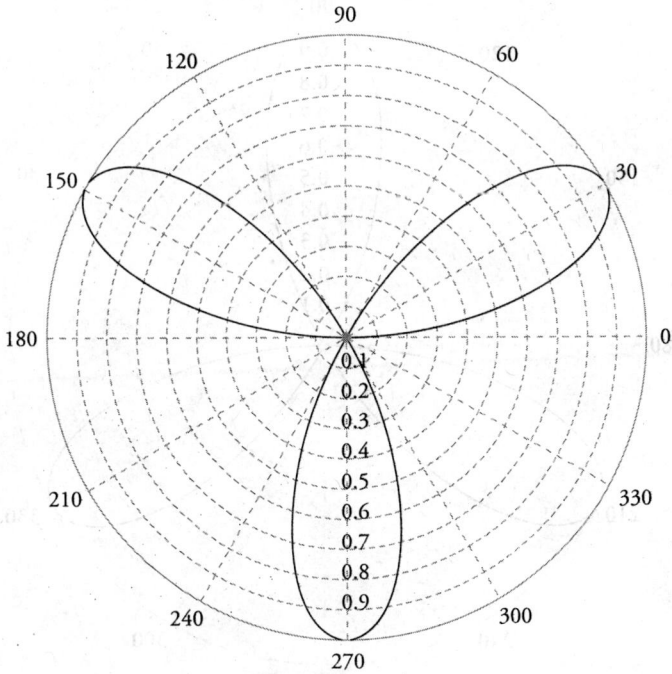

Figure 2.10   Polar plot (default) in the shape of a petalled flower

In *Figure 2.10*,

- The angle θ varies from zero to 2π radians.

- The radius vector is given by $\rho = \sin (3\theta)$

- This plot has been generated under the default Scilab settings.

The coordinate axes of the graph shown in *Figure 2.10* can be formatted using the following function.

```
function set_my_axes(fontsize, font_color, fontstyle)
a = get ("current_axes")
a.font_size=fontsize;
a.labels_font_color=font_color;
a.font_style=fontstyle;
endfunction

theta = 0:0.01:2*%pi;
rho = sin((3*theta) + %pi);
set_my_axes(2, 2, 4)
polarplot(theta, rho);
```

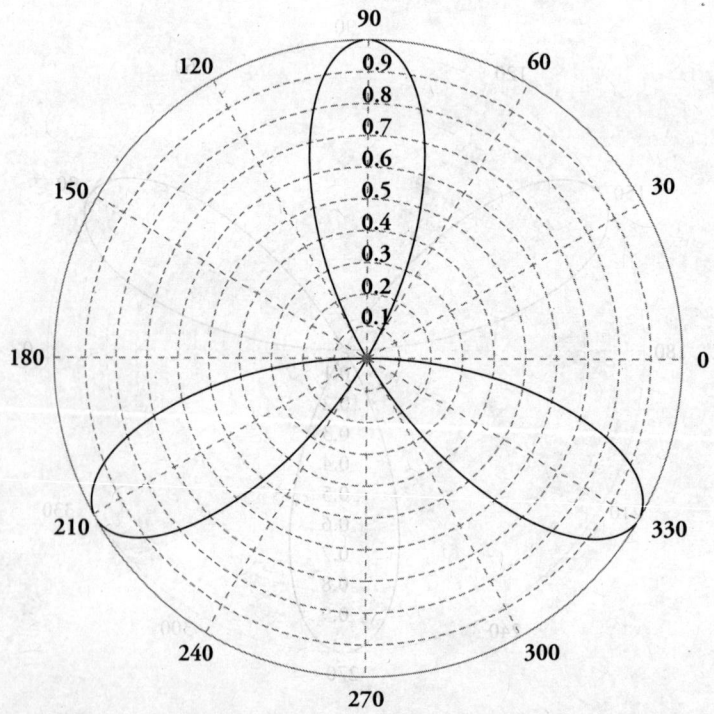

Figure 2.11   Graph showing the polar plot in the shape of a petalled flower

In *Figure 2.11*,

- The orientation of the petals of the flower has been changed through a slight modification in the relation between the angle and the radius vector.

- In this case, $\rho = \sin(3\theta + \pi)$

- The font size is more than the previous graph.

- The font colour is taken as blue and the typeface is chosen as 'Times Bold'.

2.3 Formatting of the Line Styles

This section focuses on the organization of the line styles of any curve in Scilab. Formatting of the line style mainly controls the thickness, design and colour of the curve. This aspect of the Scilab programming has been explained in the following subsections with the help of suitable examples.

2.3.1 Thickness

Consider the same data set that was used in the previous section. It can be generated in Scilab using the following commands.

```
x = -10:1:10;
y = -10:1:10;
```

The following Scilab program shows the use of a 'user-defined function' to draw the curve having a thickness that is four times the default value. As explained in the previous section, the instructions of the function can also be invoked directly from the command line to produce the same effect.

```
function set_my_line_styles(thickness)
e = gce();
e.children.thickness = thickness;
endfunction

plot2d(x,y)
set_my_line_styles(4);
```

In *Figure 2.12*,

- The graph on the left has been drawn under the default Scilab settings.

- The graph on the right has an increased thickness.

- The thickness of the curve can also be formatted through the command line by using the following program. However, it should be noticed that this program also increases the thickness of the coordinate axes. Moreover, 'xset' is now obsolete.

```
xset("thickness",4)
plot2d(x,y) or plot(x,y)
```

- The width of the curve can also be increased using the following command.

```
plot(x,y,'LineWidth',4)
```

2.3.2 Line style

The line style refers to the style of the curve. For example, the following user defined function generates a dashed curve instead of a solid line.

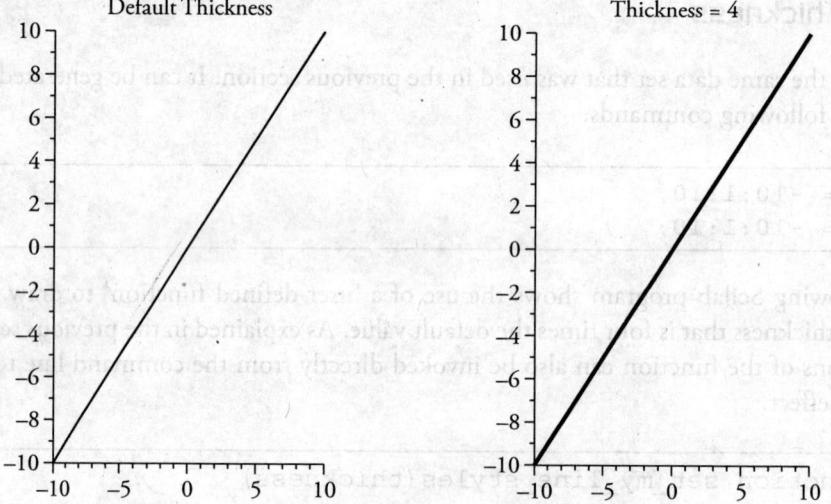

Figure 2.12   Modification of thickness of the curve

```
function set_my_line_styles(style, thickness)
e = gce();
e.children.line_style = style;
e.children.thickness = thickness;
endfunction
plot2d(x,y)
set_my_line_styles(3,4);
```

In *Figure 2.13*,

- Graph on the left has been drawn under default Scilab settings for thickness and style of the curve.

- Curve on the right has an increased thickness. The line style is a 'dashed' curve instead of the default solid line. Instructions of the function can also be invoked directly from the command line to produce the same effect.

- The direct command line instruction for dashed and thicker curve is as follows:

```
xset("line style",3)
xset("thickness",4)
plot2d(x,y)
```

- The graph on the right can also be produced using the following command.

```
plot(x,y,'--', 'LineWidth',4)
```

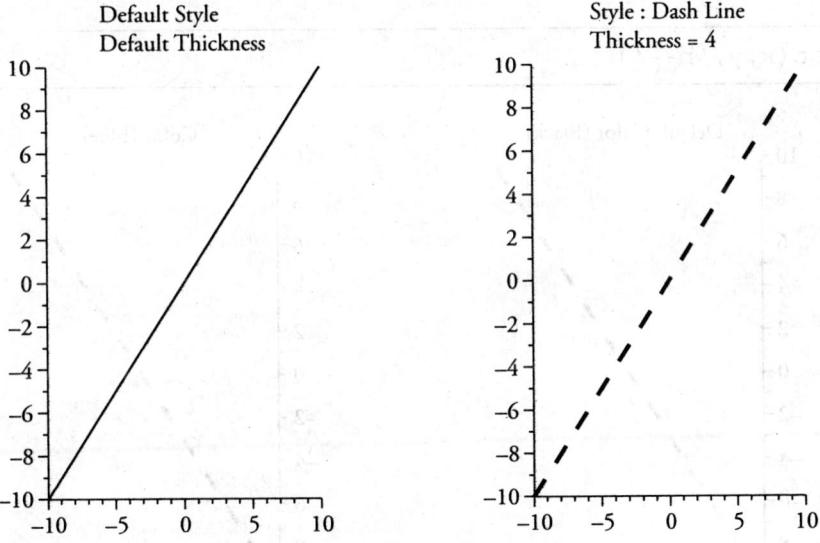

Figure 2.13   Different line styles of a curve

Numbers corresponding to some commonly used line styles are given in *Table 2.4*.

Table 2.4   Different line styles in Scilab

Line Style	Number
Solid line	1
Long dash line	2
Small dash line	3

2.3.3 Line colour

Colour of a curve can be changed directly from the command line. In this case, the number corresponding to the colour is directly written in the `plot2d` command.

```
xset("line style",3)
plot2d(x,y,2)
```

If the 'plot' command is used, then the alphabet for the colour is used.

```
plot(x,y,'b--')
```

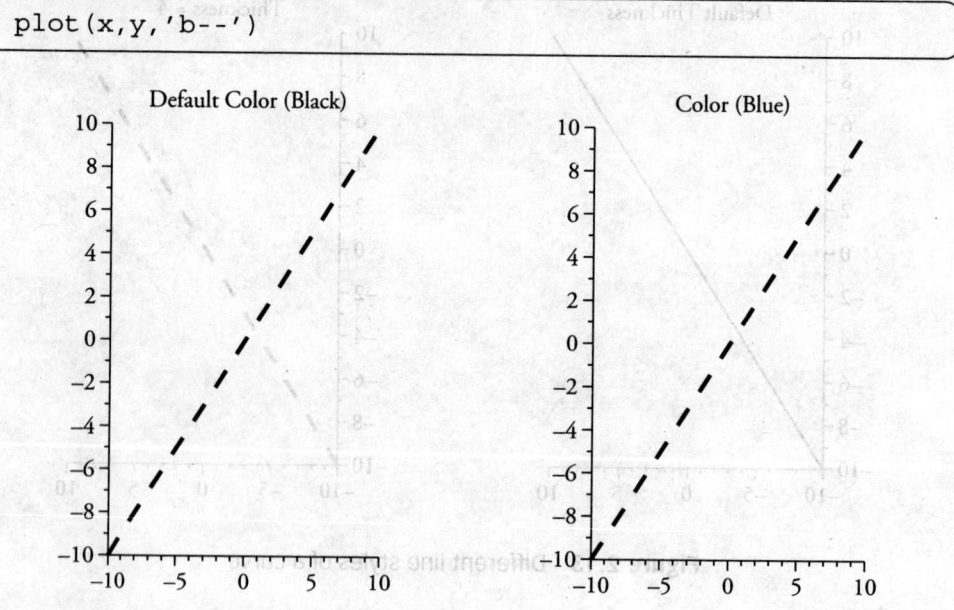

Figure 2.14   Modification of the line colour

In *Figure 2.14*,

- Graph on the left uses default colour for the curve.
- Colour of the curve is changed to 'blue' in the right panel.

2.4 Formatting of the Markers

In almost all the graphs that are made from experimental observations, it is necessary to mark the data points on the plot. This is done by making optimum use of 'markers', which are explained in the following sub-sections with suitable examples.

2.4.1 Marker style

In the following Scilab program, a 'circle' is used as the marker.

```
x = -10:1:10;
y = -10:1:10;
function set_my_line_styles_mark(markstyle)
e = gce();
e.children.mark_style = markstyle;
endfunction
plot2d(x,y)
set_my_line_styles_mark(9)
```

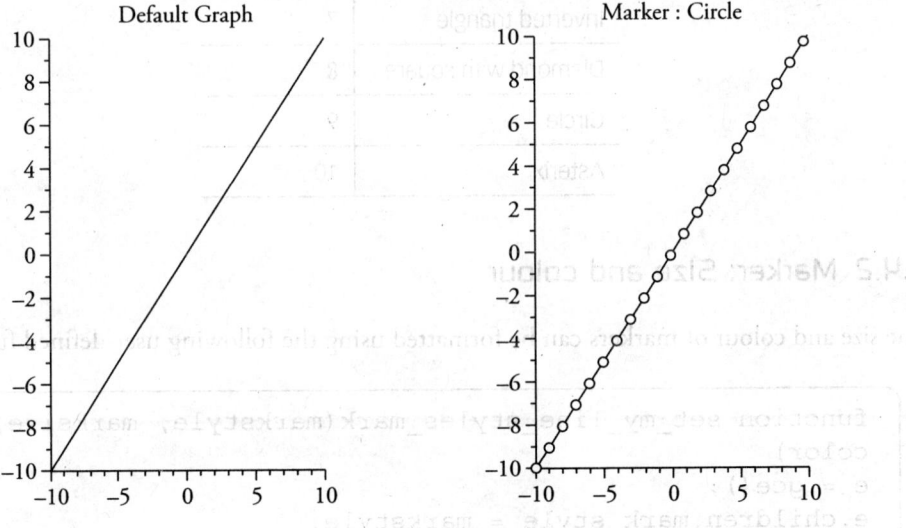

Figure 2.15   Graph showing the use of markers

In *Figure 2.15*,

- Graph on the left is generated using the default Scilab settings. This is obviously without any markers.

- Graph on the right makes use of circular markers for plotting the data points. The number '9' corresponds to a circular marker in Scilab.

- Graph on the right can also be generated by invoking the following instruction directly from the command line,

```
plot(x,y,'o-')
```

The numbers corresponding to some commonly used markers are given in *Table 2.5*.

Table 2.5   Commonly used markers in Scilab

Marker	Number
Plus	1
Cross	2
Circle with square	3
Filled diamond	4
Empty diamond	5
Triangle	6
Inverted triangle	7
Diamond with square	8
Circle	9
Asterix	10

2.4.2 Marker: Size and colour

The size and colour of markers can be formatted using the following user-defined function.

```
function set_my_line_styles_mark(markstyle, marksize, color)
e = gce();
e.children.mark_style = markstyle;
e.children.mark_size = marksize;
e.children.mark_foreground = color;
endfunction

plot2d(x,y)
set_my_line_styles_mark(9, 3, 5)
```

In *Figure 2.16*,

- Graph on the left uses a circular marker with a default size and colour.
- Size of the marker has been increased three times and the colour has been changed to 'red' in the graph on the right.
- Graph on the right can also be generated by writing the following code,

```
plot(x,y,'ro-','markersize',3)
```

2.4.3 Thickness and line mode

It is possible to change the thickness of markers in Scilab. Connecting/disconnecting a marker from its adjacent markers is also possible. This is shown in the following user-defined function.

```
function set_my_line_styles_mark(markstyle, thickness,
color)
e = gce();
e.children.thickness = thickness;
e.children.mark_style = markstyle;
e.children.mark_foreground = color;
e.children.line_mode = 'off';
endfunction
plot2d(x,y)
set_my_line_styles_mark(9, 3, 5)
```

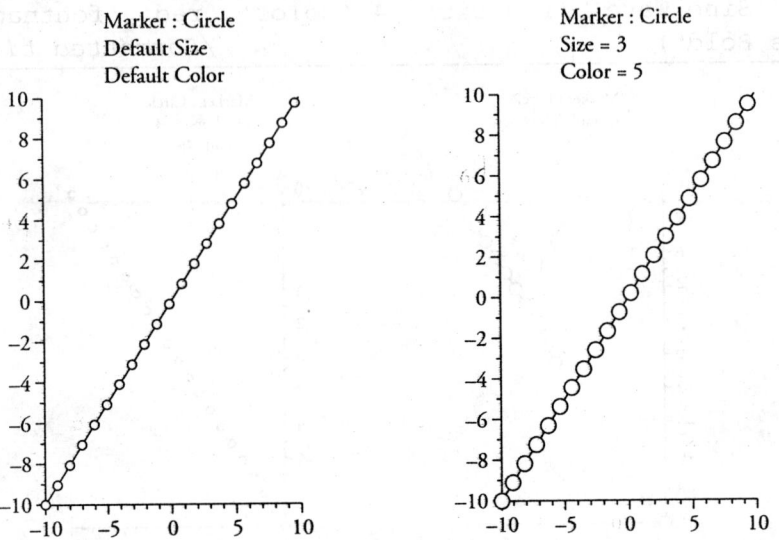

Figure 2.16   Modification of size and colour of the marker

In *Figure 2.17*,

- Graph on the left is the same as that in *Figure 2.16* (right panel).

- Thickness of the marker has been increased in the graph on the right. In this case, the size of the marker is the default size.

- Markers are not connected by a line in the graph on the right.

2.5 Formatting of the Title

Title is an important element of a well-formatted graph because it summarizes the contents of the graph. The appearance of the title is also equally important. It is usually centre aligned and positioned on top of the graph. The Scilab configuration for placement of titles is shown in the following with the help of an example.

```
//Load the *.sci file that contains plotting functions
exec('plot.sci',-1);
x = 0:0.01:2*%pi;                    //Independent variable
y = sin(x);                          //Dependent variable

subplot(121)
plot2d(x,y,2)
title('Sine Wave');                       //Default title

subplot(122)
plot2d(x,y,2)
title('Sine Wave','fontsize',4,'color','red','fontname',
'Times Bold')                             //Formatted title
```

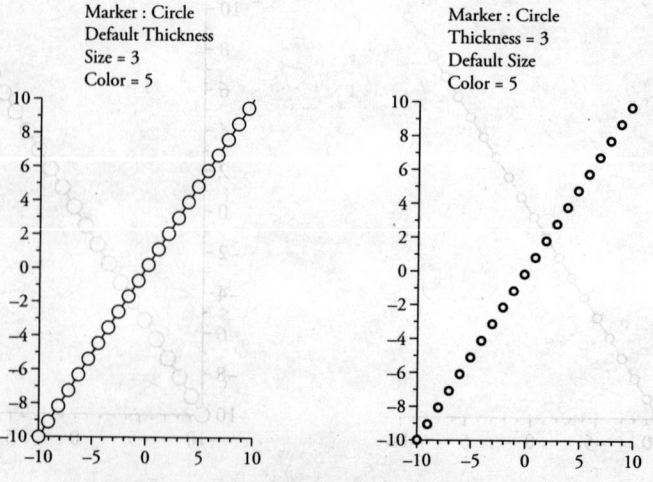

Figure 2.17   Different thickness of the marker

In *Figure 2.18*,

- Graph on the left has default settings for the title.
- Title of the graph on the right has been formatted. In this graph,
    - The font size can be increased/decreased (`'fontsize',4`).
    - The colour of the text can be changed (`'colour','red'`).
    - The font style can also be controlled using the desired font name (`'fontname','Times Bold'`).
- It is possible to put a background colour and colour the edge of the title text.

It is also possible to create a multiple line title in Scilab. For the same aforementioned example, the following formatting command will create a two line title. The graph is shown in *Figure 2.19*.

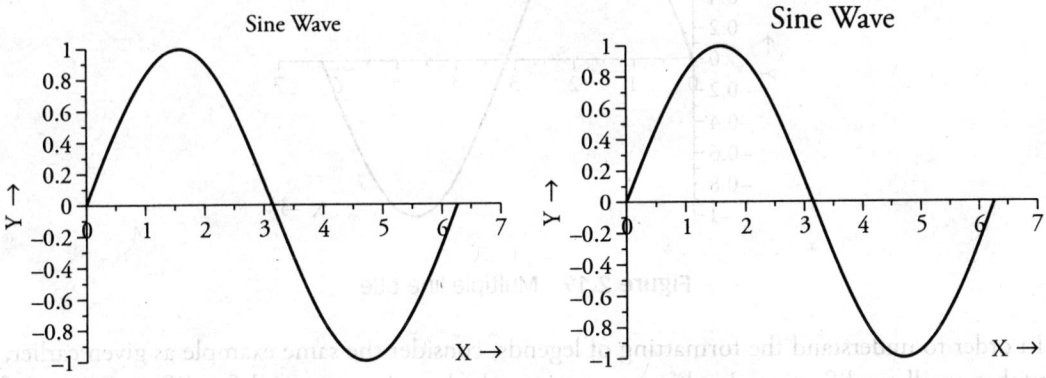

Figure 2.18   Formatting of the title of the graph

```
title({'Sine Wave';'Y =
sin(X)'},'fontsize',4,'color','black','fontname','Courri
er');
```

At various places in the subsequent chapters, titles have been generated using LaTeX syntax to show expressions and equations. However, details of LaTeX editor and typesetting is beyond the scope of this book. Therefore, the formatting of titles in subsequent chapters has been omitted to avoid confusion.

2.6 Formatting of the Legend

Graph legend is a text object that is used as key reference to the data being presented by the graph. The following user-defined function has been written to control common properties of the legend. The use of this function has been explained as follows.

```
function set_my_legend(size, style, color)
a=get("current_axes");
legend=a.children(1);
legend.font_size = size;
legend.font_style = style;
legend.font_color = color;
legend.line_mode = "off"
legend.legend_location = "in_lower_left"
endfunction
```

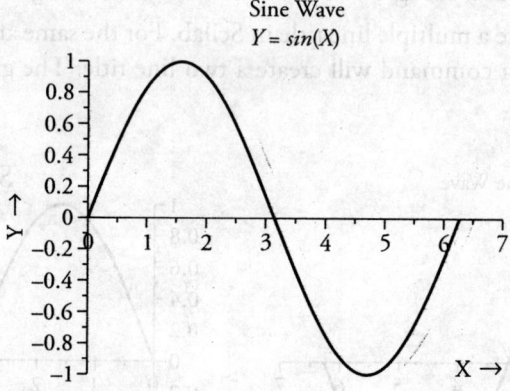

Figure 2.19   Multiple line title

In order to understand the formatting of legends, consider the same example as given earlier, with a small modification. In this case, two graphs have been created for different value of amplitude of the function. The Scilab program is written as follows.

```
//Load the *.sci file that contains plotting functions
exec('plot.sci',-1);

subplot(121)
a = 1;
x = 0:0.01:2*%pi;
y = a*sin(x);
plot2d(x,y,2)

subplot(121)
a = 2;
y = a*sin(x);
plot2d(x,y,5)

legend(['a = 1';'a = 2']);
```

```
subplot(122)
a = 1;
y = a*sin(x);
plot2d(x,y,2)

subplot(122)
a = 2;
y = a*sin(x);
plot2d(x,y,5)

legend(['a = 1';'a = 2']);
set_my_legend(5,4,13);
```

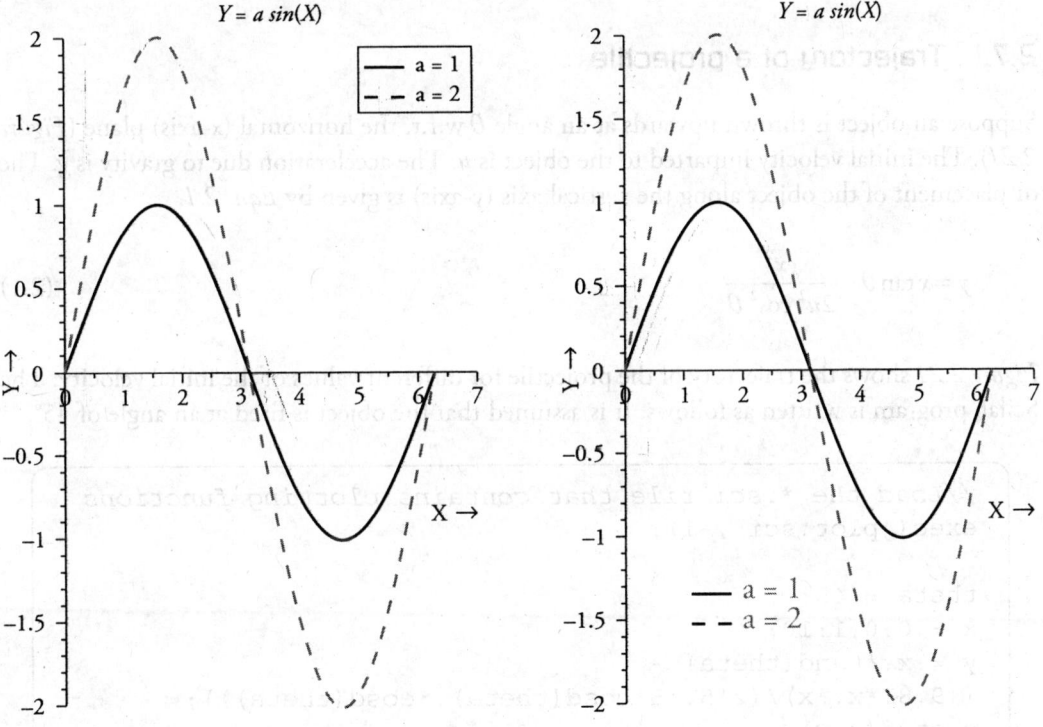

Figure 2.20 Formatting of the legends of the graph

In *Figure 2.20*,

- Graph on the left has the legend under the default configuration.
- Graph on the right has a formatted legend. It is possible to
  - Remove the border around the legend (`line_mode = "off"`).
  - Change the colour used for the text of the legend (`font_colour = colour`).

- Control the font size of the legend (`font_size = size`).
- Change the font style of the legend (`font_style = style`).
- Change the placement of the legend (`legend_location="in_lower_left"`).

2.7 Applications

The objective of this section is to apply the plotting skills developed in previous sections to various problems of physics. In the following sub-sections, various graphs depicting the experimental results and the advanced problems of physics have been shown. The reader is encouraged to reproduce these graphs using the formatting tools discussed in the previous sections.

2.7.1 Trajectory of a projectile

Suppose an object is thrown upwards at an angle θ w.r.t. the horizontal (x-axis) plane (*Figure 2.21*). The initial velocity imparted to the object is u. The acceleration due to gravity is g. The displacement of the object along the vertical axis (y-axis) is given by *Eqn. 2.1*.

$$y = x\tan\theta - \frac{gx^2}{2u^2\cos^2\theta} \qquad (2.1)$$

Figure 2.21 shows the trajectory of the projectile for different values of the initial velocity. The Scilab program is written as follows. It is assumed that the object is fired at an angle of 45°.

```
//Load the *.sci file that contains plotting functions
exec('plot.sci',-1);

theta = 45;
x = 0:0.1:10;
y = x.*tand(theta) -
((9.81*x.*x)/(2*5.*5*cosd(theta).*cosd(theta)));
plot2d(x,y)

y = x.*tand(theta) -
((9.81*x.*x)/(2*7.*7*cosd(theta).*cosd(theta)));
plot2d(x,y)

y = x.*tand(theta) -
((9.81*x.*x)/(2*9.*9*cosd(theta).*cosd(theta)));
plot2d(x,y)
```

2.7.2 Superposition of collinear harmonic oscillations

Consider that a particle is subjected to two collinear simple harmonic oscillations having the same frequency simultaneously. The displacement of the particle due to these collinear oscillations is given by *Eqns 2.2–2.3.*

$$x_1 = A_1 \cos\left(\omega t + \varphi_1\right) \tag{2.2}$$

$$x_2 = A_2 \cos\left(\omega t + \varphi_2\right) \tag{2.3}$$

From the superposition principle, the resultant displacement of the particle will be given by *Eqns. 2.4–2.5.*

$$x = A_1 \cos\left(\omega t + \varphi_1\right) + A_2 \cos\left(\omega t + \varphi_2\right) \tag{2.4}$$

$$x = A\cos\left(\omega t + \delta\right) \tag{2.5}$$

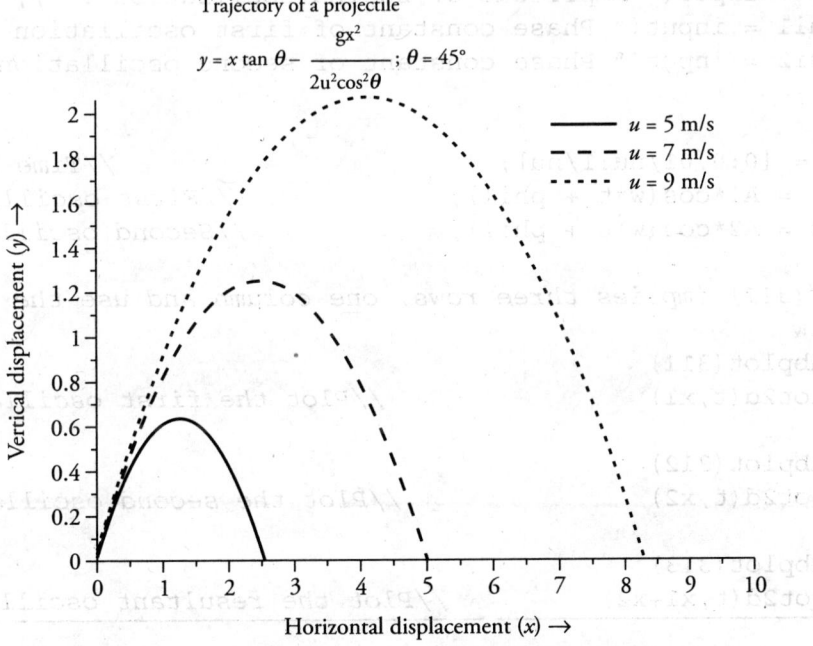

Figure 2.21 Trajectory of a projectile motion

In *Eqn. 2.5,*

- The amplitude is given by *Eqn. 2.6.*

$$A = \sqrt{A_1^2 + A_2^2 + 2A_1 A_2 \cos\left(\varphi_2 - \varphi_1\right)} \tag{2.6}$$

- The phase angle is given by *Eqn. 2.7.*

$$\delta = \tan^{-1}\left(\frac{A_1 \sin\varphi_1 + A_2 \sin\varphi_2}{A_1 \cos\varphi_1 + A_2 \cos\varphi_2} \right) \tag{2.7}$$

The following Scilab program plots the resultant motion of the particle assuming that,

- The phase difference between two oscillations is $\varphi_2 - \varphi_1 = \pi$

- As a result, the maximum amplitude of the resultant displacement will be equal to $A_1 - A_2$.

```
//Load the *.sci file that contains plotting functions
exec('plot.sci',-1);


nu = input(" Enter frequency (in Hertz) : ");
w = 2*%pi*nu;
A1 = input(" Amplitude of first oscillation : ");
A2 = input(" Amplitude of second oscillation : ");
phi1 = input(" Phase constant of first oscillation : ");
phi2 = input(" Phase constant of second oscillation :
");

t = [0:0.01/nu:1/nu];                    //Time range
x1 = A1*cos(w*t + phi1);         //First oscillation
x2 = A2*cos(w*t + phi2);         //Second oscillation

//(311) implies three rows, one column and use the first
row
subplot(311)
plot2d(t,x1)                     //Plot the first oscillation

subplot(312)
plot2d(t,x2)                     //Plot the second oscillation

subplot(313)
plot2d(t,x1+x2)          //Plot the resultant oscillation
```

The input parameters of this program are as follows.

```
Enter frequency (in Hertz): 500;
Amplitude of first oscillation: 5;
Amplitude of second oscillation: 2;
Phase constant of first oscillation: 0;
Phase constant of second oscillation: %pi
```

Figure 2.22 shows the two harmonic oscillations and the resultant oscillation of the particle.

2.7.3 Beats

A particle is simultaneously subjected to two collinear simple harmonic oscillations having different frequencies. Suppose the displacement of the particle due to these collinear oscillations is given by *Eqn. 2.8* and *Eqn. 2.9*.

$$x_1 = A_1 \cos \omega_1 t \tag{2.8}$$

$$x_2 = A_2 \cos \omega_2 t \tag{2.9}$$

It is assumed that $\omega_2 > \omega_1$ and $A_2 > A_1$. From the superposition principle, the resultant displacement of the particle will be given by *Eqn. 2.10*.

$$x = x_1 + x_2 = A_1 \cos \omega_1 t + A_2 \cos \omega_2 t \tag{2.10}$$

Substituting the following notation,

- Average frequency $= \omega_a = \dfrac{1}{2}(\omega_2 + \omega_1)$

- Modulation frequency $= \omega_m = \dfrac{1}{2}(\omega_2 - \omega_1)$

- $\omega_1 = \omega_a - \omega_m$

- $\omega_2 = \omega_a + \omega_m$

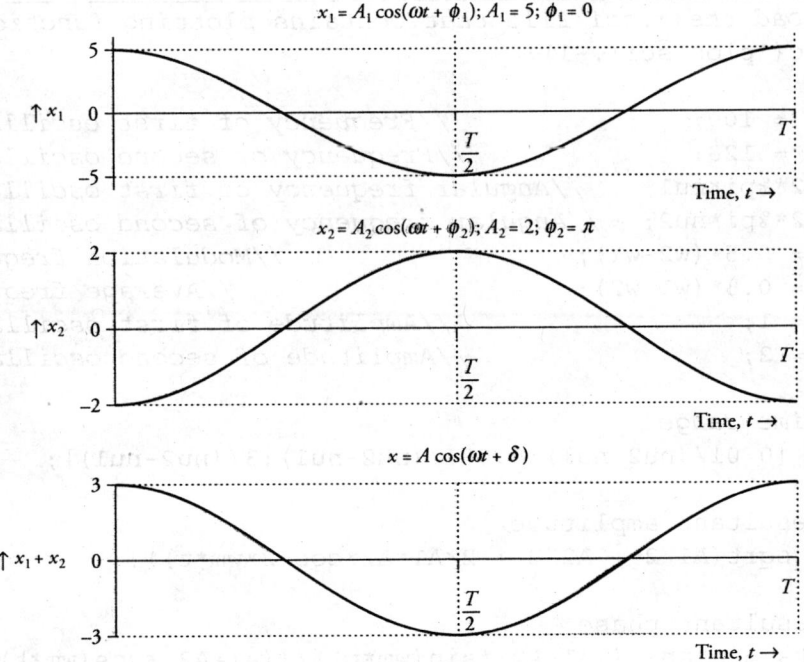

Figure 2.22 Superposition of collinear harmonic oscillations

The resultant displacement will be given by *Eqn. 2.11*,

$$x = A_1 \cos\left(\omega_a - \omega_m\right)t + A_2 \cos\left(\omega_a + \omega_m\right)t$$

$$x = \left(A_1 + A_2\right)\cos\omega_m t \cos\omega_a t + \left(A_1 - A_2\right)\sin\omega_m t \sin\omega_a t$$

$$x = A_m \cos\left(\omega_a t + \delta_m\right) \tag{2.11}$$

In *Eqn. 2.11*,

- The amplitude is given by (2.12)

$$A_m = \sqrt[2]{A_1^2 + A_2^2 + 2A_1 A_2 \cos\left(2\omega_m t\right)} \tag{2.12}$$

- The phase angle is (*Eqn. 2.13*).

$$\delta_m = \tan^{-1}\left[\frac{\left(A_1 - A_2\right)}{\left(A_1 + A_2\right)}\tan\omega_m t\right] \tag{2.13}$$

The following Scilab program plots the resultant oscillation of the particle. The graph is shown in *Figure 2.23*.

```
//Load the *.sci file that contains plotting functions
exec('plot.sci',-1);


nu1 = 100;                  //Frequency of first oscillation
nu2 = 120;                  //Frequency of second oscillation
w1=2*%pi*nu1;     //Angular frequency of first oscillation
w2=2*%pi*nu2;     //Angular frequency of second oscillation
wm = 0.5*(w2-w1);                   //Modulation frequency
wa = 0.5*(w1+w2);                      //Average frequency
A1 = 1;                  //Amplitude of first oscillation
A2 = 2;                  //Amplitude of second oscillation

//Time range
t = [0.01/(nu2-nu1):0.001/(nu2-nu1):3/(nu2-nu1)];

//Resultant amplitude
A = sqrt(A1^2 + A2^2 + 2*A1*A2*cos(2*wm*t));

//Resultant phase
delta = atan((((A1-A2)*sin(wm*t))/((A1+A2)*cos(wm*t))));

x1 = A1*cos(w1*t);
```

```
x2 = A2*cos(w2*t);
x = A.*cos((wm*t) + delta);

subplot(311)
plot2d(t,x1)

subplot(312)
plot2d(t,x2)

subplot(313)
plot2d(t,A)
plot2d(t,-A)
plot2d(t,x1+x2)
```

It should be noticed in *Figure 2.23* that

- The phenomena of 'beats' is exhibited when the frequencies of two collinear oscillations are nearly equal.

- The resultant oscillation is periodic having angular frequency ω_a.

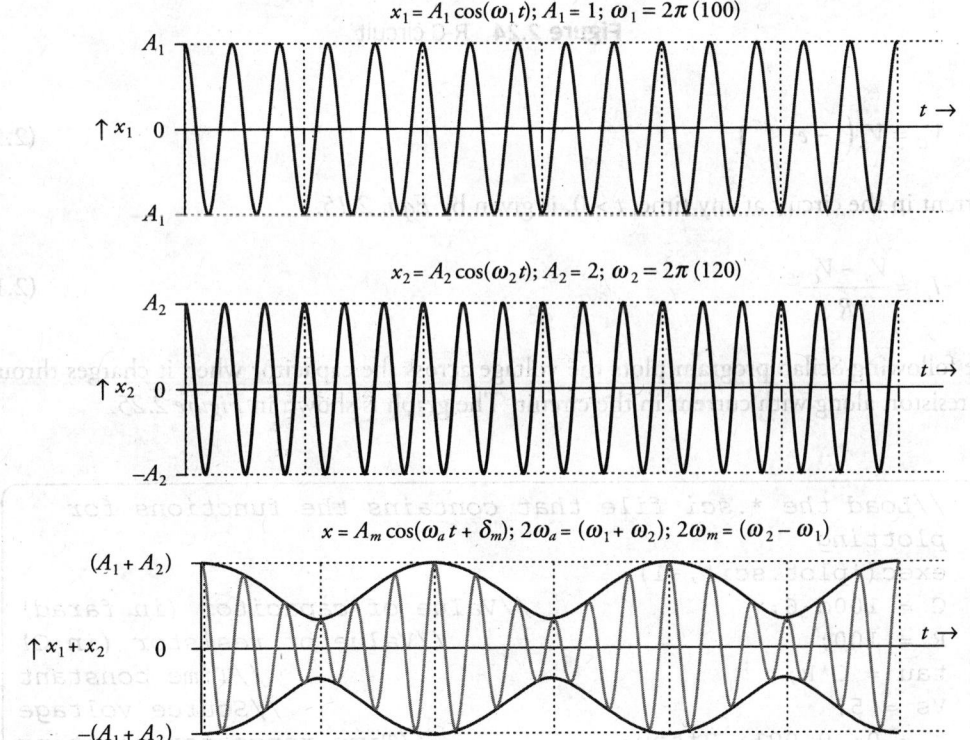

Figure 2.23 Graph showing the phenomena of beats

- The amplitude of the resultant oscillation is maximum ($= A_1 + A_2$)when two interfering waves are in the same phase, i.e., their phase differ by an even multiple of π.

- The amplitude of the resultant oscillation is minimum ($= A_2 - A_1$)when the two interfering waves are opposite in phase, i.e., their phase differ by an odd multiple of π.

2.7.4 R-C Circuit

An interesting application of simple plotting in Scilab is to plot the voltage wave for charging the capacitor in a resistor–capacitor circuit. *Figure 2.24* shows a series RC circuit comprising one resistor and one capacitor. They are driven by a d.c. voltage source. In this circuit, the capacitor (C) charges through resistor (R) when a voltage (V_s) is applied. The charging continues till the voltage across the capacitor reaches the source voltage. Assuming that the capacitor is fully discharged at time $t = 0$, the voltage across it at any time $t > 0$, is given by *Eqn. 2.14*.

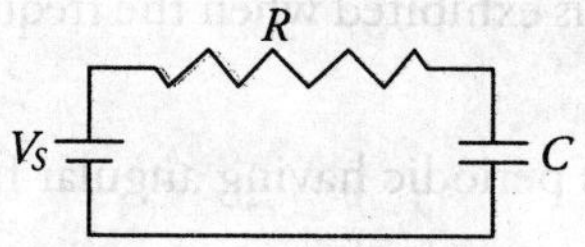

Figure 2.24   R-C circuit

$$V_C = V_S \left(1 - e^{-t/RC}\right) \tag{2.14}$$

Current in the circuit at any time, $t > 0$, is given by *Eqn. 2.15*.

$$I_C = \frac{V_S - V_C}{R} \tag{2.15}$$

The following Scilab program plots the voltage across the capacitor when it charges through the resistor, along with current in the circuit. The graph is shown in *Figure 2.25*.

```
//Load the *.sci file that contains the functions for
plotting
exec('plot.sci',-1);
C = 100d-6;                   //Value of capacitor (in farad)
R = 100;                      //Value of resistor (in Ω)
tau = C*R;                         //Time constant
Vs = 5;                            //Source voltage
t = 0: 0.001: 7*tau;          //Time range for plotting
```

```
V = Vs * (1 - exp(-t/tau));    //Voltage across capacitor
i = (Vs-V)/R;                  //Current in the circuit
subplot(211)
plot2d(t/tau,V/Vs);

subplot(212)
plot2d(t/tau,i/max(i));
```

It should be noticed that

- The rate of charging of the capacitor is faster in the beginning of the curve.

- One time constant ($\tau = RC$) is the time in which the capacitor voltage reaches 63.2% of the source voltage. In this time, the current in the circuit reduces to 36.7% of its peak value.

- The capacitor is never fully charged due to the energy stored in it; however, for all practical purposes, it is considered to be fully charged after a time equal to approximately 5τ. The current at this point is almost zero.

2.7.5 R-L Circuit

Consider the diagram in *Figure 2.26*. At time $t = 0$, the nodes '1' and '2' are connected and current flows through the inductor until $t = 2$. The inductor charges during this period through the resistor. The current flowing through the inductor during this period is (*Eqn. 2.16*),

$$I = \frac{V}{200}\left(1 - e^{-0.4t}\right) \tag{2.16}$$

At time $t = 2$, the connection between '1' and '2' is opened. The nodes '2' and '3' are now joined. The inductor discharges through the resistor. The current flowing through the inductor is (*Eqn. 2.17*),

$$I = I_{max}e^{-0.5t} \tag{2.17}$$

The following Scilab program plots the charging and discharging cycle of the inductor. The graph is shown in *Figure 2.27*.

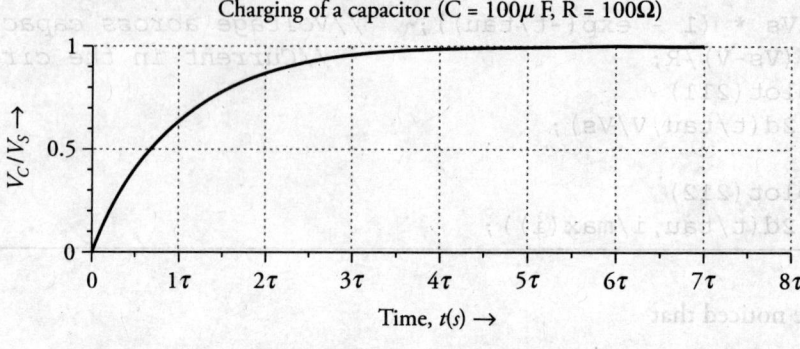

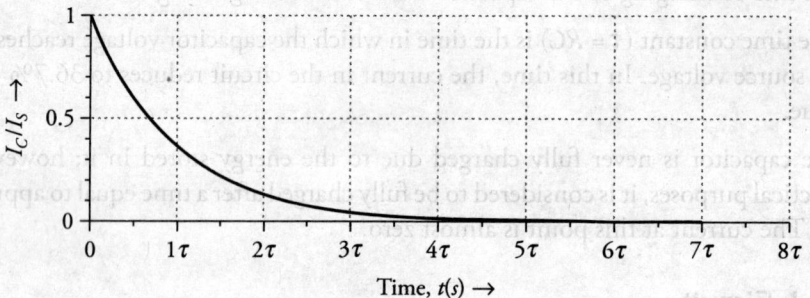

Figure 2.25 Charging of a capacitor

```
//Load the *.sci file that contains plotting functions
exec('plot.sci',-1)

R1 = 200;                       //Resistance during 0<t<2
R2 = 250;                       //Resistance during t>2
L = 500;                        //Value of inductance
tau_1 = R1/L;                   //Time constant during 0<t<2
tau_2 = R2/L;                   //Time constant during t>2
Vs = 20;                        //Source voltage
t1 = 0: 0.1: 2;                 //Time duration 0<t<2
i1 = (Vs/R1)*(1-exp(-tau_1*t1));    //Charging current
plot2d(t1,i1);

t2 = 2.1:0.1:8;                 //Time duration 2<t<8
i2=max(i1)*exp(-tau_2*(t2-max(t1)));//Discharge current
plot2d(t2,i2);
```

2.7.6 Maximum power transfer theorem

This theorem presents an interesting method to analyse electrical circuits. It has several applications such as impedance matching between audio amplifiers and loud speakers. The method has been explained as follows.

Consider the circuit in *Figure 2.28*.

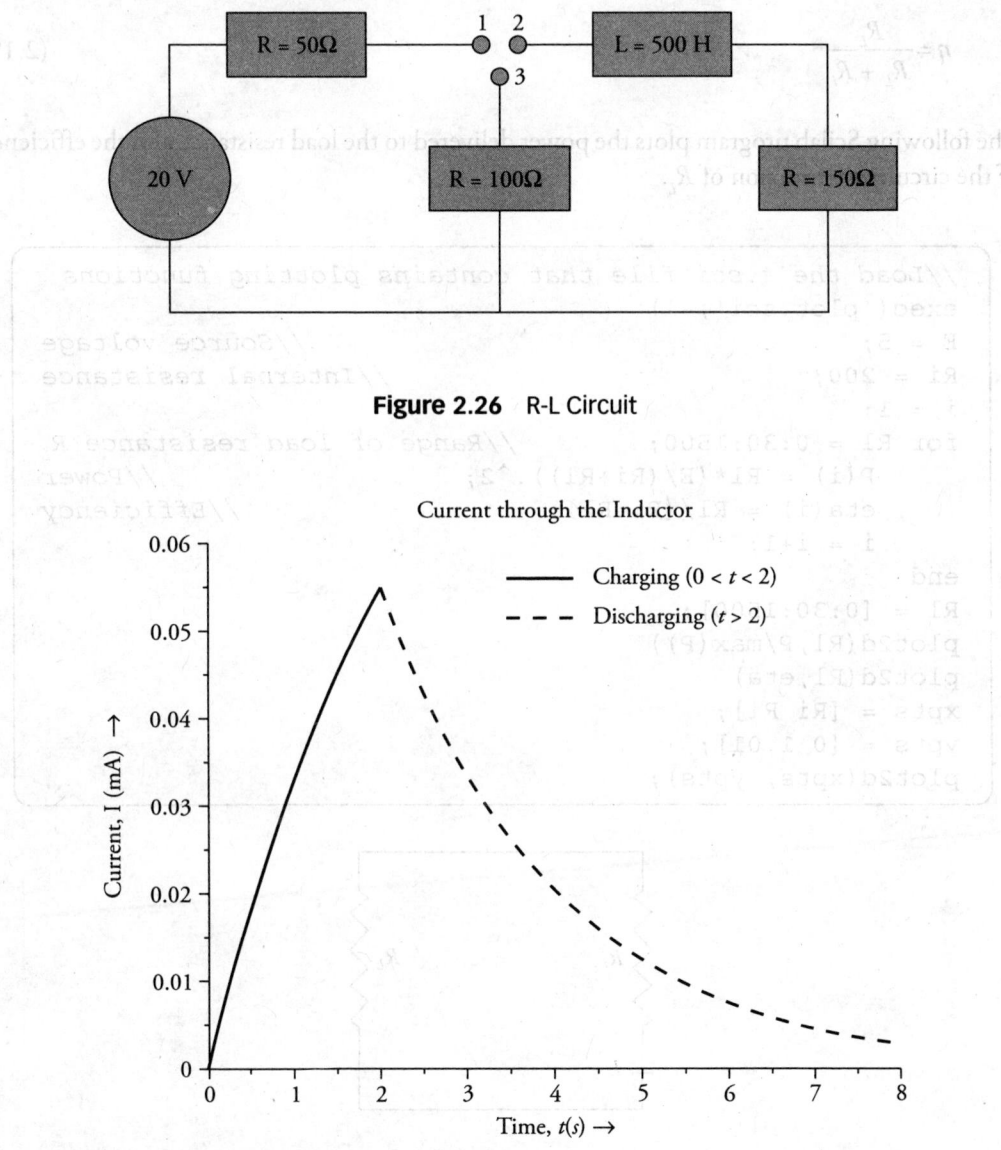

Figure 2.26   R-L Circuit

Figure 2.27   Charging and discharging of an inductor

Power delivered to the load resistance is given by *Eqn. 2.18*.

$$P = \left(\frac{E}{R_L + R_i}\right)^2 R_L \tag{2.18}$$

Efficiency of the circuit is given by *Eqn. 2.19*.

$$\eta = \frac{R_L}{R_L + R_i} \tag{2.19}$$

The following Scilab program plots the power delivered to the load resistance and the efficiency of the circuit as a function of R_L.

```
//Load the *.sci file that contains plotting functions
exec('plot.sci',-1)
E = 5;                          //Source voltage
Ri = 200;                       //Internal resistance
i = 1;
for Rl = 0:30:1500;      //Range of load resistance R_L
    P(i) = Rl*(E/(Ri+Rl)).^2;           //Power
    eta(i) = Rl/(Ri+Rl);            //Efficiency
    i = i+1;
end
Rl = [0:30:1500];
plot2d(Rl,P/max(P))
plot2d(Rl,eta)
xpts = [Ri Ri];
ypts = [0 1.01];
plot2d(xpts, ypts);
```

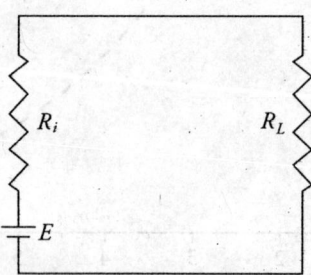

Figure 2.28 Circuit for maximum power transfer theorem

It should be noticed in *Figure 2.29* that

- Power delivered to R_L is maximum when $R_L = R_i$.
- Efficiency of the circuit is 50% when $R_L = R_i$.
- Efficiency of the circuit is maximum when R_L is effectively infinite (or R_i is zero).

2.7.7 Diode characteristics

The diode equation is given by *Eqn. 2.20*.

$$I = I_s \left(e^{(V/\eta V_T)} - 1 \right) \tag{2.20}$$

In *Eqn. 2.20*, the reverse saturation current is I_s. Its dependence on temperature is given by *Eqn. 2.21*.

$$I_s(T_2) = I_s(T_1) e^{\{k_s(T_2 - T_1)\}} \tag{2.21}$$

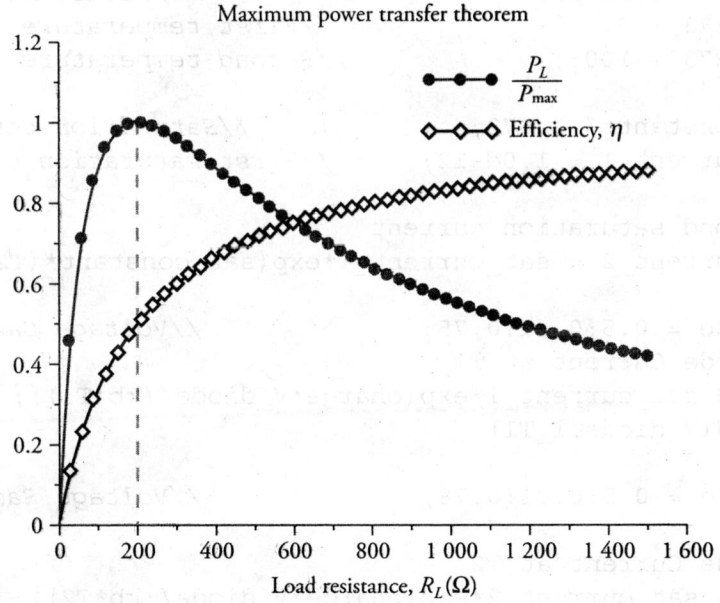

Figure 2.29   Graph for maximum power transfer theorem

It is assumed that

- Saturation constant = $k_s = 0.072/°C$

- Empirical constant is η. It is assumed to be 1.

- Thermal voltage $= V_T = \dfrac{kT}{q}$.

- Boltzmann constant $= k$

- Electronic Coulomb charge $= q$

- Temperature (in Kelvin) $= T$

The following Scilab program plots the diode characteristic for two different temperatures. The graph is shown in *Figure 2.30*.

```
//Load the *.sci file that contains plotting functions
exec('plot.sci',-1)

kb = 1.38d-23;                        //Boltzmann constant
charge = 1.6d-19;                     //Coulomb charge
T1 = 273 + 0;              //First temperature (in K)
T2 = 273 + 100;           //Second temperature (in K)

sat_constant = 0.072;               //Saturation constant
sat_current_1 = 1.0d-12;      //First saturation current

//Second saturation current
sat_current_2 = sat_current_1*exp(sat_constant*(T2-T1));

V_diode = 0.5:0.01:0.75;            //Voltage Sampling
// Diode Current at T1
l_T1 = sat_current_1*exp(charge*V_diode/(kb*T1));
plot2d(V_diode,l_T1)

V_diode = 0.5:0.01:0.78;            //Voltage Sampling

//Diode Current at T2
l_T2 = sat_current_2*exp(charge*V_diode/(kb*T2));
plot2d(V_diode,l_T2)
```

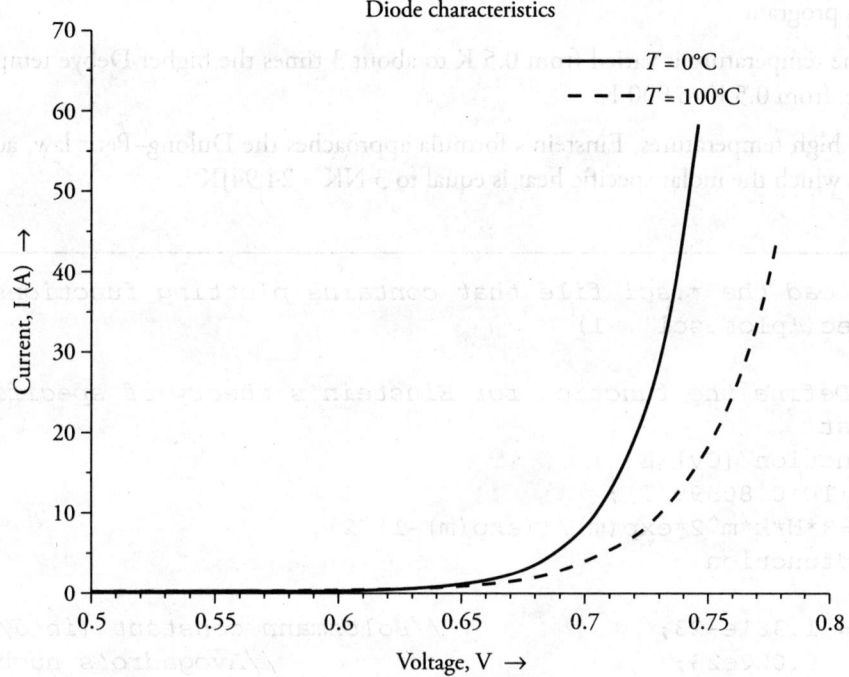

Figure 2.30   Diode characteristics at different temperatures

2.7.8 Specific heat of solids

According to Einstein's theory of specific heat, the molar specific heat of a solid is given by *Eqn. 2.22* and *Eqn. 2.23*.

$$C_v = 3Nk\left(\frac{\varepsilon}{kT}\right)^2 \frac{e^{\frac{\varepsilon}{kT}}}{\left(e^{\frac{\varepsilon}{kT}} - 1\right)^2} \tag{2.22}$$

$$C_v = 3Nk\left(\frac{T_E}{T}\right)^2 \frac{e^{\frac{T_E}{T}}}{\left(e^{\frac{T_E}{T}} - 1\right)^2} \tag{2.23}$$

The following Scilab program has been written to plot the variation of molar specific heat (C_v) of copper and sodium as a function of temperature. The graph is shown in *Figure 2.31*. It is given that the Debye's temperature (T_D) for copper and sodium is 340 K and 157 K, respectively.

In this program

- The temperature is varied from 0.5 K to about 3 times the higher Debye temperature, i.e. from 0.5 K to 900 K.

- At high temperatures, Einstein's formula approaches the Dulong–Petit law, according to which the molar specific heat is equal to 3 NK = $24.94 \mathrm{JK}^{-1}$.

```
//Load the *.sci file that contains plotting functions
exec('plot.sci',-1)

//Define the function for Einstein's theory of specific
heat
function [Cv]=E(T)
m=(TD*0.80599/T);
Cv=3*N*k*m^2*exp(m)/((exp(m)-1)^2);
endfunction


k = 1.381e-23;              //Boltzmann constant (in J/K)
N = 6.022e23;                      //Avogadro's number
n =input("Enter the number of elements for the graph : ")
for i =1:n;
    element = input("Name of the element : ","string");
    TD = input("Debye temperature (in Kelvin)? ");
    x =[0.5 : 0.1 : 900.0];
    fplot2d(x,E);
    A(i)=string(element);
end
legend(A);
```

The input parameters of this program are written as follows:

```
Enter the number of elements for the graph: 2
Name of the element: Copper
Debye temperature (in Kelvin)? 340
Name of the element: Sodium
Debye temperature (in Kelvin)?: 157
```

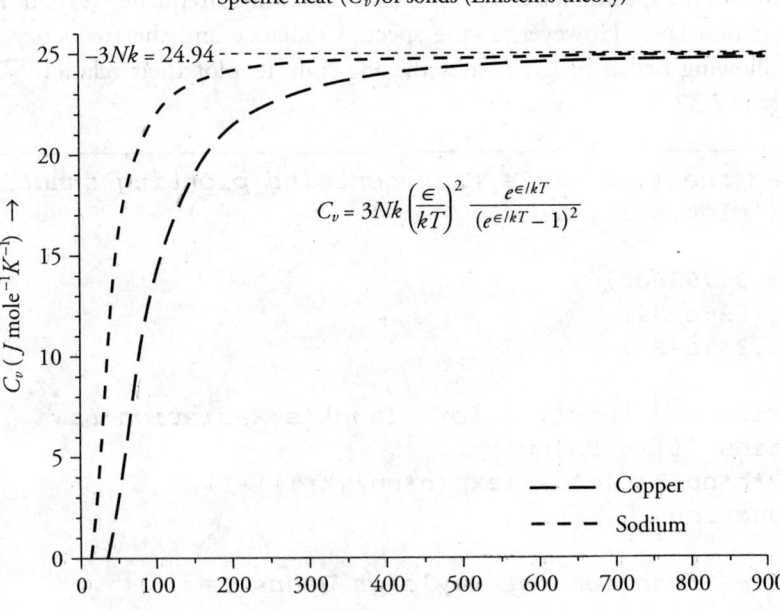

Figure 2.31   Specific heat of solids

2.7.9 Spectral radiance of a blackbody radiation

The expression for spectral radiance of a blackbody is given by *Eqns. 2.24–2.26.*

- Planck's law

$$S_{P-L}\left(v\right)=\frac{2hv^{3}}{c^{2}}\frac{1}{exp\left(\dfrac{hv}{kT}\right)-1}$$

(2.24)

- Rayleigh Jeans law

$$S_{R-L}\left(v\right)=\frac{2v^{2}kT}{c^{2}}$$

(2.25)

- Wien's law

$$S_{Wien's}\left(v\right)=\frac{2hv^{3}}{c^{2}}\frac{1}{exp\left(\dfrac{hv}{kT}\right)}$$

(2.26)

It is desired to plot the spectral radiance ($S(v)$) (as a function of frequency (v) and thus compare the three radiation laws. However, as the spectral radiance and the frequency span a long range, the following Scilab program uses the log-scale to plot their relation. The graph is shown in *Figure 2.32*.

```
//Load the *.sci file that contains plotting functions
exec('plot.sci',-1)

c = 2.997925d8;
h = 6.626e-34;
k = 1.381e-23;

//Define the function for Planck's Radiation Law
function [f] = PL(nu)
f = 2*h*nu^3/((c*c)*(exp(h*nu/(k*T))-1))
endfunction

//Define function for Rayleigh Jeans Law
function [f1] = RL(nu)
f1 = 2*nu*nu*k*T/(c*c)
endfunction

//Define the function for Wien's law
function [f2] = WL(nu)
f2 = 2*h*nu^3/(c*c*exp(h*nu/(k*T)))
endfunction

T = 2500;
x = [0.01*c*T/0.0029 : 0.01*c*T/0.0029 : 3*c*T/0.0029];

fplot2d(x,PL,style=[color("red")],logflag="ll");
fplot2d(x,RL,style=[color("green")],logflag="ll");
fplot2d(x,WL,style=[color("blue")],logflag="ll");
```

2.7.10 Miller indices

Miller indices are a convention used to describe the orientation of a plane or a family of planes within a lattice in relation to the unit cell. This is useful to quantitatively analyse problems in material science. However, students find it difficult to imagine the orientation of lattice planes of symmetry in a unit cell.

The following Scilab program gives examples of how to plot lattice planes in a unit cell. The corresponding plots are shown in *Figures 2.33–2.35*. The reader is advised to change the coordinates and construct planes of their choice.

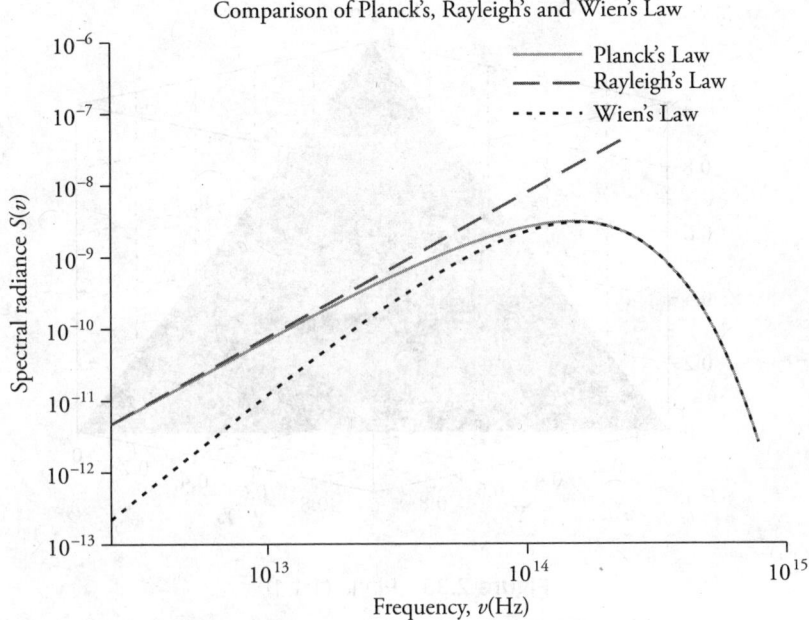

Figure 2.32   Spectral radiance of a black body

- ***Example 1***: Plane (1 1 1)

```
//Define the x-, y-, z- coordinates of the corners
x = [1 0 0 1];
y = [0 1 0 0];
z = [0 0 1 0];
param3d(x,y,z,alpha=75);
a =get("hdl");
a.foreground = 5;
a.thickness=5;
a.polyline_style = 5;

a = gca();
a.grid = [2 2 2];
a.font_size = 3;
a.labels_font_color = 1;
a.thickness = 2;
a.grid = [2 2 2];
a.grid_thickness = [2,2,2];
```

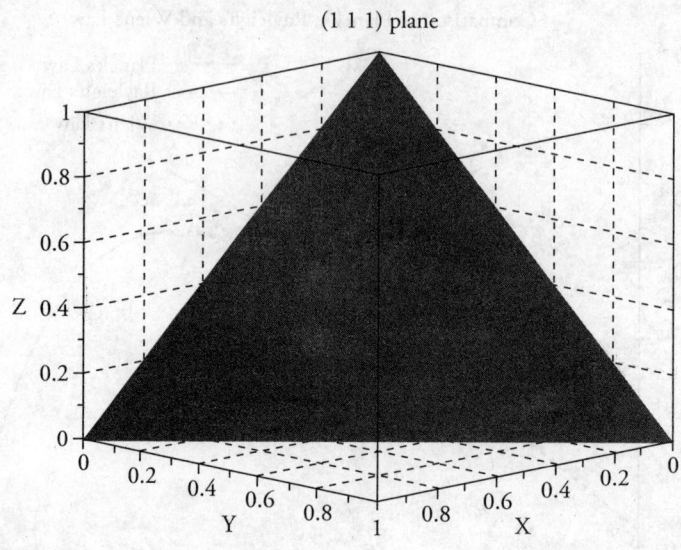

Figure 2.33   Plane (1 1 1)

- *Example 2*: Plane (1 0 0)

```
//Define the x-, y-, z- coordinates of the corners
x = [1 1 1 1 1];
y = [1 0 0 1 1];
z = [0 0 1 1 0];
param3d(x,y,z,alpha=75);
a = get("hdl");
a.foreground = 5;
a.thickness=5;
a.polyline_style = 5;

a = gca();
a.grid = [2 2 2];
a.font_size = 3;
a.labels_font_color = 1;
a.thickness = 2;
a.grid = [2 2 2];
a.grid_thickness = [2,2,2];
```

- *Example 3*: Plane (0 1 1)

```
//Define the x-, y-, z- coordinates of the corners
x = [0 1 1 0 0];
```

```
y = [0 0 1 1 0];
z = [1 1 0 0 1];
param3d(x,y,z,alpha=75);
a = get("hdl");
a.foreground = 5;
a.thickness = 5;
a.polyline_style = 5;

a = gca();
a.grid = [2 2 2];
a = get("current_axes")
a.font_size = 3;
a.labels_font_color = 1;
a.thickness = 2;
a.grid = [2 2 2];
a.grid_thickness = [2,2,2];
```

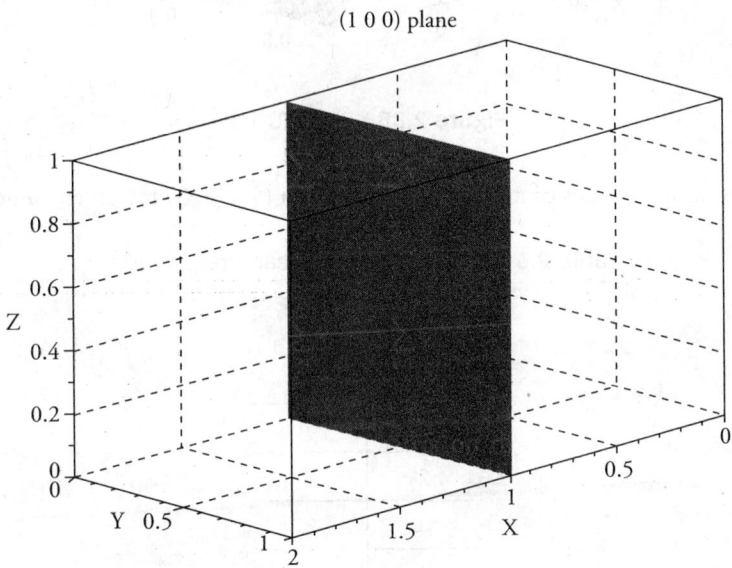

Figure 2.34 Plane (1 0 0)

2.7.11 Linear interpolation

In physics, one generally deals with data set spread over a range of variables. It represents the values of a function at certain discrete values of the independent variable. It is often desired to estimate the value of the function at some intermediate value. Interpolation is used to construct new data points within a pair of known discrete data points. It is an easy way to fill in the gaps in a data set. Linear interpolation uses the method of curve fitting using linear polynomials.

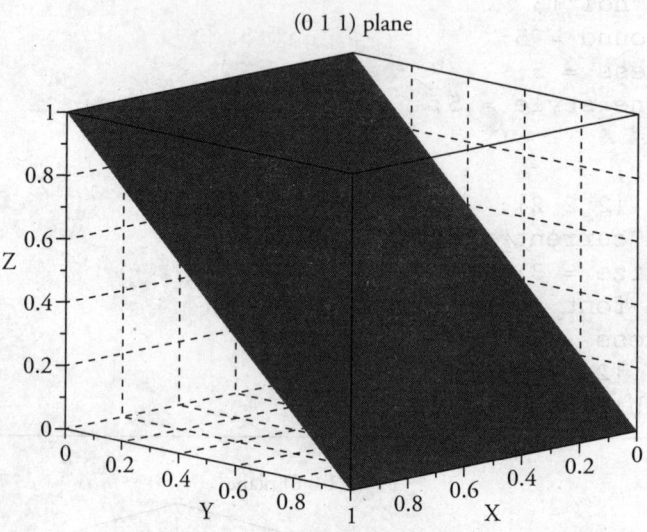

Figure 2.35   Plane (0 1 1)

Table 2.6 gives sample values of an unknown function $(Y = f(X))$ for given values of 'X'.

Table 2.6   Sample data for linear interpolation

X	Y
0	0
1	1
2	3
3	3
4	3
5	4
6	5.1
7	5.1
8	8.5
9	9
10	10

In physics, one generally deals with discrete values of a range of variables. It represents the value of a function at certain discrete values of the said variable. It is often useful to estimate the value of the function at some interval. Interpolation is used to

It is desired to determine the value of the function at the following values of 'X',

[1.2.2.5.4.7 6.0 8.8]

The following Scilab program plots the given data. The graph is shown in *Figure 2.36* where the data points have been marked with 'filled diamonds'. It also determines the values of the dependent variable at in-between desired points using the built-in Scilab function 'interpln' and plots them on the same graph using a different marker (circle).

```
x = 0:10;
y = [0 1 3 3 3 4 5.1 5.1 8.5 9 10];

data = [x;y]                         //Given data set
x_new=[1.2 2.5 4.7 6.0 8.8]; //New values of X-variable
y_new = interpln(data,x_new);        //Interpolated values

plot2d(data(1,:),data(2,:))          //Plot original data
plot2d(x_new,y_new)                  //Plot interpolated data
```

The interpolated values of the function at given points are

[1.4 3 3.7 5.1 8.9]

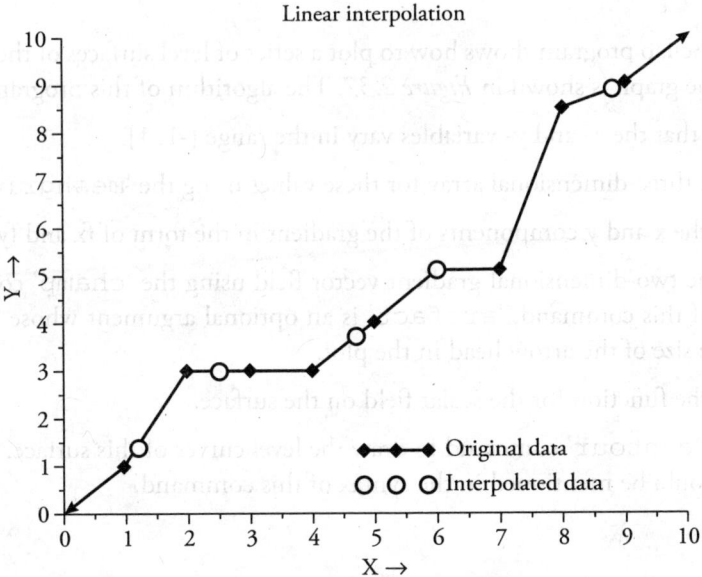

Figure 2.36 Linear interpolation

2.7.12 Gradient of a scalar field

It is a vector whose magnitude is the rate of change of the scalar field at the point of gradient. This vector field points in the direction of the maximum rate of increase of the scalar field. There are several applications of gradient in physics. For example,

- Determination of rate of change of electric field due to a point charge, as one moves away from the charge (or towards the charge).

- Determination of rate of change of potential with respect to the displacement.

- Determination of velocity field of a fluid flowing through a pipe.

- Determination of the rate at which temperature (a scalar field) changes as one moves away from a fireplace placed at one corner of a room.

Scilab is a convenient tool to represent the scalar field with a series of level surfaces (such as equipotential and isothermal surface) each having a constant value. It can also be used to visualize the magnitude and direction of rate of change of the scalar field at different points on this surface. For example, consider the two-dimensional scalar field given in *Eqn. 2.27*.

$$f(x, y) = x^2 + y^2 \tag{2.27}$$

The gradient of this scalar function in rectangular coordinates is given by *Eqn. 2.28*.

$$\nabla f(x, y) = \frac{\partial f(x, y)}{\partial x} \hat{i} + \frac{\partial f(x, y)}{\partial y} \hat{j} = 2x \hat{i} + 2y \hat{j} \tag{2.28}$$

The following Scilab program shows how to plot a series of level surfaces of the scalar field and its gradient. The graph is shown in *Figure 2.37*. The algorithm of this program is as follows.

- Assume that the x- and y- variables vary in the range [-1, 1].

- Create a three-dimensional array for these values using the 'meshgrid' command.

- Define the x and y components of the gradient in the form of fx and fy, respectively.

- Draw the two-dimensional gradient vector field using the 'champ' command. In the syntax of this command, 'artfact' is an optional argument whose value is used to scale the size of the arrow head in the plot.

- Define the function for the scalar field on the surface.

- Use the 'contour' command to draw the level curves of this surface. The number of levels should be mentioned in the syntax of this command.

```
x = -1:0.1:1;
y = -1:0.1:1;
[X,Y] = meshgrid(x,y);
fx = 2.*X;
fy = 2.*Y;
champ(x,y,fx',fy',arfact=2);

function z=my_surface(x, y)
z = (x^2+y^2)
endfunction

contour(x,y,my_surface,5)
```

The gradient vector field, $\nabla f(x, y)$
$f(x, y) = x^2 + y^2$

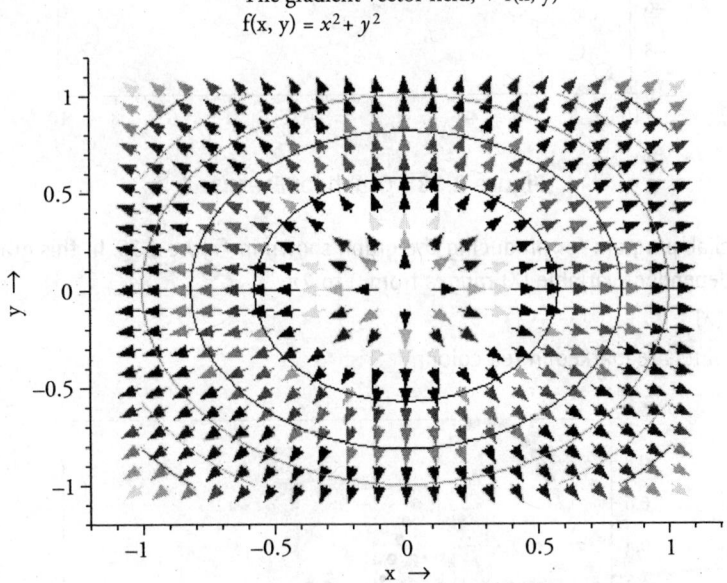

Figure 2.37 Gradient and contour of a scalar field

2.8 Exercises

1. Write a command line Scilab program for producing the graph shown in *Figure 2.38*, where data points are marked with yellow coloured markers outlined with a different colour.

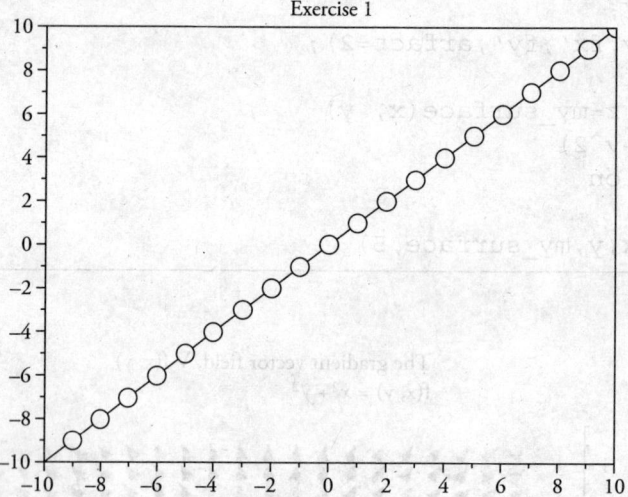

Figure 2.38 Graph for *Exercise 1*

2. Write a Scilab program for producing the graph shown in *Figure 2.39*. In this graph,
    *   The independent variable (X) ranges from 0 to 2π
    * $Y = \sin(X)$
    *   Data points are marked in red colour

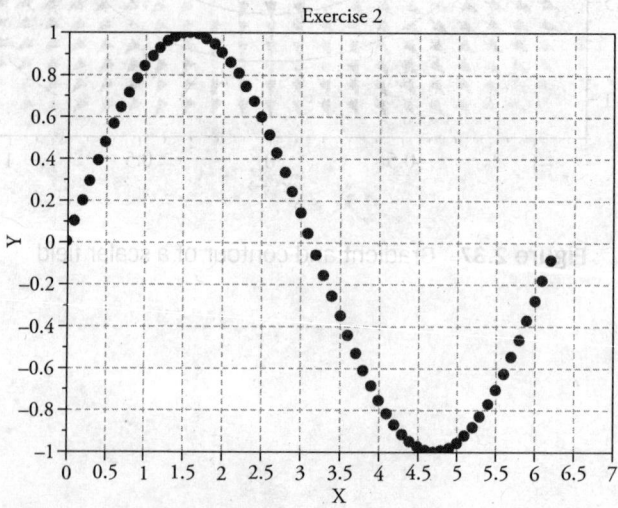

Figure 2.39 Graph for *Exercise 2*

3. Write a Scilab program for producing the graph shown in *Figure 2.40*. In this graph,
   - The independent variable (X) ranges from 0 to 2π
   - $Y = \sin(X)$
   - Data points are marked in red colour outlined with a different colour

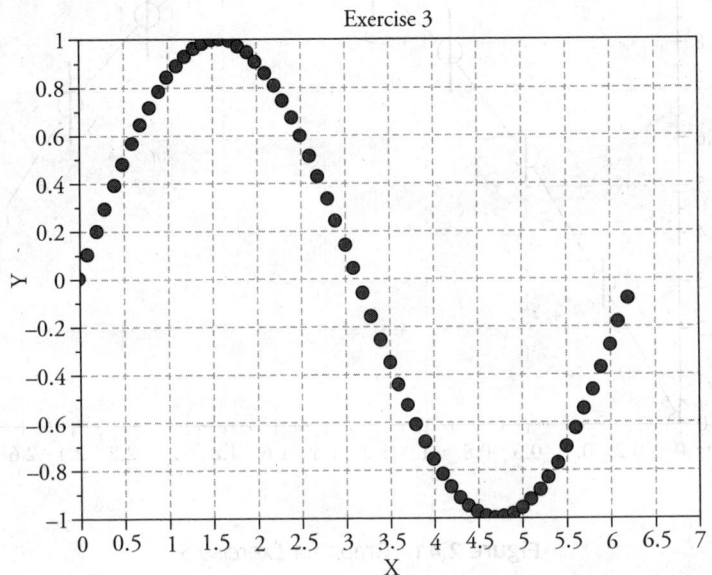

Figure 2.40 Graph for *Exercise 3*

4. Write a Scilab program for producing a histogram for generation of 1000 integer numbers between -5 and 5. Assume that the number generation obeys normal distribution. Mark the frequency of each number on the graph.

5. Write a Scilab program for producing the graph shown in *Figure 2.41*. In this graph,
   - The independent variable (X) ranges from 0 to π in steps size of 0.3
   - $Y = \sin(X)$
   - The error on the dependent variable (Y) is equal to $\sqrt{Y}\,/10$

6. Write a Scilab program to plot the data set given in *Table 2.7*.

   Mark the data points on the graph. Determine the values of the dependent variable (Y) at X = [0.25 0.7 2.3 3.3 4]. Mark the interpolated values on the same graph with a different marker.

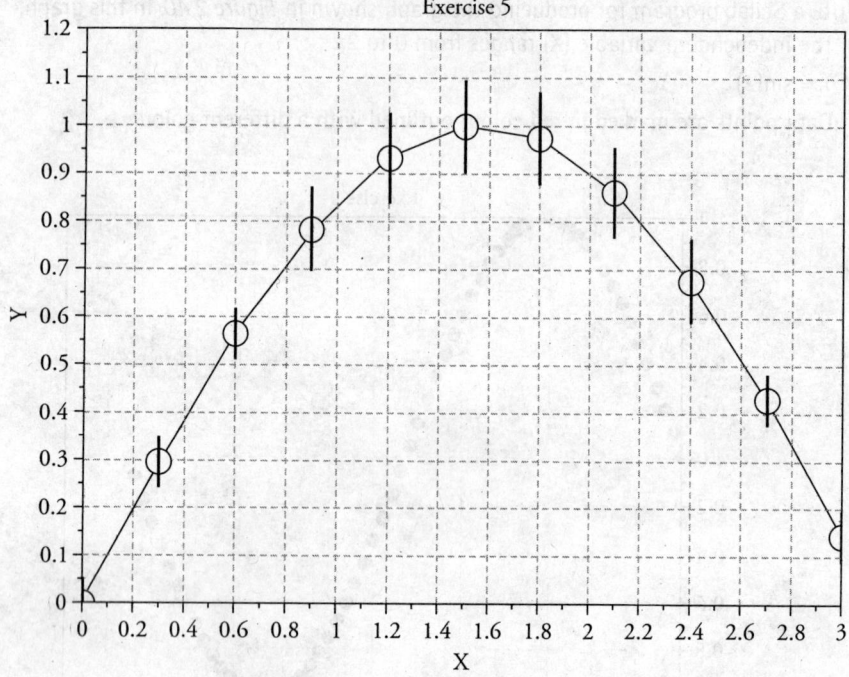

Figure 2.41 Graph for *Exercise 5*

7. Write a Scilab program to plot the data points given in *Table 2.8*. Connect the data points with a straight line as well as with a smooth curve going through them.

8. This question is with reference to *Exercise 4*. For the same data configuration, write a Scilab program to draw the normalized histogram. Mark the frequency of each number and join them with a smooth curve.

Table 2.7 Sample data for *Exercise 6*

X	Y
0	0
0.5	0.51
1.0	0.51
1.0	0.2
2.0	-0.31
3.0	-0.25
3.5	0.2
4.51	0.25

Table 2.8 Sample data for *Exercise 7*

X	Y
0	0
1.2	0.51
2.1	0.51
2.9	0.93
3.5	0.31
4.9	0.15
6	0.4
7.1	0.25

9. Write a Scilab program for producing the graph shown in *Figure 2.42*.

10. An object is fired at an angle θ w.r.t. the horizontal axis with an initial velocity of 10 ms⁻¹. Trace the path of the object for different values of θ (30° ,45° ,75°).

11. A particle is subjected to two collinear simple harmonic oscillations simultaneously. Suppose the displacement of the particle due to these collinear oscillations is given by

$$x_1 = A_1 \cos\left(\omega t + \varphi_1\right)$$

$$x_2 = A_2 \cos\left(\omega t + \varphi_2\right)$$

Write a Scilab program to plot the resultant oscillation of the particle assuming that the phase difference between the two oscillations is equal to 2π.

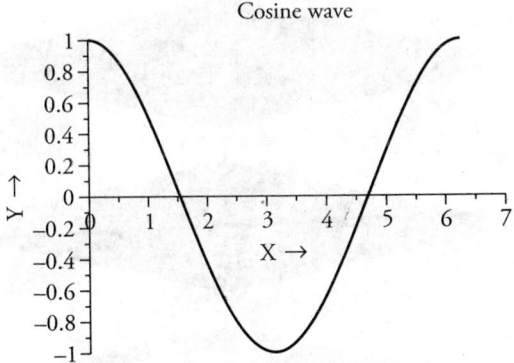

Figure 2.42   Graph for *Exercise 9*

12. In *Section 2.7.4*, the charging of a capacitor through a resistor was explained. Suppose the voltage source is disconnected and is replaced by a short circuit. The capacitor will gradually discharge its current through the resistor. Write a Scilab program to plot the discharge curve of the capacitor voltage along with the current in the circuit.

13. Write a Scilab program to plot the following function. Show that as the standard deviation (σ) becomes smaller, the function approaches the Dirac delta representation.

$$f(x) = \frac{1}{\sqrt{2\pi\sigma^2}} \exp\left(-\frac{(x-a)^2}{2\sigma^2}\right)$$

14. Write a Scilab program to plot the spectral radiance of a blackbody radiation as a function of wavelength (in μm). Study the effect of temperature by taking three different temperatures (2500 K, 3000 K and 3500 K). The law is given by the following equation,

$$f(\lambda) = \frac{2hc^2}{\lambda^5} \frac{1}{\exp\left(\frac{hc}{\lambda kT}\right) - 1}$$

15. Write a Scilab program to generate the family of lattice planes shown in *Figure 2.43*.

16. The Maxwell Boltzmann speed distribution function is given as follows. Write a Scilab program to plot the variation of $f(v)$ as a function of v for an oxygen molecule at temperatures 100 K, 500 K and 1000 K.

$$f(v) = \sqrt{\left(\frac{m}{2\pi kT}\right)^3} 4\pi v^2 e^{-\left(mv^2/2kT\right)}$$

17. Repeat the previous question for three different gas molecules (Helium, Neon, Argon) at 300 K.

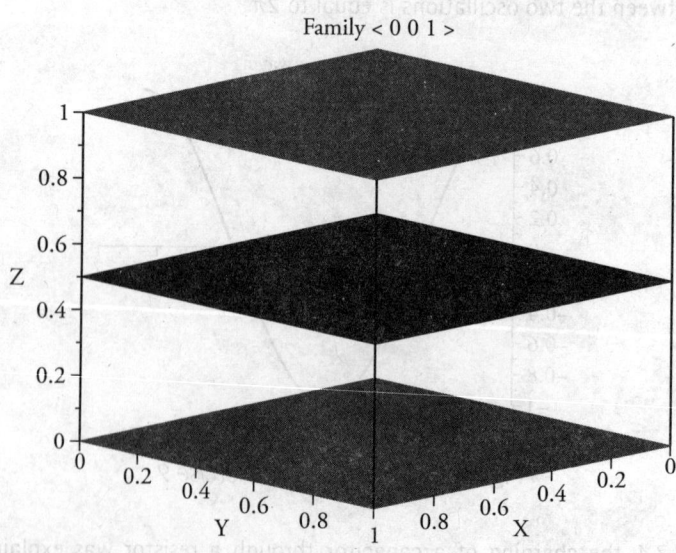

Figure 2.43 Plot for *Exercise 15*

18. Write a Scilab program to plot the following Fermi–Dirac energy distribution function. Take $m = 2$ eV and temperature, $T = 0$ K, 2000 K and 5000 K. In the case of 0K, it will be sufficient to use a small temperature, say 0.1 K, just to avoid error due to division by zero.

$$f(E) = \frac{1}{1 + \exp\left(\dfrac{E - m}{kT}\right)}$$

19. Write a Scilab program to plot the following Bose–Einstein energy distribution function. Take $m = 0$ eV and temperature, $T = 1000$ K, 2000 K, 5000 K .

$$f(E) = \frac{1}{\exp\left(\dfrac{E - m}{kT}\right) - 1}$$

20. Compare the Fermi–Dirac, Maxwell–Boltzmann and Bose–Einstein energy distribution curves at 500 K. Take $m = 0$.

21. Write a Scilab program to draw the level surfaces and gradient of the scalar field, $f(x, y) = x^2 - y^2$

22. Write a Scilab program to draw the level surfaces and gradient of the following scalar field.

$$f(x,y) = \frac{x^2}{2} - \frac{x^4}{4} - \frac{y^2}{2}$$

23. Write a Scilab program to draw the level surfaces and gradient of the scalar field, $f(x,y) = xe^{-x^2 - y^2}$

24. Write a Scilab program to determine the divergence of the following vectors at the mentioned points.
    a) $\vec{A} = xy\,\hat{i} - y^3\,\hat{j} + (x - 1)z\hat{k}$ at (5,3,8)
    b) $\vec{A} = x^2\,\hat{i} + y^2\,\hat{j} + z^2\,\hat{k}$ at (2.1.3)

25. Write a Scilab program to determine the curl of the following vectors at (2,1,3).
    a) $\vec{A} = xy^2\,\hat{i} - xyz\,\hat{j} + x^2z^2\,\hat{k}$
    b) $\vec{A} = y^2z^2\,\hat{i} - xz\,\hat{j} + xy\,\hat{k}$

Least Square Curve Fitting

After going through this chapter, the reader should be able to

◊   Understand the theory of least square fitting of linear data.

◊   Fit data whose variables are linearly related.

◊   Understand the theory of least square fitting of non-linear data.

◊   Fit data whose variables have a non-linear relation.

◊   Understand the theory of least square fitting of a polynomial function.

◊   Perform polynomial curve fitting and determine the roots of a polynomial.

◊   Apply the curve fitting rules to various problems in physics.

3.1 Introduction

Curve fitting refers to the process of generating a curve that fits to a set of data points. For example, consider a series of data points that consists of 'n' data points $[(x_1,y_1),(x_2,y_2)\dots(x_n,y_n)]$. The variables y_i are related to their respective x_i through a function $f(x_i)$.

In this configuration, the residual (R_i) of the ith data pair is defined as the difference between the theoretical value and the observed value (*Eqn. 3.1*).

$$R_i = f(x_i) - y_i \tag{3.1}$$

The aim of curve fitting is to minimize R_i, i.e., to minimize the difference between expected values and the observed values. From *Eqn. 3.1*, it is evident that the value of R_i can be positive for some values of 'i' and it can be negative for others. Therefore, the objective of 'least square' in curve fitting is to minimize the sum of square of all the residuals.

In the following sections, the least square fitting of different types of data sets will be discussed. The curve fitting for a linear and a non-linear data set is described in *Sections 3.2* and *3.3* respectively. This is followed by polynomial fitting in *Section 3.4*. The next section explains the fitting procedure using built-in Scilab function. Applications of these methods have been discussed in *Section 3.6*.

3.2 Fitting of Linear Data

Consider a data set having 'n' data points $[(x_1,y_1),(x_2,y_2),\dots (x_n,y_n)]$. The linear relation between x_i and y_i is given by *Eqn. 3.2*.

$$f(x_i) = y_i = mx_i + c \tag{3.2}$$

Eqn. 3.2 represents a straight line whose slope is 'm' and the intercept on y-axis is 'c'. The residual (R_i) is given by *Eqn. 3.3*.

$$R_i = f(x_i) - y_i = (mx_i + c) - y_i \tag{3.3}$$

Eqn. 3.4 gives the sum of squares of all the residuals.

$$S = \sum_{i=1}^{n}(R_i)^2 = \sum_{i=1}^{n}\{(mx_i + c) - y_i\}^2 \tag{3.4}$$

The objective of *least square fitting* method is to find the values of m and c, such that they minimize S. The minimum value of S can be determined by differentiating it w.r.t. the slope and the constant and then equating the differential to zero, i.e.

$$\frac{\partial S}{\partial m} = \frac{\partial S}{\partial c} = 0 \tag{3.5}$$

In *Eqn. 3.5*,

   a.   The first term, $\dfrac{\partial S}{\partial m} = 0$ implies,

$$\sum_{i=1}^{n} 2x_i\{(mx_i + c) - y_i\} = 0 \tag{3.6}$$

$$m\sum_{i=1}^{n}x_i^2 + c\sum_{i=1}^{n}x_i = \sum_{i=1}^{n}x_i y_i \tag{3.7}$$

b.   The second term, $\dfrac{\partial S}{\partial c} = 0$ implies,

$$\sum_{i=1}^{n}\left\{(mx_i + c) - y_i\right\} = 0 \tag{3.8}$$

$$m\sum_{i=1}^{n}x_i + nc = \sum_{i=1}^{n}y_i \tag{3.9}$$

Simultaneous solution of *Eqns. 3.7* and *3.9* will give,

$$m = \frac{n\sum_{i=1}^{n}x_i y_i - \sum_{i=1}^{n}x_i \sum_{i=1}^{n}y_i}{n\sum_{i=1}^{n}x_i^2 - \left(\sum_{i=1}^{n}x_i\right)^2} \tag{3.10}$$

$$c = \frac{\sum_{i=1}^{n}x_i^2 \sum_{i=1}^{n}y_i - \sum_{i=1}^{n}x_i \sum_{i=1}^{n}x_i y_i}{n\sum_{i=1}^{n}x_i^2 - \left(\sum_{i=1}^{n}x_i\right)^2} \tag{3.11}$$

Eqns. 3.7 and *3.9* can also be written in the matrix notation, i.e.

$$\begin{bmatrix} \sum_{i=1}^{n}x_i^2 & \sum_{i=1}^{n}x_i \\ \sum_{i=1}^{n}x_i & n \end{bmatrix} \begin{bmatrix} m \\ c \end{bmatrix} = \begin{bmatrix} \sum_{i=1}^{n}x_i y_i \\ \sum_{i=1}^{n}y_i \end{bmatrix} \tag{3.12}$$

Eqn. 3.12 implies that

$$[A][C] = [B] \tag{3.13}$$

$$[C] = \frac{[B]}{[A]} \tag{3.14}$$

In *Eqn. 3.14*, the first element of the matrix C will give the slope of the best fit curve. The second element of this matrix will give the intercept on the y-axis.

The values of slope and intercept can be determined using either the matrix method (*Eqns. 3.12–3.14*) or from their direct expression (*Eqns. 3.10–3.11*). The following example explains the least square fitting of a linear data set by making use of a small Scilab program for doing the direct calculation of the slope and the constant.

Consider a set of 10 data points (x, y) following a linear relation (*Figure 3.1*). These have been generated using a simple Scilab program.

```
x = linspace(1,10,10);
n = length(x);                              //n = 10
y = rand(1,n).*x          //Generate random y-values
```

As shown in *Figure 3.1*, the 10 values of '*x*' and '*y*' show a zigzag pattern superimposed on a linear trend. These data points have been connected with a straight line. The following Scilab program fits a straight line curve to the entire data set and also determines the value of the slope and constant. The result is plotted in *Figure 3.2*.

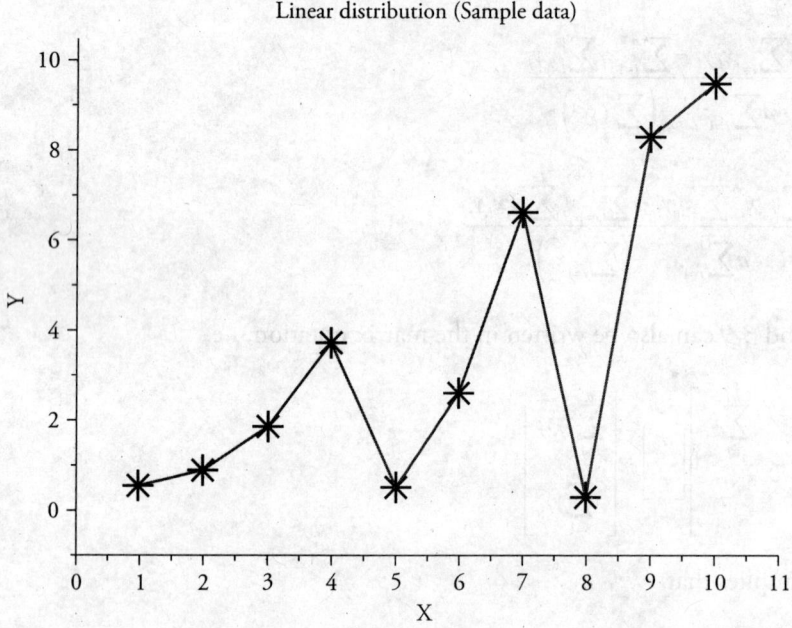

Figure 3.1   Generation of a data set obeying a linear relation

```
function f = bestfit(a)        //Define a linear function
f = m*a+c;
endfunction

x = linspace(1,10,10);
n = length(x);                                    //n = 10
y = rand(1,n).*x //Generate random y-values for each 'x'
plot2d(x,y)                               //Plot the data

//Steps to determine the value of slope and constant

x1 = sum(x);                                   // $\sum_{i=1}^{n} x_i$
```

```
x2   = sum(x.*x);                                    // $\sum\limits_{i=1}^{n} x_i^2$

x1y1 = sum(x.*y);                                    // $\sum\limits_{i=1}^{n} x_i y_i$

y1   = sum(y);                                        // $\sum\limits_{i=1}^{n} y_i$

m = (n*x1y1 - (x1*y1))/ (n*x2 - (x1)^2);      //Slope
c = (x2*y1 - (x1*x1y1))/ (n*x2-(x1)^2);       //Constant
z = x(1)/2:0.01:x(n)+x(1)/2.0;

fplot2d(z,bestfit);                    //Plot the best fit curve
```

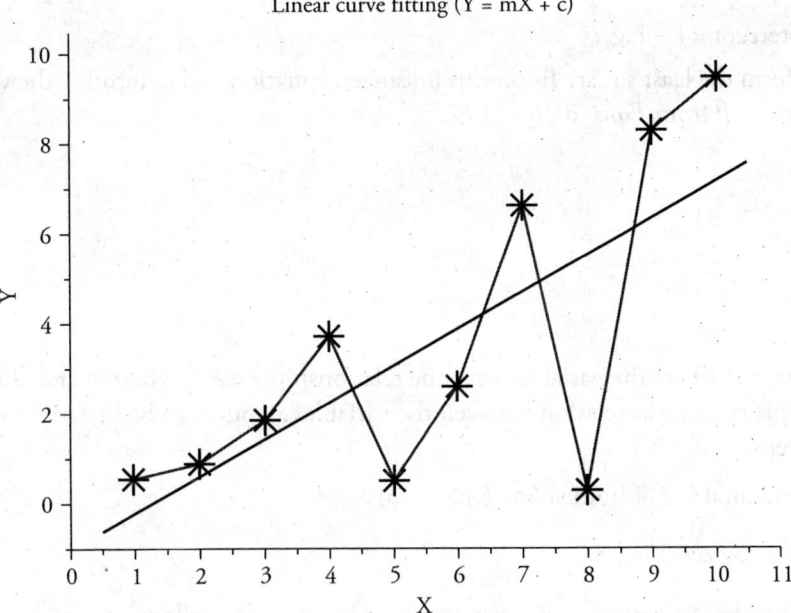

Linear curve fitting (Y = mX + c)

Figure 3.2 Least square fitting of a linear data

3.3 Fitting of Non-Linear Data

Sometimes the variables (x_i, y_i) of the experimental data do not follow a linear relation. However, it is possible to linearize their relation by making some minor modifications. Two such cases have been discussed here.

Case (I)

Consider the case where the variables have the relationship $y = ae^{\beta x}$, where α and β are constant numbers. This non-linear relation between the variables x and y can be linearized through the following steps.

- Take natural log of both sides (*Eqn. 3.15*)

$$\log_e y = \log_e \alpha + \beta x \tag{3.15}$$

- This will be the equation of $(Y = mX + c)$ as long as
  - $Y = \log_e y$
  - $X = x$
  - Slope $= m = \beta$
  - Intercept $= c = \log_e \alpha$

Hence, perform the least square fit on this linearized equation and determine the values of the constants α and β from *Eqns. 3.16 – 3.17*.

$$\alpha = e^c \tag{3.16}$$

$$\beta = m \tag{3.17}$$

Case (II)

Consider the case where the variables have the relationship $y = ax^\beta$, where α and β are constant numbers. This non-linear relation between the variables x and y can be linearized through the following steps.

- Take natural log of both sides (*Eqn. 3.18*)

$$\log_e y = \log_e \alpha + \beta \log_e x \tag{3.18}$$

- This will be the equation of a straight line $(Y = mX + c)$ as long as
  - $Y = \log_e y$
  - $X = \log_e x$
  - Slope $= m = \beta$
  - Intercept $= c = \log_e \alpha$

Hence, perform the least square fit on this linearized equation and determine the values of the constants α and β from *Eqns. 3.19–3.20*.

$$\alpha = e^c \tag{3.19}$$

$$\tag{3.20}$$

The least square fitting of a non-linear data (both the cases) has been explained below with the help of two examples.

Example 1

Consider a set of 10 data points (x, y) having an exponential relation (*Figure 3.3*). The data set has been generated using the following Scilab program.

```
x = linspace(0.1,1,10);
n = length(x);                           //n=10
y = %e^(3*x) + (3*rand(1,n)); //Generate random y-values
```

This data set shows a zigzag type pattern superimposed on an exponential trend. The best fit curve can be obtained using the following Scilab program. The result is shown in *Figure 3.4*.

```
//Define a function for the exponential fitting
function [f,m,c] = exponential_fit(x,y)
n = length(x);
for i=1:n
    z(i)=log(y(i));
end

x1 = sum(x);
x2 = sum(x.*x);
x1y1 = sum(x'.*z);
y1 = sum(z);

A = [x2 x1; x1 n];
B = [x1y1; y1];
C = A\B;
m = C(1);
c = C(2);
c = exp(c);

for i = 1:n
    f(i) = c*exp(m*x(i));
end
endfunction

//Generate a random data set around
x = linspace(0.1,1,10);
n = length(x);
y = %e^(3*x) + (3*rand(1,n));
plot2d(x,y)
```

```
//Determine best fit parameters and plot best fit curve
[bestfit,m,c] = exponential_fit(x,y)

plot2d(x,bestfit)
```

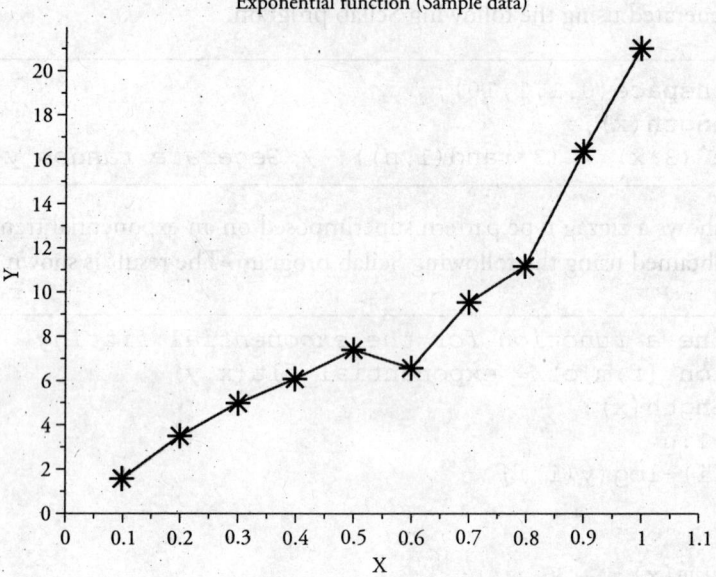

Figure 3.3   Generation of a data set obeying an exponential relation

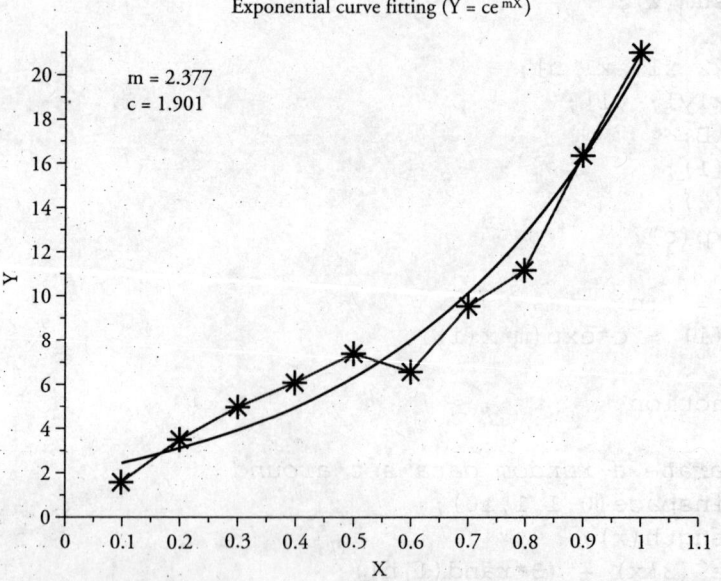

Figure 3.4   Best fit exponential function

Example 2

This example explains how to perform the method of least square fitting for Case (II). Consider a set of 10 data points (x_i, y_i) having an exponential relation $y \sim x^4$ (*Figure 3.5*). These have been generated using the following Scilab program.

```
x = linspace(0.1,1,10);
n = length(x);                           //n = 10
a = x.^4
y = a + a.*rand(1,n);          //Generate random y-values
```

In *Figure 3.5*, the data set shows a zigzag type pattern superposed on an exponential trend. The least square fitting can be applied using the following Scilab program. The best fit curve is shown in *Figure 3.6*.

```
//Define the function for the best fit
function f = bestfit(a)
f = c*(a^m);
endfunction

x = linspace(0.1,1,10);
n = length(x);
a = x.^4
y = a + a.*rand(1,n);

plot2d(x,y)
for i = 1:n
    Y(i) = log(y(i));  //Y = log_e y
    X(i) = log(x(i));  //X = log_e x
end

x1 = sum(X);
x2 = sum(X.*X);
x1y1 = sum(X.*Y);
y1 = sum(Y);

A = [x2 x1; x1 n];
B = [x1y1; y1];
C = A\B;

m = C(1);                                //β = m
c = C(2);
c = %e^c;                                //α = e^c

z= x(1):(x(2)-x(1))*0.1:x(n)+x(1);

fplot2d(z,bestfit)           //Plot the best fit curve
```

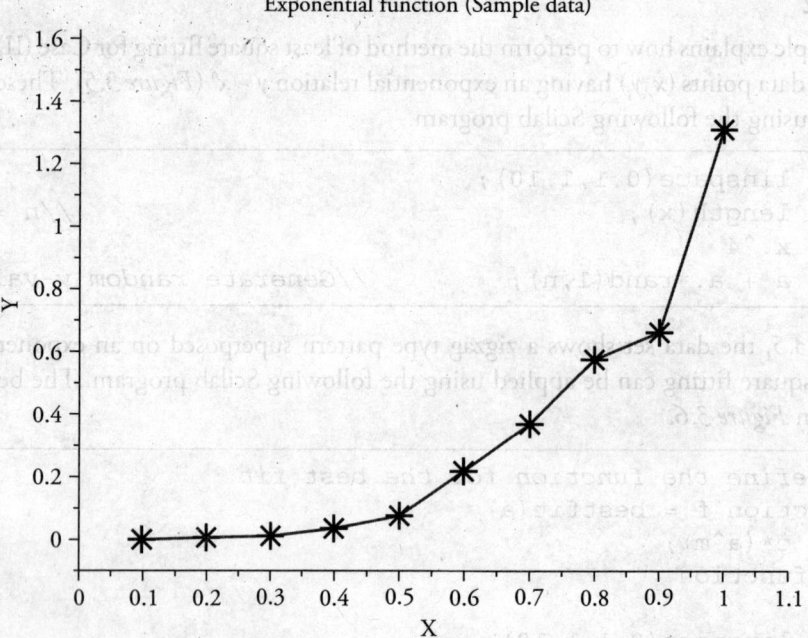

Figure 3.5   Generation of a data set obeying an exponential relation

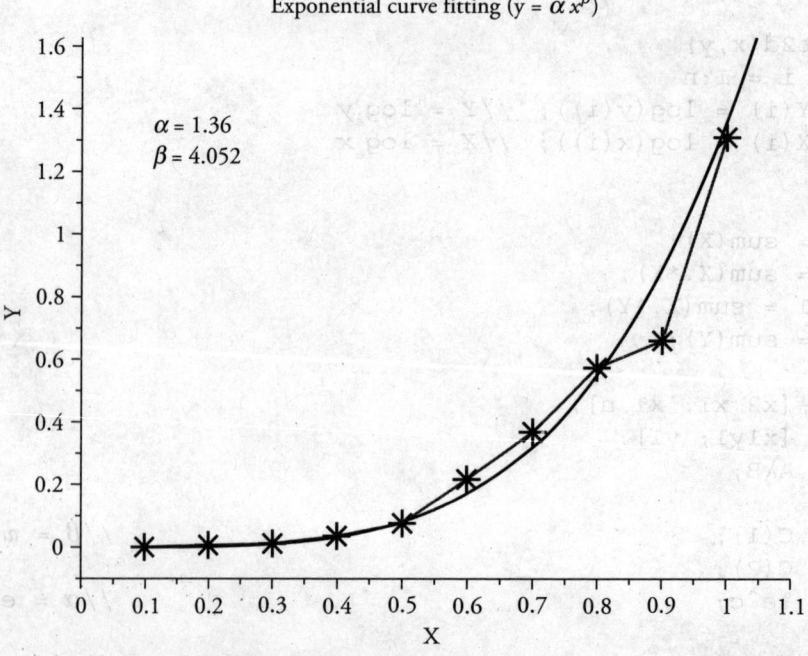

Figure 3.6   Best fit exponential function

3.4 Polynomial Fitting

A polynomial is a mathematical expression that contains terms summed over the product of powers of the variable and coefficients. In general, any k^{th} order polynomial in variable x can be written as follows:

$$y = a_0 + a_1 x + a_2 x^2 + \ldots + a_k x^k \tag{3.21}$$

The residual for this polynomial is given by

$$R \equiv \sum_{i=1}^{n} \left[y_i - \left\{ a_0 + a_1 x_i + a_2 x_i^2 + \ldots + a_k x_i^k \right\} \right]^2 \tag{3.22}$$

The objective of least square fitting of polynomials is to determine the coefficients, $a_0, a_1, \ldots, a_k$ such that they minimize the residual R. This is done by taking the partial derivative of R w.r.t. each coefficient and equating the derivative to zero.

The partial derivatives are given by the following constituent equations:

$$\frac{\partial R}{\partial a_0} = -2 \sum_{i=1}^{n} \left[y_i - \left\{ a_0 + a_1 x_i + a_2 x_i^2 + \ldots + a_k x_i^k \right\} \right] = 0$$

$$\frac{\partial R}{\partial a_1} = -2 \sum_{i=1}^{n} \left[y_i - \left\{ a_0 + a_1 x_i + a_2 x_i^2 + \ldots + a_k x_i^k \right\} \right] x_i = 0$$

$$\frac{\partial R}{\partial a_2} = -2 \sum_{i=1}^{n} \left[y_i - \left\{ a_0 + a_1 x_i + a_2 x_i^2 + \ldots + a_k x_i^k \right\} \right] x_i^2 = 0$$

$$\vdots$$

$$\vdots$$

$$\frac{\partial R}{\partial a_k} = -2 \sum_{i=1}^{n} \left[y_i - \left\{ a_0 + a_1 x_i + a_2 x_i^2 + \ldots + a_k x_i^k \right\} \right] x_i^k = 0 \tag{3.23}$$

This implies that

$$a_0 n + a_1 \sum_{i=1}^{n} x_i + a_2 \sum_{i=1}^{n} x_i^2 + \ldots + a_k \sum_{i=1}^{n} x_i^k = \sum_{i=1}^{n} y_i$$

$$a_0 \sum_{i=1}^{n} x_i + a_1 \sum_{i=1}^{n} x_i^2 + \ldots + a_k \sum_{i=1}^{n} x_i^{k+1} = \sum_{i=1}^{n} x_i y_i$$

$$a_0 \sum_{i=1}^{n} x_i^2 + a_1 \sum_{i=1}^{n} x_i^3 + \ldots + a_k \sum_{i=1}^{n} x_i^{k+2} = \sum_{i=1}^{n} x_i^2 y_i$$

$$\vdots$$

$$\vdots$$

$$a_0 \sum_{i=1}^{n} x_i^k + a_1 \sum_{i=1}^{n} x_i^{k+1} + \ldots + a_k \sum_{i=1}^{n} x_i^{2k} = \sum_{i=1}^{n} x_i^k y_i \tag{3.24}$$

Eqn. 3.24 can be written as

$$
\begin{bmatrix}
n & \sum_{i=1}^{n} x_i & \sum_{i=1}^{n} x_i^2 & \cdots & \sum_{i=1}^{n} x_i^k \\
\sum_{i=1}^{n} x_i & \sum_{i=1}^{n} x_i^2 & \sum_{i=1}^{n} x_i^3 & \cdots & \sum_{i=1}^{n} x_i^{k+1} \\
\sum_{i=1}^{n} x_i^2 & \sum_{i=1}^{n} x_i^3 & \sum_{i=1}^{n} x_i^4 & \cdots & \sum_{i=1}^{n} x_i^{k+2} \\
& & \vdots & & \\
& & \vdots & & \\
\sum_{i=1}^{n} x_i^k & \sum_{i=1}^{n} x_i^{k+1} & \sum_{i=1}^{n} x_i^{k+2} & \cdots & \sum_{i=1}^{n} x_i^{2k}
\end{bmatrix}
\begin{bmatrix}
a_0 \\ a_1 \\ \vdots \\ \vdots \\ a_k
\end{bmatrix}
=
\begin{bmatrix}
\sum_{i=1}^{n} y_i \\
\sum_{i=1}^{n} x_i y_i \\
\sum_{i=1}^{n} x_i^2 y_i \\
\vdots \\
\vdots \\
\sum_{i=1}^{n} x_i^k y_i
\end{bmatrix}
\tag{3.25}
$$

Therefore, if the first matrix on the left side of *Eqn. 3.25* is denoted by matrix 'A' and the matrix on the right side is denoted by matrix 'B', then the coefficient matrix will be given by

$$
\begin{bmatrix}
a_0 \\ a_1 \\ \vdots \\ \vdots \\ a_k
\end{bmatrix}
= B/A
$$

Here,

$$
A = \begin{bmatrix}
n & \sum\limits_{i=1}^{n} x_i & \sum\limits_{i=1}^{n} x_i^2 & \cdots & \sum\limits_{i=1}^{n} x_i^k \\[2ex]
\sum\limits_{i=1}^{n} x_i & \sum\limits_{i=1}^{n} x_i^2 & \sum\limits_{i=1}^{n} x_i^3 & \cdots & \sum\limits_{i=1}^{n} x_i^{k+1} \\[2ex]
\sum\limits_{i=1}^{n} x_i^2 & \sum\limits_{i=1}^{n} x_i^3 & \sum\limits_{i=1}^{n} x_i^4 & \cdots & \sum\limits_{i=1}^{n} x_i^{k+2} \\[2ex]
\vdots & & & & \\[1ex]
\sum\limits_{i=1}^{n} x_i^k & \sum\limits_{i=1}^{n} x_i^{k+1} & \sum\limits_{i=1}^{n} x_i^{k+2} & \cdots & \sum\limits_{i=1}^{n} x_i^{2k}
\end{bmatrix}
$$

$$
B = \begin{bmatrix}
\sum\limits_{i=1}^{n} y_i \\[2ex]
\sum\limits_{i=1}^{n} x_i y_i \\[2ex]
\sum\limits_{i=1}^{n} x_i^2 y_i \\[1ex]
\vdots \\[1ex]
\sum\limits_{i=1}^{n} x_i^k y_i
\end{bmatrix}
$$

The following example will show the use of the matrix method to determine the coefficients $a_0, a_1, \ldots, a_k$. Consider the data written in *Table 3.1*.

Table 3.1   Data for polynomial curve fitting

x	y
-2	25
-1	3
0	-3
1	1
2	9
3	15

x	y
4	13
5	-3

This data set can be easily plotted in Scilab using the following commands. The result is shown in *Figure 3.7*.

```
x = [-2 -1 0 1 2·3 4 5];
y = [25 3 -3 1 9 15 13 -3];
plot2d(x,y)
```

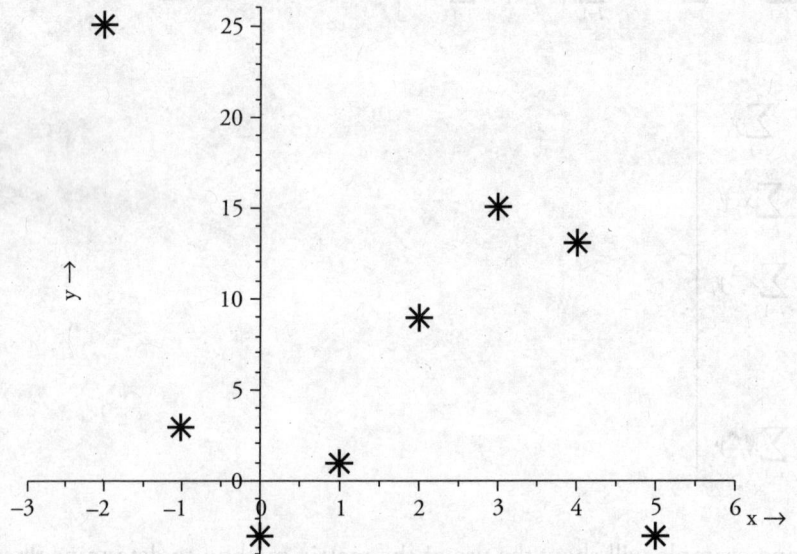

Figure 3.7   Plot of data points

In order to fit a polynomial to the given data set, one should make a quick estimate of the order of the polynomial. Looking at the data, it is obvious that the order of best fit polynomial for this data set should be 3. Accordingly, the following Scilab function can be written for determining the coefficient matrix.

```
function alpha = polynomial_fit(x,y,k)
format(6);
n = length(x);
X(1) = n;
```

```
//Calculate sum of all powers (power 1 to) of all the x-
```
data, till $\sum_{i=1}^{n} x_i^{2k}$

```
for i = 1:2*k                    //k is order of the polynomial
    X(i+1) = sum(x.^i);
end

//Generate matrix A
for i = 1:k+1
    for j = 1:k+1
        A(i,j) = X(i+j-1);
    end
end

//Generate the matrix B
for i = 0:k
    B(i+1) = sum(x.^i.*y);
end
B = -B;

//Determine coefficient matrix using linear equation
solver and return the value of coefficients.
alpha = linsolve(A,B);
endfunction

k = 3;                           //Guess order of the polynomial

//Call the function
coefficient = polynomial_fit(x,y,k);

//Determine and plot the best fit polynomial
j = 1;
for x = -2:0.1:5;
    clear d
    for i = 1:k+1;
        d(i) = coefficient(i)*x.^(i-1);
    end
    xfit(j) = x;
    yfit(j) = sum(d);
    j = j+1;
end
plot2d(xfit,yfit)
```

The last step in the polynomial fitting is to determine the roots of the polynomial and to plot them on the best fit polynomial. This is done using the following Scilab program. The roots are shown in *Figure 3.9*.

```
//Determine the roots of the polynomial and plot them
p = poly([coefficient],'x','c');
result = roots(p)

for i = 1:k;
    xroot(i) = result(i);
    yroot(i) = 0;
end

plot2d(xroot,yroot)
```

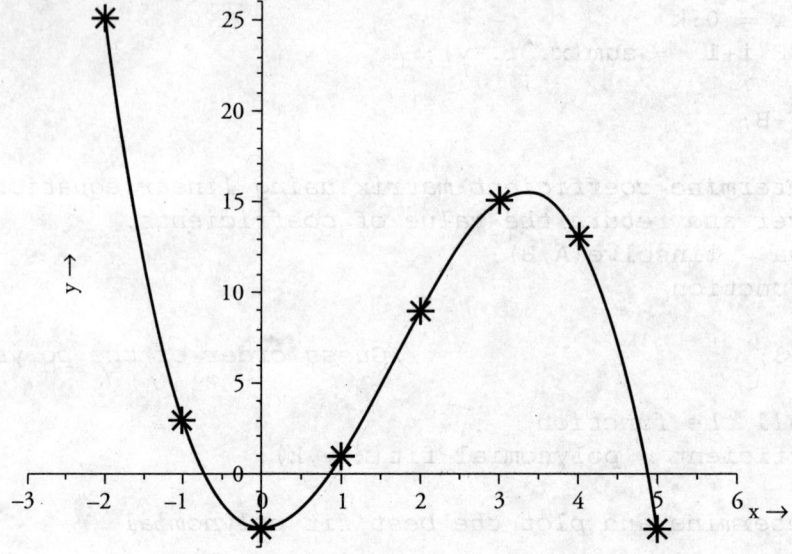

Figure 3.8   Fitting of the polynomial

k^{th} order polynomial fitting
$y = (-1)x^3 + (5)x^2 + (-0.00)x^1 + (-3)x^0$
(**Roots at** $x = -0.72, 0.85, 4.87$)

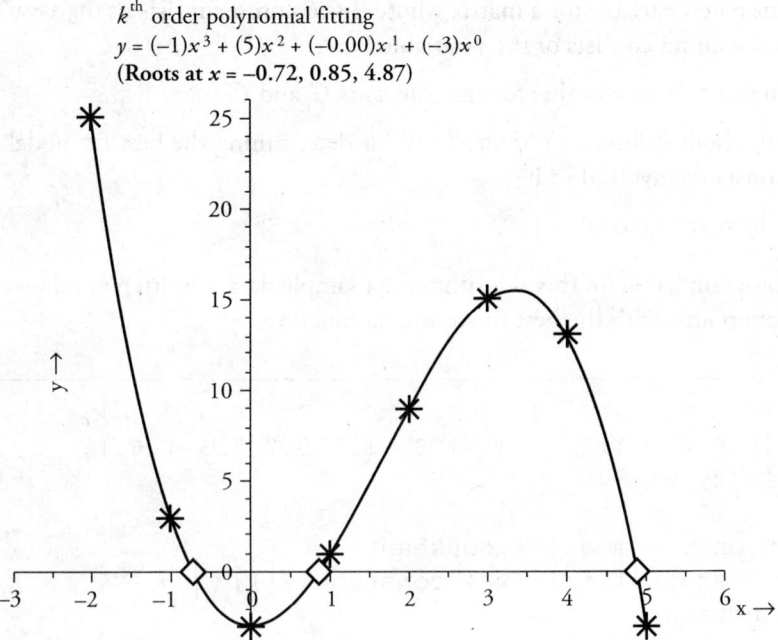

Figure 3.9   Plotting the roots of best fit polynomial

3.5 Fitting with Built-in Scilab Function – 'datafit'

The built-in Scilab function 'datafit' makes use of an error function to fit the data. The error function is the difference between the given values of the dependent variable and that obtained from the model. The function 'datafit' determines the best fit model by minimizing this error.

The algorithm for using this method to fit a dataset having a sinusoidal variation is as follows.

- Define and mark the data points on the graph.

- Define a function for the expected model that may fit the data set. For example, in this case, model for a sinusoidal function is defined in the following manner.

$$C_1 \sin\left(x + C_2\right)$$

- Here, value of the constants C_1 and C_2 are determined by the best fit model.

- Define another function for error estimation in the model.

- Define a new variable for a matrix whose first column consists of the x-variable and the second column consists of the y-variable.
- Define a trial (guess) value for the constants C_1 and C_2.
- Call the built-in function 'datafit' for determining the best fit model and value of the constants involved in it.
- Plot the best fit model.

The Scilab program based on this algorithm for a sample data is written as follows. *Figure 3.10* shows the data points with the best fit sinusoidal function.

```
x = [1    2    3    4    5    6    7    8    9    ];
y = [0.8 2.1 1.2 -0.4 -1.8 -1.5 0.2 1.9 1.6 ];
plot2d(x, y)

function y = model(x,constant)
y = constant(1)*sin(x + constant(2));
endfunction

function err = model_error(constant,z)
err = z(2) - model(z(1),constant);
endfunction

z = [x ; y];
constant_trial = [0 0]';

[best_fit_constant, err] = datafit(model_error, z, con-
stant_trial);

x = linspace(0, 10, 100);
y = model(x,best_fit_constant);
plot2d(x, y)
```

3.6 Applications

There are numerous physical situations that require curve fitting to evaluate the best possible solution of the phenomena. Some common applications have been discussed in the following sub-sections.

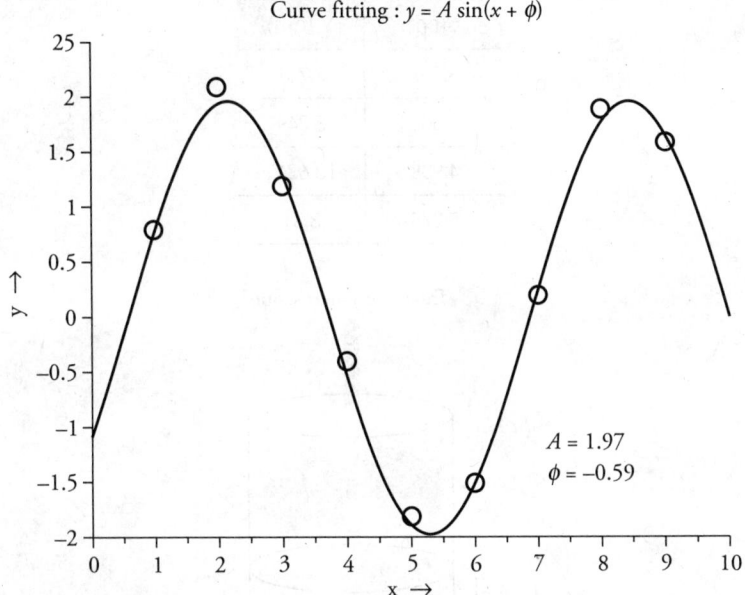

Curve fitting : $y = A \sin(x + \phi)$

$A = 1.97$
$\phi = -0.59$

Figure 3.10   Curve fitting using 'datafit'

3.6.1 Refractive index of water

The refractive index of water can be calculated with the help of a travelling microscope, as shown in *Figure 3.11*. The travelling microscope is focused on the target, which is fixed in a beaker. When some water is poured in the beaker, the position of the target appears to rise. The travelling microscope therefore has to be focused again on the target.

The refractive index of water is given by *Eqn. 3.26*.

$$\text{Refractive Index} = \frac{\text{Real Depth}(X)}{\text{Apparent Depth}(Y)} \qquad (3.26)$$

Suppose the readings given in *Table 3.2* are obtained for different levels of water.

Table 3.2   Sample readings for *Section 3.6.1*

X (cm)	Y (cm)
1.871	1.421
2.225	1.651
2.705	2.009
3.259	2.418

X (cm)	Y (cm)
3.707	2.757
4.501	3.349
4.908	3.634
5.275	3.81

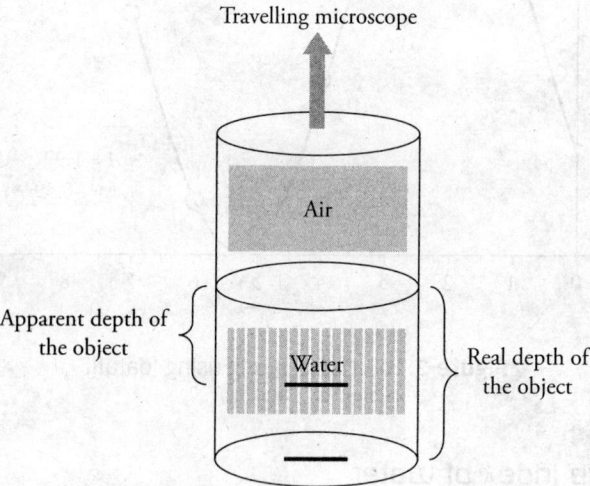

Figure 3.11   Schematic diagram for determining refractive index of water

The following Scilab program plots these data points, determines the best fit curve and calculates the refractive index of the liquid. The result is shown in *Figure 3.12*.

```
//Define a function for the best straight line fit
function func = bestfit(a)
func = m*a+c;
endfunction

n = input("How many data points ? ")
for i=1:n
    x(i)= input("Enter x(" + string(i)+ ") = ")
end

for i=1:n
    y(i)= input("Enter y(" + string(i)+ ") = ")
end

plot2d(y,x)                                    //Plot the data
```

```
//Determination of the refractive index

x1 = sum(x);                                //Calculate $\sum\limits_{i=1}^{n} x_i$

x2 = sum(x.*x);                             //Calculate $\sum\limits_{i=1}^{n} x_i^2$

x1y1 = sum(x.*y);                           //Calculate $\sum\limits_{i=1}^{n} x_i y_i$

y1 = sum(y);                                //Calculate $\sum\limits_{i=1}^{n} y_i$

//Calculate value of slope
m = (n*x1y1 - (x1*y1))/ (n*x2 - (x1)^2);

//Calculate value of intercept on the y-axis
c = (x2*y1 - (x1*x1y1))/ (n*x2-(x1)^2);

//Range of x-axis and plot the best fit curve
z = 0:0.01:x(n)+x(1)/2.0;
fplot2d(z,bestfit)

//Determination of slope and constant (matrix method)
A = [x2 x1; x1 8]
B = [x1y1; y1]
C = A\B;
m = C(1);
c = C(2);
```

3.6.2 Spring constant

If a mass 'm' is hung from one end of a spring, then from Hooke's law,

$$F = -kx \tag{3.27}$$

In *Eqn. 3.27*,

- F is force
- k is spring constant
- x is the extension produced in the spring

This implies that

$$k = \frac{F}{x} = \frac{mg}{x} = \frac{g}{\left(\frac{x}{m}\right)} \tag{3.28}$$

Figure 3.13 shows the extension (*x*) produced in a spring when a mass '*m*' is hung from one of its ends. Suppose the extension produced in the spring due to different loads is given in *Table 3.3*.

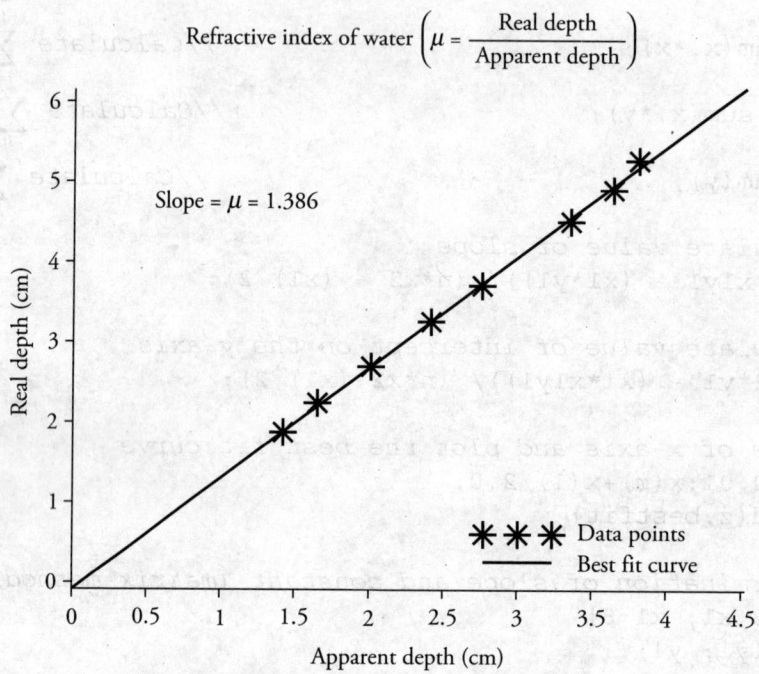

Refractive index of water $\left(\mu = \dfrac{\text{Real depth}}{\text{Apparent depth}}\right)$

Slope = μ = 1.386

Data points
Best fit curve

Figure 3.12   Graph for refractive index of water

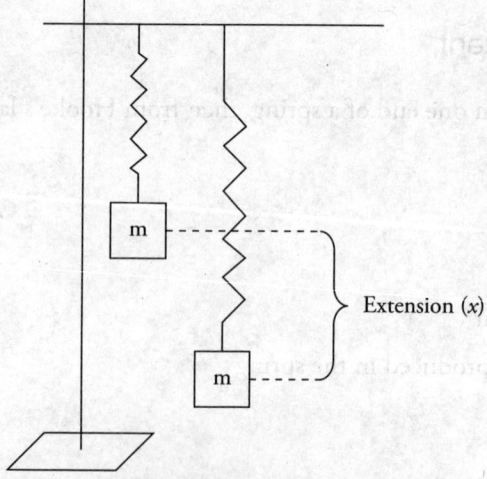

Figure 3.13   Schematic diagram for determination of spring constant

Table 3.3 Data for determining spring constant

Load (gram)	Extension (cm)
50	0.30
100	0.70
150	0.95
200	1.20
250	1.55
300	1.80
350	2.00

The following Scilab program determines the spring constant and fits the data points with the best fit straight line curve. The best fit curve is shown in *Figure 3.14*.

```
//Define a function for best fit straight line
function func = bestfit(a)
func = m*a+c;
endfunction

n = input("How many data points? ")
for i=1:n
    x(i)= input("Enter x(" + string(i)+ ") = ")
end

for i=1:n
    y(i)= input("Enter y(" + string(i)+ ") = ")
end

plot2d(x,y)                              //Plot the data

//Determination of the spring constant
x1 = sum(x);              //Calculate $\sum\limits_{i=1}^{n} x_i$
x2 = sum(x.*x);           //Calculate $\sum\limits_{i=1}^{n} x_i^2$
x1y1 = sum(x.*y);         //Calculate $\sum\limits_{i=1}^{n} x_i y_i$
y1 = sum(y);              //Calculate $\sum\limits_{i=1}^{n} y_i$

//Calculate value of slope
m = (n*x1y1 - (x1*y1))/ (n*x2 - (x1)^2);
```

```
//Calculate the value of intercept on y-axis
c = (x2*y1 - (x1*x1y1))/ (n*x2-(x1)^2);

//Range of x-axis and plot the best fit curve
z= x(1)/2:0.01:x(n)+x(1)/2.0;
fplot2d(z,bestfit)
```

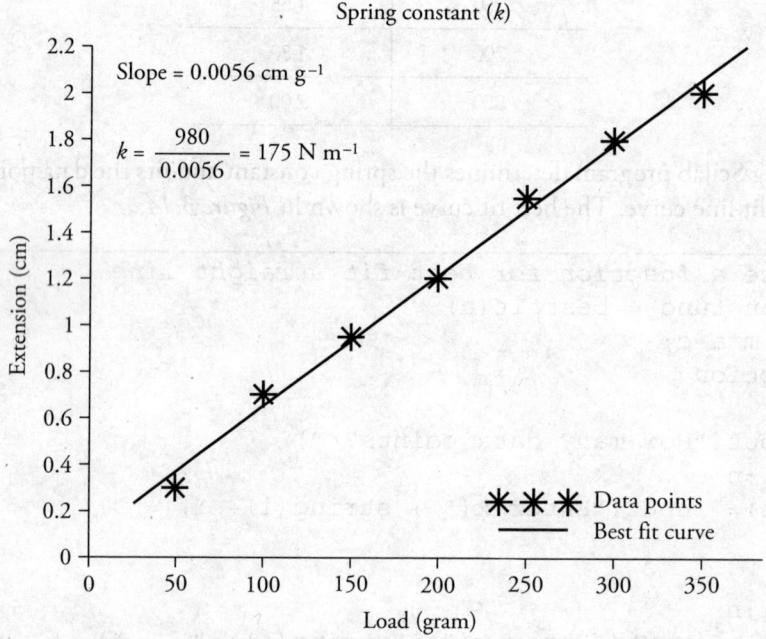

Figure 3.14 Graph for determination of spring constant

3.6.3 Cauchy's constant of a prism

Cauchy's equation is given by

$$\mu = A + \frac{B}{\lambda^2}$$

(3.29)

In *Eqn. 3.29*,

- The refractive index of the transparent material is μ and wavelength of light is λ
- Cauchy's constants are A and B

Consider *Table 3.4* for a sample data.

Table 3.4   Data for determining the Cauchy's constants

$\frac{1}{\lambda^2}\left(\times 10^{12}\ \text{m}^{-2}\right)$	Refractive Index
2.07	1.655
2.57	1.66
2.98	1.663
3.0	1.666
3.353	1.67
5.13	1.69
5.29	1.6925
6.01	1.7016
6.12	1.702

The following Scilab program calculates the Cauchy's constants and draws best fit straight line curve which is shown in *Figure 3.15*.

```
//Define function for the best fit straight line
function func = bestfit(a)
func = m*a+c;
endfunction

//Enter data points
x =[2.07   2.57   2.98   3.0   3.353   5.13   5.29   6.01
6.12]*10^12;
y =[1.655   1.66   1.663   1.666   1.67   1.69   1.6925
1.7016   1.702];
plot2d(x,y)
n = length(x)

//Determination of Cauchy's constants
x1 = sum(x);
x2 = sum(x.*x);
x1y1 = sum(x.*y);
y1 = sum(y);

//Calculate value of Cauchy's constant (B)
m = (n*x1y1 - (x1*y1))/ (n*x2 - (x1)^2);
```

```
//Calculate value of Cauchy's constant (A)
c = (x2*y1 - (x1*x1y1))/ (n*x2-(x1)^2);

//Range of x-axis and plot the best fit curve
z = 0:0.1e12:7e12;
fplot2d(z,bestfit)
```

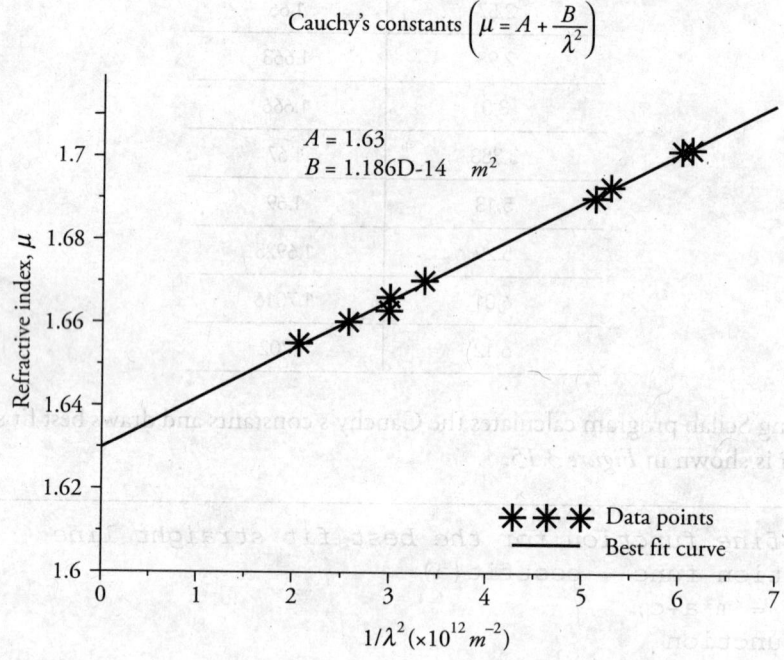

$$\text{Cauchy's constants}\left(\mu = A + \frac{B}{\lambda^2}\right)$$

$A = 1.63$
$B = 1.186D\text{-}14 \quad m^2$

Figure 3.15 Graph for determination of Cauchy's constants

3.6.4 RC Time constant

RC time constant (τ) is the time required to discharge a capacitor through a resistor from initial charge voltage to 36.8% of the initial charge voltage. The voltage across the capacitor is

$$V = V_o e^{-t/\tau} \tag{3.30}$$

Table 3.5 gives the value of voltage across a capacitor at different times. The variables $V(\equiv y)$ and time $t(\equiv x)$ are related by the relation $y = \alpha e^{\beta x}$.

Table 3.5   Data for determining the RC time constant

Time (s)	Voltage across Capacitor (V)
0.0	11.5
1.0	7.0
2.0	3.5
3.0	2.0
4.0	1.5
5.0	0.8
6.0	0.5
7.0	0.2
8.0	0.15
9.0	0.08
10.0	0.06

The following Scilab program calculates the time constant for above data. The best fit graph is drawn in *Figure 3.16*.

```
//Define the function for best fit curve
function func = bestfit(a)
func = c*exp(m*a);
endfunction

//Enter the data points and plot them
x = [0.0 1.0 2.0 3.0 4.0 5.0 6.0 7.0 8.0 9.0 10.0];
y = [11.5 7.0 3.5 2.0 1.5 0.8 0.5 0.2 0.15 0.08 0.06];
plot2d(x,y)

n = length(x);

//Change of variable
for i = 1:n
    z(i) = log(y(i));
end
//Calculation of time constant
x1 = sum(x);
```

```
x2 = sum(x.*x);

x1y1 = sum(x'.*z);                  //Calculate $\sum_{i=1}^{n} x_i \log(y(i))$
y1 = sum(z);                        //Calculate $\sum_{i=1}^{n} \log(y(i))$

//Use of matrix method to determine time constant
A = [x2 x1; x1 n];
B = [x1y1; y1];
C = A\B;
m = C(1);
tau = -1.0/m;
c = C(2);
c = exp(c);

//Range for calculation of best fit curve
z= 0:0.1:x(n);
fplot2d(z,bestfit)                  //Plot the curve
```

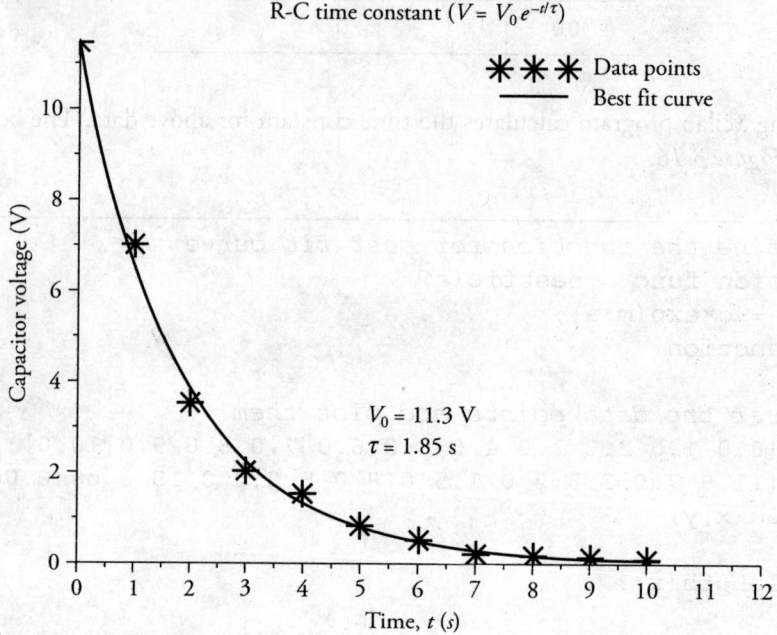

R-C time constant ($V = V_0 e^{-t/\tau}$)

$V_0 = 11.3\ \text{V}$
$\tau = 1.85\ \text{s}$

Figure 3.16   Graph for determination of RC-time constant

3.6.5 Coefficient of electronic heat capacity and Debye's temperature

The specific heat data of an element at different temperatures is given in *Table 3.6*. From Debye's model of specific heat, at low temperatures, the heat capacity of metals is the sum of contribution from electrons (linear dependence on temperature) and phonons (cubic dependence on temperature). This dependence is given in *Eqn. 3.31*.

Table 3.6   Data for variation of specific heat of an element

Temperature, T (K)	Specific Heat, C ($\times 10^{-4}$ Jmole^{-1}K^{-1})
1.0	1.0
1.5	2.0
2.0	4.0
2.5	8.0
3.0	13.5
3.5	21.5
4.0	32.0

$$C = \alpha T + \beta T^3 \tag{3.31}$$

Here

- α is called the coefficient of electron heat capacity

$$\beta = \frac{12\pi^4 Nk}{5T_D{}^3} \quad \Rightarrow \quad T_D = \sqrt[3]{\frac{12\pi^4 Nk}{5\beta}}$$

- T_D is the Debye's temperature

The following Scilab program plots the given data such that $\dfrac{C}{T}$ is along the y-axis and T^2 is along the x-axis. A straight line is fit to this data, such that,

$$y = mx + c \Leftrightarrow \frac{C}{T} = \beta T^2 + \alpha$$

The intercept of the best fit curve gives the value of α. The slope of the best fit curve is β from which Debye's temperature can be easily calculated. The graph in *Figure 3.17* shows the best fit curve for the determination of electronic heat capacity.

```
n = input("How many observations? ")
disp("Enter values of temperature: ")
for i = 1:n
    x(i) = input("")
    X(i) = x(i)*x(i);
end
disp("Enter corresponding values of specific heat: ")
for i = 1:n
    y(i) = (10^-4)*input("")
    Y(i) = y(i)/x(i);
end

A(1,1) = sum(X.*X);
A(1,2) = sum(X);
A(2,1) = sum(X);
A(2,2) = n;
B(1) = sum(X.*Y);
B(2) = sum(Y);
Z = A\B
mprintf("\n Electronic heat capacity coefficient is
%f\n",Z(2))

TD = nthroot(12*(%pi^4)*8.31/(5*Z(1)),3);
mprintf("\n Debye''s temperature is %f K\n",TD)

F = Z(1)*X+Z(2);
plot2d(X,F)                     //Plot the best fit curve
plot(X,Y)                       //Plot the data points
```

The input parameters are as follows.

```
How many observations? 7
Enter values of temperature:
1
1.5
2
2.5
3
3.5
4
```

```
Enter corresponding values of specific heat:
1.0
2.0
4.0
8.0
13.5
21.5
32.0
```

Answer is as follows.

```
Electronic heat capacity coefficient is 0.000028
Debye's temperature is 344.066422 K
```

3.6.6 Lennard–Jones potential

The Lennard–Jones potential describes the potential energy of interaction between a pair of molecules or neutral atoms. *Eqn. 3.32* gives the expression for the Lennard–Jones potential.

$$V = \varepsilon\left[\left(\frac{\sigma}{r}\right)^{12} - 2\left(\frac{\sigma}{r}\right)^{6}\right] \qquad (3.32)$$

In *Eqn. 3.32*,

- ε is the depth of the potential well
- r is the distance between particles
- σ is the distance at which the potential becomes minimum.
- Value of the potential at $r = \sigma$ is equal to $-\varepsilon$.
- Finite distance at which the inter-particle potential is zero is given by $\sigma / \sqrt[6]{2}$.
- This distance is also called Van der Waal's radius.
- The first term on the right (r^{-12} term) corresponds to Pauli's repulsion at short range.
- The second on the right (r^{-6} term) corresponds to the long range attraction.

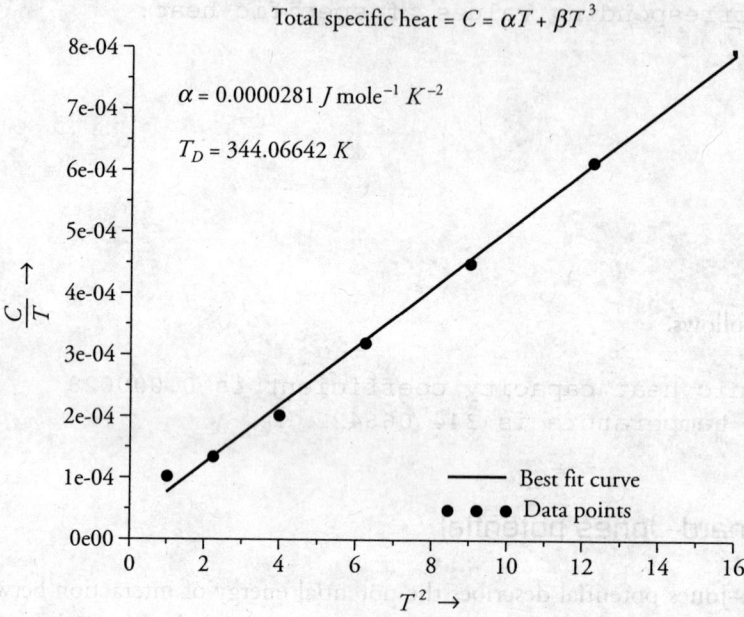

Figure 3.17   Graph for electronic heat capacity

The concept of least square curve fitting can be used to fit the data set given in *Table 3.7* and determine the distance at which the inter-particle potential is minimum.

Table 3.7   Data set for Lennard–Jones Potential

Inter-particle distance (r)	V	$\dfrac{V}{\varepsilon}$
3.2	10	2
3.3	5	1
3.35	−2.5	−0.5
4.2	−4	−0.8
5.2	−1	−0.2
6.3	−0.5	−0.1
8.3	0	0
10.2	0	0

The following Scilab program fits the given data set with the Lennard–Jones potential and determines the value of sigma.

```
function y = LJP(x)
y = ((sigma/x).^12 - 2*((sigma/x).^6));
endfunction

x = [3.2 3.3 3.35 4.2 5.2 6.3 8.3 10.2];
y = [10 5 -2.5 -4 -1 -0.5 0 0]/5;

mindif = 1D30;
sigma_min = 0;

for sigma = 2:0.1:5;
    funcprot(0);
    diff = 0;

    for i = 1:length(x);
        diff = diff + (LJP(x(i)) - y(i)).^2;
    end
    if diff<mindif then
        mindif = diff;
        tmin = sigma;
    end
end

sigma = tmin;
x1 = [min(x):0.01:max(x)];
fplot2d(x1,LJP);
plot2d(x,y);
```

3.6.7 Spectral radiance of blackbody radiation

Planck's law for blackbody radiation states that the spectral radiance (amount of energy emitted per unit surface area of the blackbody, per unit time, at temperature T, per unit solid angle over which the radiation is measured, per unit wavelength) is given by *Eqn. 3.33*. *Table 3.8* gives the set of data points that follow Planck's radiation law.

$$f(\lambda) = \frac{2hc^2}{\lambda^5} \frac{1}{\exp\left(\dfrac{hc}{\lambda kT}\right) - 1} \tag{3.33}$$

Table 3.8   Data for determining the temperature of blackbody radiation

Wavelength (nm)	Flux density ($\times 10^{11}$ Watt m^{-2})
600	1.67
1200	4.37
1700	3.04
2300	1.77
2800	1.04
3400	0.63

Lennard-Jones potential $\left[V = \epsilon \left\{ \left(\frac{\sigma}{r} \right)^{12} - 2 \left(\frac{\sigma^6}{r} \right) \right\} \right]$

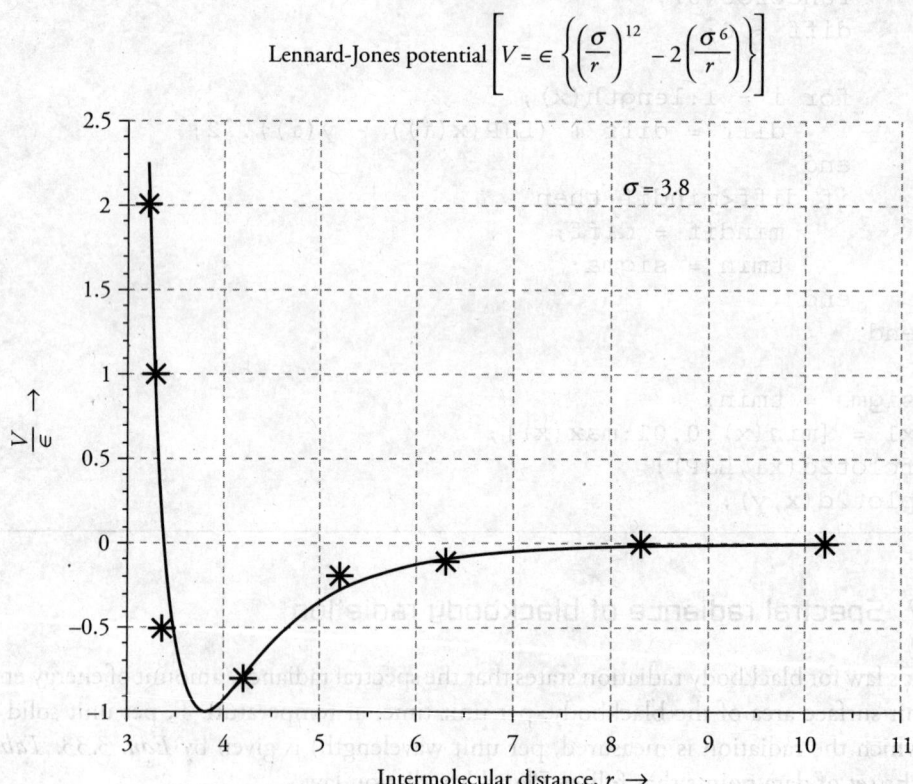

Figure 3.18   Lennard–Jones potential

It is desired to write a Scilab program for determining the temperature at which Planck's radiation law fits best to these data points. It is clear from the data that the peak radiation occurs around wavelength 1200 nm. Therefore from Wien's law ($\lambda_m T = 0.002898$), the best fit temperature should have a value of approximately 2415 K.

The following Scilab program performs least square fitting to find the temperature that minimizes the residue. The best fit curve is given in *Figure 3.19*.

```
c = 2.997925e8;
k = 1.381e-23;
h = 6.626e-34;

//Function for Planck's Radiation Law
function y = PL(lamda)
y = 2*h*c*c/((lamda^5)*(exp(h*c/(lamda*k*T))-1))
endfunction

//Enter the data points
data_lamda = [600,1200,1700,2300,2800,3400]*1e-9;
data_radiance = [1.67,4.37,3.04,1.77,1.04,0.63]*1e11;

mindif = 1D30;          //A large number for comparison
tmin = 0;               //Initialize temperature to zero

//Give the temperature range around the expected value
for T = 2000:1:3000
funcprot(0);   //Clear the function for every temperature
diff = 0;               //Initialize the difference to zero
    for i = 1:6
        diff = diff + (PL(data_lamda(i)) - data_
radiance(i))^2;
    end
    if diff < mindif then
        mindif = diff;
        tmin = T;
    end
end

T = tmin;
lamda = [5e-7:5e-9:3.5e-6];

//Plot Planck's radiation curve corresponding to the best
fit temperature
fplot2d(lamda,PL);

//Plot the data points
plot(data_lamda,data_radiance)
```

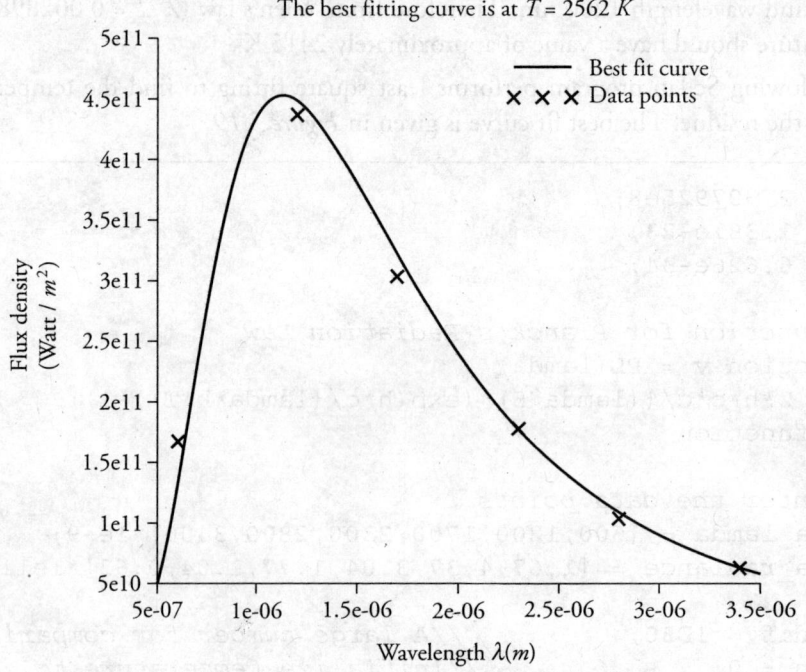

The best fitting curve is at $T = 2562\ K$

Figure 3.19 Least square curve fitting for Planck's radiation law

3.7 Exercises

1. Suppose an independent variable 'x' varies between 1 and 10. Write a Scilab program to generate random y-values such that $y \sim \div x$. Perform least square curve fitting (matrix method) on this random trend. Plot the best fit curve.

2. The following observations were made while performing an experiment for the determination of spring constant by dynamical method. Write a Scilab program to fit this data set and determine the value of the spring constant. Moreover, determine mass of the spring.

Load (Mass in gram)	Time taken for 50 oscillations (in seconds)
300	15.28
400	16.94
500	19.02
600	20.16
700	21.69

3. Suppose an independent variable 'x' varies between 0 and 1. Write a Scilab program to generate random y-values such that

$$y \sim e^{-3x}$$

Perform least square curve fitting on this trend and plot the best fit curve.

4. Plot the following data points and determine the best fit. Assume that the data points follow the relationship,

$$y_i = \alpha x_i^\beta$$

x	y
10	55
20	210
30	440
40	794
50	1205
60	1812
70	2451
80	3172
90	4022
100	5020

5. Write a Scilab program for fitting a polynomial to the following.

x	y
-3	23
-2	10
-1	5
0	1
1	3
2	8
3	15

6. The amplitude of oscillation (A) of a vibrating spring decays exponentially such that

$$A = A_0 \, e^{-\alpha t}$$

The following data gives the amplitude at various instants of time. Write a Scilab program to determine the value of decay constant (α).

Time, t (s)	Amplitude, A (cm)
0	25
10	14
20	7.5
30	4.2
40	2.3
50	1.3

7. The following set of data points follow Planck's radiation law. Write a Scilab program for determining the temperature at which Planck's radiation law fits best to these data points.

Wavelength (µm)	Spectral Radiance ($\times 10^6$ Watt m^{-2})
4	1.5
6	5.2
8	9.1
10	9.9
12	8.9
14	7.4
16	6.1
18	4.6
20	3.6
22	3.0
24	2.5
26	1.9
28	1.5

8. Write a Scilab program to fit the following data set using the 'datafit' Scilab function. Take the model to be

$$C_1 x + e^{-C_2 x} \sin(x + C_3)$$

x	y
2	0
4	0.2
6	1.1
8	1.2
10	1
12	2
14	1.9
16	1.8
18	2.5

9. Write a Scilab program to fit an eclipse profile to the following data set by using the 'datafit' Scilab function.

x	y
1	4.1
2	3.8
3	3
4	1.1
5	0.9
6	1.1
7	1.1
8	1
9	2
10	4.8
11	5.1
12	5

10. Write a Scilab program to fit the data given in *Exercises 1 – 7* using the 'datafit' Scilab function.

Ordinary Differential Equation

After going through this chapter, the reader should be able to

◊ Write Scilab programs to solve initial value problems involving first and second order linear ordinary differential equations using Euler's method, the modified Euler's method and the Runge–Kutta methods.

◊ Understand the difference between these numerical methods for determining the approximate solution of differential equations.

◊ Appreciate the significance of 'step size' in these numerical methods.

◊ Write Scilab programs for solving boundary value problems involving first and second order linear differential equations using the finite difference method.

◊ Apply the aforementioned skills to solve advanced problems, such as problems based on,

  ◊ Radioactive decay

  ◊ Orthogonal trajectories

  ◊ Motion of a freely falling object

  ◊ Atwood's machine

  ◊ Simple harmonic motion

  ◊ Damped motion

  ◊ Forced motion

  ◊ Schrödinger equation

  ◊ Lagrangian mechanics

4.1 Introduction

Differential equations are key mathematical tools for modelling physics problems. They are frequently used in all branches of physics while expressing the variation in one quantity w.r.t. the other. There are various kinds of differential equations, such as:

- Ordinary differential equations with initial and boundary value problems.

- Partial differential equations involving functions of multiple independent variables and their partial derivatives.

This chapter introduces the necessary numerical tools for determining approximate solutions of ordinary differential equations. It focuses only on initial and boundary value problems involving first and second order ordinary linear differential equations. These equations contain functions of one independent variable, and derivatives in that variable.

There are several numerical techniques for determining the solutions of differential equations. In this chapter, some commonly used methods have been explained. *Section 4.2* shows the use of Euler's method to determine the solution of a differential equation. This is followed by modified Euler's method in *Section 4.3*, Runge–Kutta second order method in *Section 4.4* and Runge–Kutta fourth order method in *Section 4.5*. A graphical comparison of these four methods is presented in *Section 4.6*. In *Section 4.7*, a quick review of the finite difference method has been provided for second order boundary value problems. Some advanced application problems of physics involving the first and second order differential equations have been discussed in *Section 4.8*.

This chapter uses the plotting skills developed in the second chapter. The reader is encouraged to refine their understanding of the plotting techniques.

4.2 Euler's Method

This is the most basic method of numerical integration. It is a first order method for approximating solutions of differential equations. This method uses the initial value as the starting point and approximates the next point of the solution curve using a tangent line to that point. The accuracy crucially depends on the step size used to approximate the subsequent point on the solution curve.

The algorithm for writing a Scilab program based on Euler's method is explained in *Section 4.2.1* (first order) and in *Section 4.2.2* (second order) with the help of suitable examples.

4.2.1 First order differential equation

Consider the general form of a first order differential equation as given in *Eqn. 4.1*.

$$\frac{dy}{dx} = f(x, y) \tag{4.1}$$

The initial condition is given in *Eqn. 4.2*.

$$y(x_0) = y_0 \qquad (4.2)$$

The aim of any numerical method is to solve this differential equation and determine the value of dependent variable (*y*) at a given value of independent variable (*x*). In Euler's method, the slope or derivative of '*y*' at the initial point is used to extrapolate the solution of the differential equation at the next step-point. According to this method, the general formula for solving the differential equation is given by *Eqn. 4.3*.

$$y_{n+1} = y_n + hf(x_n, y_n) \qquad (4.3)$$

In *Eqn. 4.3*,

- *n* corresponds to the nth solution of the differential equation, starting from y_0.
- *h* is the step size

Euler's method is a slow process for determining the solutions of differential equations and its accuracy depends on the value of *h*. To get a reasonably accurate solution, the value of *h* should be as small as possible. However, a smaller value implies a slower speed of calculation.

The following example shows how to write a Scilab program for using Euler's method to determine the solution of a first order differential equation. The significance of the value of *h* has also been highlighted.

Consider the differential equation given in *Eqn. 4.4*.

$$\frac{dy}{dx} = -y \qquad (4.4)$$

For *Eqn. 4.4*,

- The initial condition is *y*(at *x* = 0) = 1.
- The aim is to determine the solution at *x* = 0.1, 0.2, 0.3, 0.4.

The step wise algorithm for writing the Scilab program is as follows.

- **Step 1:** Write a user-defined function for Euler's formula given in *Eqn. 4.3*.

```
function [x,y] = euler(initial_x,initial_y,h,final);
i = 1;
x(i) = initial_x;
y(i) = initial_y;
while(x(i) < final);
    x(i+1) = x(i) + h;
    y(i+1) = y(i) + (f(x(i),y(i)).*h);
```

```
        disp('Euler_Method : Solution at x = ' +string(x(i))+
' is : ' +string(y(i)));
    i = i+1;
end
endfunction
```

- **Step 2:** Write a user-defined function for the differential equation given in *Eqn. 4.4*.

```
function dy = f(x,y);
dy = -y;
endfunction
```

- **Step 3:** Define the initial values, step size and the final value of the independent variable

```
x(1)=input('Enter initial value of independent variable : ');
y(1)=input('Enter initial value of dependent variable : ');
final=input('Enter final value of independent variable : ');
h = input('Enter step size : ');
```

- **Step 4:** Call the function for Euler's formula

```
[x,y] = euler(x(1),y(1),h,final);
```

- **Step 5:** For comparison, the built-in Scilab function 'ode' can also be used to determine the solution of the differential equation. The simplest syntax for the 'ode' command is

$$y = ode(y0,t0,t,f)$$

Here,

- The argument, $y0$, represents the initial value of the dependent variable
- The argument, $t0$, represents the initial value of the independent variable
- The argument, t, is a vector of points at which the solution y has to be computed
- The argument, f, corresponds to the right side function of the first order differential equation.

The Scilab program based on the use of 'ode' is written as follows.

```
j = x(1);
while(j < max(x));
    ode_result = ode(y(1),x(1),j,f);
    disp('ode : Solution at x = ' +string(j)+ ' is : '
```

```
                +string(ode_result));
      j = j+h;
   end
```

The input parameters used for this example are

```
x(1) = 0;
y(1) = 1;
final = 0.5;
h = 0.1;
```

The result of the Scilab code from Euler's formula is as follows.

```
Euler_Method : Solution at x = 0 is : 1
Euler_Method : Solution at x = 0.1 is : 0.9
Euler_Method : Solution at x = 0.2 is : 0.81
Euler_Method : Solution at x = 0.3 is : 0.729
Euler_Method : Solution at x = 0.4 is : 0.6561
```

The result of the Scilab code from the built-in function is as follows.

```
ode : Solution at x = 0 is : 1
ode : Solution at x = 0.1 is : 0.9048
ode : Solution at x = 0.2 is : 0.8187
ode : Solution at x = 0.3 is : 0.7408
ode : Solution at x = 0.4 is : 0.6703
```

The analytical solution is

$$y = e^{-x}$$

Therefore, the direct results from the analytical solution are

```
exp(-0.1) = 0.9048;
exp(-0.2) = 0.8187;
exp(-0.3) = 0.7408;
exp(-0.4) = 0.6703;
```

It is clear from the aforementioned steps that

- The built-in 'ode' function gives a more accurate result.
- As compared to the analytical solution, results from Euler's formula are less accurate.

- It can be verified that for smaller step sizes ($h = 0.001$), the analytical solution provides a much better match as shown in the following.

```
Euler_Method : Solution at x = 0.1 is : 0.9048
Euler_Method : Solution at x = 0.2 is : 0.8186
Euler_Method : Solution at x = 0.3 is : 0.7407
Euler_Method : Solution at x = 0.4 is : 0.6702
```

4.2.2 Second order differential equation

Euler's method for solving second order differential equations has been explained for differential equation (*Eqn. 4.5*) and initial conditions (*Eqn. 4.6*).

$$\frac{d^2y}{dx^2} + 4\frac{dy}{dx} + 3y = 0 \qquad (4.5)$$

$$y(\text{at } x = 0) = 1$$

$$\left.\frac{dy}{dx}\right|_{x=0} = 5 \qquad (4.6)$$

Eqn. 4.7 gives the analytical solution of *Eqn. 4.5*

$$y = 4e^{-x} - 3e^{-3x} \qquad (4.7)$$

In this method, the second order differential equation is reduced to two first order equations that can be solved simultaneously. This is done in the following manner.

Let

$$\frac{dy}{dx} = z \qquad (4.8)$$

This implies that

$$\frac{dz}{dx} = \frac{d^2y}{dx^2} = -4\frac{dy}{dx} - 3y = -4z - 3y \qquad (4.9)$$

The following Scilab program uses Euler's method to solve these two first order differential equations (*Eqn. 4.8* and *Eqn. 4.9*) simultaneously.

```
function [t,x,y] = euler2(initial_t,initial_x,initial_y,
h,final);
i = 1;
t(i) = initial_t;
x(i) = initial_x;
y(i) = initial_y;
while (t(i) < final);
    t(i+1) = t(i) + h;
    x(i+1) = x(i) + (f1(t(i), x(i), y(i)).*h);
    y(i+1) = y(i) + (f2(t(i), x(i), y(i)).*h);
    i = i+1;
end
endfunction
```

As compared to the previous Euler's program, this program has three variables (t, x, y) and two functions $(f_1$ and $f_2)$, one each for the first order differential equations.

The following steps demonstrate the usefulness of this program.

- **Step 1:** Define user-defined Scilab functions for both the first order differential equations.

```
function yprime = f1(x,y,z)
yprime = z;
endfunction

function zprime = f2(x,y,z)
zprime = (-4*z) + (-3*y);
endfunction
```

- **Step 2:** Define the initial values, step size and the final value of independent variable.

```
x(1) = 0;                                    //Initial x
y(1) = 1;                                    //Initial y
z(1) = 5;                                    //Initial z
final = 3;                                     //Final x
h = 0.1;                                     //Step size
```

- **Step 3:** Call the function for Euler's formula and plot the result

```
[x,y,z] = euler2(x(1),y(1),z(1),h,final);
plot2d(x,y);
```

- **Step 4:** Euler's formula can be invoked again for a different step size. A smaller step size gives a more accurate solution curve.

- **Step 5:** For comparison, the analytical solution of the second order differential equation has also been generated (*Eqn. 4.7*) and plotted in *Figure 4.1* using the following commands.

```
x = 0:0.1:3;
plot2d(x,4*exp(-x)-3*exp(-3*x));
```

4.3 Modified Euler's Method

As seen in the previous section, Euler's method can provide reasonably accurate solutions of ordinary differential equations only if the step size is small. The modified Euler's method is an improvement over this scheme. For the same step size as in Euler's method, the modified version uses the ordinary differential equation to evaluate the slope of the solution curve at the starting point and at the subsequent point. It then averages the two slopes to determine an improved step size. This averaging scheme is made clear with the help of an appropriate example.

Consider the general form of the first order differential equation given in *Eqn. 4.1* and the initial condition given in *Eqn. 4.2*. According to the modified Euler's method, the solution of this differential equation is given by *Eqn. 4.10*.

$$y_1^{(n+1)} = y_0 + \frac{h}{2}\left[f\left(x_0, y_0\right) + f\left(x_1, y_1^{(n)}\right)\right] \tag{4.10}$$

$$\frac{d^2 y}{dx^2} + 4\frac{dy}{dx} + 3y = 0$$

$$y(0) = 1 \; ; \; \left.\frac{dy}{dx}\right|_0 = 5$$

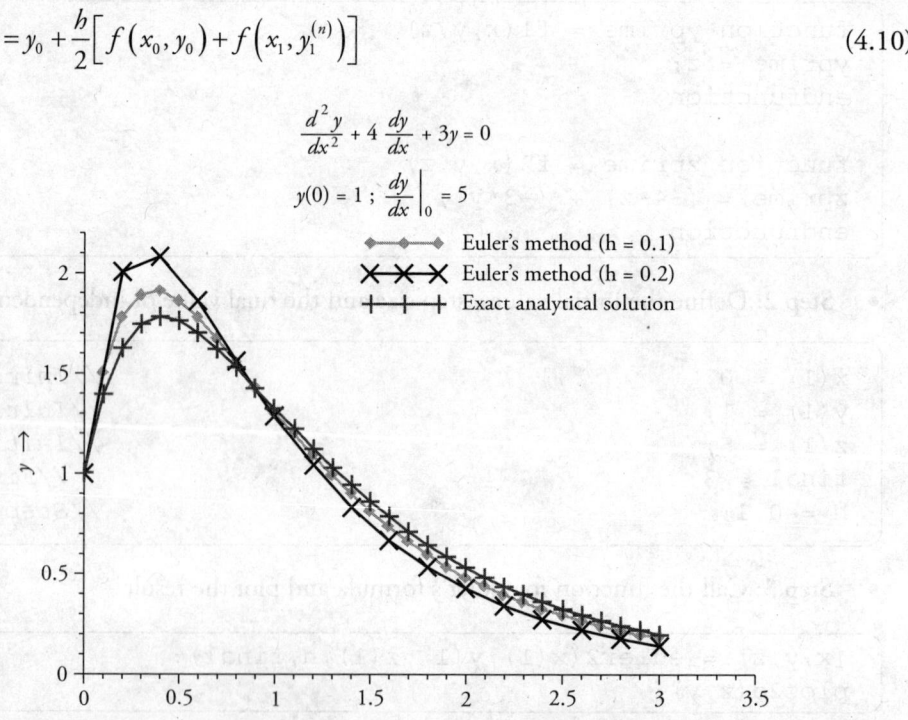

Figure 4.1   Solution of the second order differential equation

In *Eqn. 4.10,*

- n corresponds to the nth approximation to y_1.
- h is the step size
- The starting value of y_1 is generally taken from Euler's method. It is given by *Eqn. 4.11.*

$$y_1^{(0)} = y_0 + h f\left(x_0, y_0\right) \tag{4.11}$$

The following example shows how to write a Scilab program for using modified Euler's method to determine the solution of a first order differential equation.

Consider the differential equation given in *Eqn. 4.4* along with its initial condition. The aim is to determine the solution at x = 0.1, 0.2, 0.3, 0.4, 0.5. The step wise procedure for writing the Scilab program is as follows.

- **Step 1:** Write a user-defined function for the modified Euler's formula given in *Eqn. 4.10.*

```
function[x,y] = modeuler(initial_x,initial_y,h,final)
i = 1;
x(i) = initial_x;
y(i) = initial_y;
while(x(i) < final);
    x(i+1) = x(i) + h;
    y(i+1) = y(i) + (f(x(i),y(i)).*h);
    z = y(i) + (h/2).*(f(x(i),y(i)) + f(x(i+1),y(i+1)));
        disp('Modified Euler_Method : Solution at x = '
        +string(x(i+1))+ ' is : ' +string(z));
    i = i+1;
end
endfunction
```

- **Step 2:** Define a function for the given differential equation

```
function dy = f(x,y);
dy = -y;
endfunction
```

- **Step 3:** Define the initial values, step size and final value of the independent variable.

```
x(1)=input('Enter initial value of independent variable : ');
y(1)=input('Enter initial value of dependent variable : ');
final=input('Enter final value of independent variable : ');
h=input('Enter step size : ');
```

- **Step 4:** Call the function for the modified Euler's formula

```
[x,y] = modeuler(x(1),y(1),h,final);
```

The input parameters used for this example are

```
x(1) = 0;
y(1) = 1;
final = 0.5;
h = 0.1;
```

The result of the Scilab program from the modified Euler's formula is as follows.

```
Modified Euler_Method : Solution at x = 0.1 is : 0.905
Modified Euler_Method : Solution at x = 0.2 is : 0.814
Modified Euler_Method : Solution at x = 0.3 is : 0.733
Modified Euler_Method : Solution at x = 0.4 is : 0.659
Modified Euler_Method : Solution at x = 0.5 is : 0.593
```

This solution is more accurate than that determined from Euler's method.

4.4 Second Order Runge–Kutta Method

It was seen in *Section 4.2*, that Euler's method is inefficient in the sense that it gives a reasonably accurate result only if the value of step size is small. This is because the estimation of the solution at any subsequent point is based on the rate of change of variable at the current point. However, a small step size results in longer computation time.

If a larger step size is used, then Euler's method is bound to give a poor estimation of the solution. In this case, the method is unable to take into account the true curvature of the solution curve.

In the Runge–Kutta method, the initial derivative at each point is used to find a point halfway across the interval up to the subsequent point. Therefore, for the same step size, the Runge–Kutta method gives better results as compared to Euler's method.

Consider the general form of a first order differential equation given in *Eqn. 4.1* and the initial condition given in *Eqn. 4.2*. According to the second order Runge–Kutta method, the general formula for the solution of this differential equation is given by *Eqn. 4.12*.

$$y_1 = y_0 + \frac{1}{2}\left[k_1 + k_2\right] \tag{4.12}$$

In *Eqn. 4.12*,

- k_1 gives an estimate of the solution that is determined from the value of the function ($f(x, y)$) at the beginning. It is used to step half way through the time step. Its value is given by *Eqn. 4.13*.

$$k_1 = h f\left(x_0, y_0\right) \tag{4.13}$$

- Similarly, k_2 gives an estimate of the slope in the second half of the interval. It is given by *Eqn. 4.14*.

$$k_2 = h f\left(x_0 + h, y_0 + k_1\right) \tag{4.14}$$

- h is the step size

The following example shows how to write a Scilab program for using the second order Runge–Kutta method for finding the solution of the first order differential equation.

Consider the differential equation given in *Eqn. 4.4* along with its initial condition. The aim is to determine the solution at $x = 0.1, 0.2, 0.3, 0.4$. The step-wise procedure for writing the Scilab program is as follows.

- **Step 1:** Write a user-defined Scilab function for the second order Runge–Kutta method given in *Eqn. 4.12*.

```
function [x,y] = rk2(initial_x,initial_y,h,final);
i = 1;
x(i) = initial_x;
y(i) = initial_y;
while (x(i) < final);
    x(i+1) = x(i) + h;
    k1 = h.*(f(x(i),y(i)));
    k2 = h.*(f(x(i)+h,y(i)+k1));
    y(i+1) = y(i)+((k1+k2)./2);
    disp('Second_Order_Runge-Kutta : Solution at x = '
        +string(x(i))+ ' is : ' +string(y(i)));
    i = i+1;
end
endfunction
```

- **Step 2:** Define a function for the given differential equation.

```
function alpha = f(x,y);
alpha = -y;
endfunction
```

- **Step 3:** Define the initial values, step size and the final value of the independent variable.

```
x(1)=input('Enter initial value of independent variable :
');
y(1)=input('Enter initial value of dependent variable : ');
final=input('Enter final value of independent variable : ');
h=input('Enter step size : ');
```

- **Step 4:** Call the function for the second order Runge–Kutta method

```
[x,y] = rk2(x(1),y(1),h,final);
```

The input parameters used for this example are as follows:

```
x(1) = 0;
y(1) = 1;
final = 0.5;
h = 0.1;
```

The result of the Scilab code from the Runge–Kutta method is as follows.

```
Second_Order_Runge-Kutta : Solution at x = 0 is : 1
Second_Order_Runge-Kutta : Solution at x = 0.1 is : 0.905
Second_Order_Runge-Kutta : Solution at x = 0.2 is : 0.8190
Second_Order_Runge-Kutta : Solution at x = 0.3 is : 0.7412
Second_Order_Runge-Kutta : Solution at x = 0.4 is : 0.6708
```

This solution is more accurate than that provided by Euler's methods discussed in *Sections 4.2* and *4.3*.

4.5 Fourth Order Runge–Kutta Method

In the previous sections, it was shown that an estimate of the slope at two points of an interval (using the second order Runge–Kutta method) gives a more accurate result as compared to a single estimation (in Euler's method). It is therefore practical to assume that more estimates of the slope will give even more accurate results. *Sections 4.5.1* and *4.5.2* respectively explain this aspect for the first and second order differential equations.

4.5.1 First order differential equation

Consider the general form of the first order differential equation (*Eqn. 4.1*) and the initial condition (*Eqn. 4.2*). According to the fourth order Runge–Kutta method, the general formula for the solution of this differential equation is given by *Eqn. 4.15*.

$$y_1 = y_0 + \frac{1}{6}\left[k_1 + 2k_2 + 2k_3 + k_4\right] \qquad (4.15)$$

In *Eqn. 4.15*,

- k_1 gives an estimate of the slope at the beginning of the time step (*Eqn. 4.16*).

$$k_1 = h f\left(x_0, y_0\right) \qquad (4.16)$$

- k_2 gives an estimate of the slope at the midpoint of the interval (*Eqn. 4.17*).

$$k_2 = h f\left(x_0 + \frac{h}{2}, y_0 + \frac{k_1}{2}\right) \qquad (4.17)$$

- k_3 also gives an estimate of the slope at the midpoint of the interval (*Eqn. 4.18*).

$$k_3 = h f\left(x_0 + \frac{h}{2}, y_0 + \frac{k_2}{2}\right) \qquad (4.18)$$

- k_4 gives an estimate of the slope at the end point (*Eqn. 4.19*).

$$k_4 = h f\left(x_0 + h, y_0 + k_3\right) \qquad (4.19)$$

- h is the step size

The following example shows how to write a Scilab program for using the fourth order Runge–Kutta method for finding the solution of the first order differential equation. Consider the differential equation given in *Eqn. 4.4* along with its initial condition. The aim is to determine the solution at $x = 0.1, 0.2, 0.3, 0.4$. The step-wise procedure for writing the Scilab program is as follows.

- **Step 1:** Write a user-defined Scilab function for the fourth order Runge–Kutta method given in *Eqn. 4.15*.

```
function [x,y] = rk4(initial_x,initial_y,h,final);
i = 1;
x(i) = initial_x;
y(i) = initial_y;
while (x(i) < final);
    x(i+1) = x(i) + h;
    k1 = h.*(f(x(i),y(i)));
    k2 = h.*(f(x(i) + (h/2),y(i) + (k1/2)));
    k3 = h.*(f(x(i) + (h/2),y(i) + (k2/2)));
    k4 = h.*(f(x(i) + h, y(i) + k3));
    y(i+1) = y(i) + ( ((k1 + (2.*(k2+k3)) + k4))./6 );
```

```
        disp('Fourth_Order_Runge-Kutta : Solution at x = `
             +string(x(i))+ ` is : ` +string(y(i)));
        i = i+1;
    end
endfunction
```

- **Step 2:** Define the function for the given differential equation

```
function alpha = f(x,y);
alpha = -y;
endfunction
```

- **Step 3:** Define initial values, step size and final value of the independent variable

```
x(1)=input('Enter initial value of independent variable
: ');
y(1)=input('Enter initial value of dependent variable :
');
final=input('Enter final value of independent variable :
');
h=input('Enter step size : ');
```

- **Step 4:** Call the function for the fourth order Runge–Kutta method

```
[x,y] = rk4(x(1),y(1),h,final);
```

The input parameters used for this example are as follows

```
x(1) = 0;
y(1) = 1;
final = 0.5;
h = 0.1;
```

The result of the Scilab program from the fourth order Runge–Kutta method is

```
Fourth_Order_Runge-Kutta : Solution at x = 0 is : 1
Fourth_Order_Runge-Kutta : Solution at x = 0.1 is : 0.9048
Fourth_Order_Runge-Kutta : Solution at x = 0.2 is : 0.8187
Fourth_Order_Runge-Kutta : Solution at x = 0.3 is : 0.7408
Fourth_Order_Runge-Kutta : Solution at x = 0.4 is : 0.6703
```

4.5.2 Second order differential equation

The Runge–Kutta method to solve the second order differential equations is explained in this section for the differential equation given in *Eqn. 4.5* and the initial conditions given in *Eqn. 4.6*. The analytical solution of this differential equation is given in *Eqn. 4.7*.

As done in the case of Euler's method, the second order differential equation is reduced to two first order equations that are then solved simultaneously (Refer to *Eqns. 4.8–4.9*). The following Scilab program shows the Runge–Kutta method for solving these two first order equations simultaneously.

```
function [t,x,y] =
rk42(initial_t,initial_x,initial_y,h,final);
i = 1;
t(i) = initial_t;
x(i) = initial_x;
y(i) = initial_y;
while (t(i) < final);
    t(i+1) = t(i) + h;
    k1 = h.*(f1(t(i),x(i),y(i)));
    l1 = h.*(f2(t(i),x(i),y(i)));

    k2 = h.*(f1(t(i) + (h/2),x(i) + (k1/2),y(i) + (l1/2)));
    l2 = h.*(f2(t(i) + (h/2),x(i) + (k1/2),y(i) + (l1/2)));

    k3 = h.*(f1(t(i) + (h/2),x(i) + (k2/2),y(i) + (l2/2)));
    l3 = h.*(f2(t(i) + (h/2),x(i) + (k2/2),y(i) + (l2/2)));

    k4 = h.*(f1(t(i) + h, x(i) + k3, y(i) + l3));
    l4 = h.*(f2(t(i) + h, x(i) + k3, y(i) + l3));

    x(i+1) = x(i) + ( ((k1 + (2.*(k2+k3)) + k4))./6 );
    y(i+1) = y(i) + ( ((l1 + (2.*(l2+l3)) + l4))./6 );

    i = i+1;
end
endfunction
```

The following steps demonstrate the usefulness of this program.

- **Step 1:** Define the functions for both the first order differential equations.

```
function yprime = f1(x,y,z)
yprime = z;
endfunction
```

```
function zprime = f2(x,y,z)
zprime = (-4*z) + (-3*y);
endfunction
```

- **Step 2:** Define the initial values, step size and the final value of independent variable.

```
x(1) = 0;
y(1) = 1;
z(1) = 5;
final = 3;
h = 0.1;
```

- **Step 3:** Call the function for the Runge–Kutta method and plot the result

```
[x,y,z] = rk42(x(1),y(1),z(1),h,final);
plot2d(x,y);
```

- **Step 4:** For comparison, the analytical solution of the second order differential equation has also been generated and plotted in *Figure 4.2* using the following commands.

```
x = 0:0.1:3;
plot2d(x,4*exp(-x) - 3*exp(-3*x));
```

$$\frac{d^2 y}{dx^2} + 4\frac{dy}{dx} + 3y = 0$$

$$y(0) = 1 \; ; \; \frac{dy}{dx}\bigg|_0 = 5$$

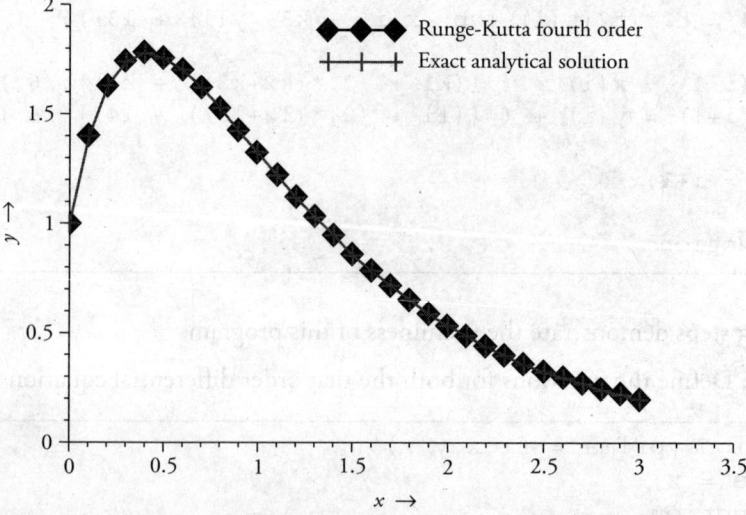

Figure 4.2 Solution of the second order differential equation

For the sake of completeness, the second order differential equation given in the following example has also been solved using the built-in 'ode' function of Scilab. The result is shown in *Figure 4.3*. For using this technique, the second order differential equation is converted into two first order differential equations as shown in *Eqn. 4.20* and *Eqn. 4.21*.

$$Y = \begin{pmatrix} y \\ \dfrac{dy}{dx} \end{pmatrix} \tag{4.20}$$

$$dY = \begin{pmatrix} \dfrac{dy}{dx} \\ \dfrac{d^2 y}{dx^2} \end{pmatrix} \tag{4.21}$$

The following program explains the algorithm.

```
//Give initial values of Y in the form of a matrix.
y0 = [1;5];
//Define both the first order equations in a common
```
function. Here, $dY(1) \equiv \dfrac{dy}{dx} = Y(2)$ and $dY(2) \equiv \dfrac{d^2 y}{dx^2}$

```
function dy = f(x,y);
dy(1) = y(2);
dy(2) = -4.*y(2) - 3.*y(1);
endfunction

x = 0:0.1:3;                    //Range of x variable
y = ode(y0,0,x,f);             //Call the ode function

//The output will have two rows. The number of columns
depends on step size and range of x variable. The first
row corresponds to the solution of the differential
equation (y in this case). The second row corresponds to
first derivative of the solution. Therefore, plot all the
columns of the first row.

plot2d(x,y(1,:));
```

For the sake of completeness, the second order differential equation given in the following example has also been solved using the built-in 'ode' function of Scilab. The result is shown in Figure 4.3. For using this technique, the second order differential equation is converted into two first order differential equations as shown in Eqn. 4.20 and Eqn. 4.21.

$$\frac{d^2 y}{dx^2} + 4\frac{dy}{dx} + 3y = 0$$

$$y(0) = 1 \; ; \; \left.\frac{dy}{dx}\right|_0 = 5$$

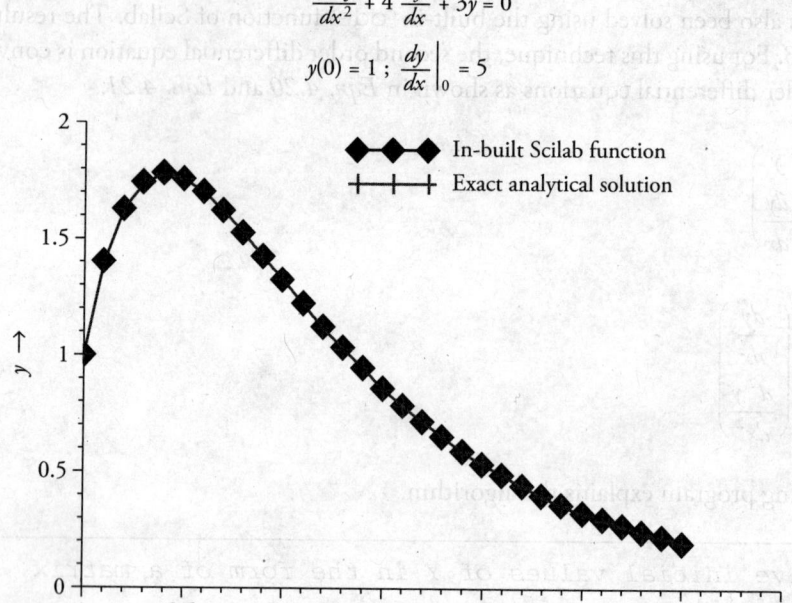

Figure 4.3 Solution of the second order differential equation

4.6 Comparison of the Four Methods

In this section, a graphical comparison of the four methods described in the previous sections has been done with the built-in 'ode' function of Scilab. The computation time required for each of these methods has also been calculated.

Consider the general form of the first order differential equation given in *Eqn. 4.1*. The function for each method can be defined in a similar manner as done in their respective sections. Each function is called with the same initial conditions.

```
x(1) = 0;
y(1) = 1;
final = 1;
h = 0.1;

[x,y] = euler(x(1),y(1),h,final);
plot2d(x,y)

[x,y] = modeuler(x(1),y(1),h,final);
plot2d(x,y)
```

```
[x,y] = rk2(x(1),y(1),h,final);
plot2d(x,y)

[x,y] = rk4(x(1),y(1),h,final);
plot2d(x,y)

j = x(1);
k = 1;

while(j <= max(x));
    ode_result(k) = ode(y(1),x(1),j,f);
    j = j+h;
    k = k+1;
end
plot2d(x,ode_result)
```

The graph showing a comparison between these methods is given in *Figure 4.4*. It is clear from the graph that the Runge–Kutta (second and fourth order) method gives a better estimate of the solution of the differential equation.

The stop watch in Scilab can be used to determine the time required by each method for computing the solution. *Table 4.1* gives the computation time required for each method.

```
tic()
[x,y] = euler(x(1),y(1),h,final);
toc()
tic()
[x,y] = modeuler(x(1),y(1),h,final);
toc()
tic()
[x,y] = rk2(x(1),y(1),h,final);
toc()
tic()
[x,y] = rk4(x(1),y(1),h,final);
toc()
```

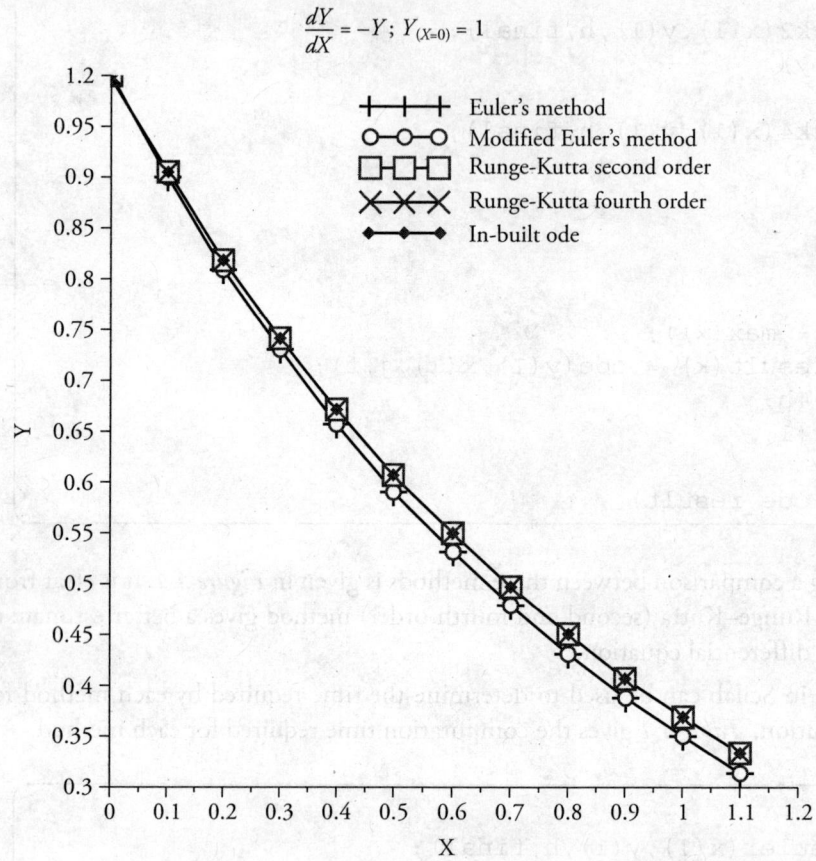

Figure 4.4 Comparative graph of various numerical methods for solving differential equations

Table 4.1 Comparison of computation time for different numerical methods

Method	Step Size	Computation Time (s)
Euler's method	0.1	0.059
	0.001	0.237
Modified Euler's method	0.1	0.064
Runge–Kutta (second order)	0.1	0.068
Runge–Kutta (fourth order)	0.1	0.057

4.7   Finite Difference Method

The finite difference method is a numerical technique that is used to solve boundary value problems of ordinary differential equations. In this method, the solution is approximated by replacing the derivatives with a linear combination of a set of weights and the function values.

Consider the general form of the second order linear differential equation given in *Eqn. 4.22*. The boundary conditions are given in *Eqn. 4.23*.

$$\frac{d^2 y}{dx^2} + f(x)\frac{dy}{dx} + g(x)y = r(x) \tag{4.22}$$

$$y_1 = y(\text{at } x = x_1 = a) = ya$$

$$y_n = y(\text{at } x = x_n = b) = yb \tag{4.23}$$

It is assumed that there are n points (or nodes) in the interval $[a, b]$ at which the solution is to be determined. The step size (or interval between successive nodes) is equal to h and its value is given by *Eqn. 4.24*.

$$h = \frac{b-a}{n-1} \tag{4.24}$$

According to the central difference method, the first order derivative of the dependent variable is given by the approximation given in *Eqn. 4.25*. The finite difference approximation of the second order derivative is given by *Eqn. 4.26*.

$$\frac{dy}{dx} \approx \lim_{h \to 0} \frac{y(x+h) - y(x-h)}{2h} \tag{4.25}$$

$$\frac{d^2 y}{dx^2} \approx \lim_{h \to 0} \frac{y(x+h) - 2y(x) + y(x-h)}{h^2} \tag{4.26}$$

Substituting *Eqns. 4.25–4.26* in *Eqn. 4.22* will give *Eqn. 4.27*.

$$\frac{y(x+h) - 2y(x) + y(x-h)}{h^2} + f(x)\frac{y(x+h) - y(x-h)}{2h} + g(x)y(x) = r(x) \tag{4.27}$$

Eqn. 4.27 implies that

$$\left(1 - \frac{h}{2}f_i\right)y_{i-1} + \left(-2 + g_i h^2\right)y_i + \left(1 + \frac{h}{2}f_i\right)y_{i+1} = r_i h^2 \tag{4.28}$$

In *Eqn. 4.28*,

- The index i depends on the number of intervals between the boundary.
- The ith index corresponds to the value of the variable at x.
- The $(i-1)^{th}$ index corresponds to the value of the variable at $x - h$.
- The $(i+1)^{th}$ index corresponds to the value of the variable at $x + h$.

The following Scilab function determines the solution of boundary value problems by using the finite difference method. The algorithm of this function is as follows.

- This function has syntax 'finite_diff(a,b,h,ya,yb,f,g,r)'. Here,
  - a and b are the boundary values for the variable x.
  - h is the step size.
  - ya and yb are the values of dependent variable y at the boundary.
  - f, g and r are pre-defined functions of x as described in *Eqn 4.22*.
- The number of nodes is determined by using *Eqn. 4.29*.

$$n = \frac{b-a}{h} + 1 \tag{4.29}$$

- A zero matrix (A) of order $(n \times n)$ is created.
- For the first node,

  $i = 1$

  $y_1 = ya$

  - For the last node,

  $i = n$

  $y_n = yb$

  - For the nodes in between, the component form of *Eqn. 4.28* gives,

$$\left(1 - \frac{h}{2}f_2\right)y_1 + \left(-2 + g_2 h^2\right)y_2 + \left(1 + \frac{h}{2}f_2\right)y_3 = r_2 h^2$$

$$\left(1 - \frac{h}{2}f_3\right)y_2 + \left(-2 + g_3 h^2\right)y_3 + \left(1 + \frac{h}{2}f_3\right)y_4 = r_3 h^2$$

and so on till the $(n-1)^{th}$ node,

$$\left(1-\frac{h}{2}f_{n-1}\right)y_{n-2}+\left(-2+g_{n-1}h^2\right)y_{n-1}+\left(1+\frac{h}{2}f_{n-1}\right)y_n=r_{n-1}h^2$$

- The matrix representation of these equations is given by

$$[A]\,[Y]=[B]$$

Here,

$$A=\begin{bmatrix} 1 & 0 & 0 & .. & .. & 0 \\ \dfrac{1}{h^2}-\dfrac{1}{2h}f & g-\dfrac{2}{h^2} & \dfrac{1}{h^2}+\dfrac{1}{2h}f & 0 & .. & 0 \\ 0 & \dfrac{1}{h^2}-\dfrac{1}{2h}f & g-\dfrac{2}{h^2} & \dfrac{1}{h^2}+\dfrac{1}{2h}f & .. & 0 \\ \ddots & \ddots & \ddots & \ddots & \ddots & \ddots \\ 0 & 0 & .. & .. & .. & 1 \end{bmatrix}$$

$$Y=\begin{bmatrix} y_1 \\ y_2 \\ \vdots \\ y_{n-1} \\ y_n \end{bmatrix}$$

$$B=\begin{bmatrix} ya \\ r(x_2) \\ r(x_3) \\ \vdots \\ r(x_{n-1}) \\ yb \end{bmatrix}$$

- The solution vector $[Y]$ is determined using the equation, $[Y]=[B]/[A]$
- This function also saves the value of 'x' variable. These values can be used for plotting the solution curve.

```
function [x, Y] = finite_diff(a,b,h,ya,yb,f,g,r)
n = ((b-a)/h)+1;
A = zeros(n,n)
A(1,1) = 1;
A(n,n) = 1;
```

```
x0 = a;
x = a;
for i = 2:n-1;
    x = x + h;
    x0 = [x0 x];
    A(i,i) = g(x) - (2/(h*h));
    A(i,i+1) = (1 + (0.5*h*f(x)))/(h*h);
    A(i,i-1) = (1 - (0.5*h*f(x)))/(h*h);
    B(i,1) = r(x);
end

B(1,1) = ya;
B(n,1) = yb;


Y = A\B;
x = [x0 b];
endfunction
```

The usefulness of this function can be understood with the help of the following example. Let the second order differential equation be given by *Eqn. 4.30*.

$$\frac{d^2 y}{dx^2} + y = 0 \qquad\qquad (4.30)$$

The values of the variable at the boundaries are given by *Eqn. 4.31*.

$$y(0) = 0 \qquad\qquad (4.31)$$

$$y\left(\frac{\pi}{2}\right) = 3$$

The following Scilab program determines the value of y between $x = 0$ and $x = \frac{\pi}{2}$.

```
function def_r = r(x)                          //Define r(x)
def_r = 0;
endfunction

function def_f = f(x)                          //Define f(x)
def_f = 0;
endfunction

function def_g = g(x)                          //Define g(x)
def_g = 1;
endfunction
```

```
//Load the *.sci file which contains the function for the
finite difference method
exec('differentiation.sci',-1);
a = 0;                                              //Initial x
b = %pi/2;                                          //Final x
ya = 0;                                             //Initial y
yb = 3;                                             //Final y
h = 0.1;                                            //Step size

//Call the function for the finite difference method
[x,y] = finite_diff(a,b,h,ya,yb,f,g,r);

//Plot y for the x range 'a' to 'b'
plot2d(x,y)
```

Figure 4.5 depicts the solution of the differential equation for the entire range of '*x*' variables.

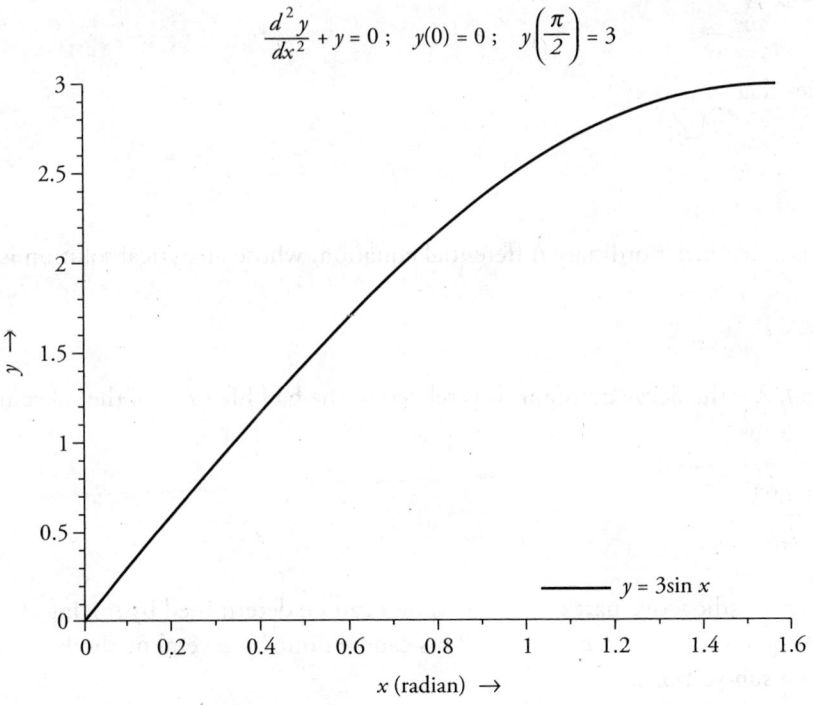

$$\frac{d^2y}{dx^2} + y = 0; \quad y(0) = 0; \quad y\left(\frac{\pi}{2}\right) = 3$$

Figure 4.5   Solution of *Eqns. 4.30–4.31*

4.8 Applications

4.8.1 Radioactive decay

Radioactive decay is a physical phenomenon through which unstable atoms/nuclei lose energy in the form of radiation, such as emission of alpha and beta particles. It is a random process at the atomic level and it is impossible to predict the time at which a particular atom will decay. It is however possible to determine the rate of decay for a collection of atoms.

The time required by a radioactive substance to decrease to 50% of its initial amount is called as the half-life of the substance. It is a characteristic constant of the substance. The rate of decay of an unstable radioactive substance can be determined from its half-life.

Suppose, N_0 is the number of radioactive atoms/nuclei at time $t = 0$, and N is number of radioactive atoms/nuclei at any given time t. In time interval dt, the number of atoms/nuclei reduces by an amount dN. The rate of change of the amount of radioactive substance is given by *Eqn. 4.32*.

$$-\frac{dN}{dt} \propto N \tag{4.32}$$

This implies that

$$\frac{dN}{dt} = -\lambda N \tag{4.33}$$

Eqn. 4.33 is a first order ordinary differential equation, whose analytical solution is given by

$$N = N_o e^{-\lambda t} \tag{4.34}$$

In *Eqn. 4.34*, λ is the decay constant. It is related to the half life $(T_{1/2})$ of the substance by *Eqn. 4.35*.

$$\lambda = \frac{0.693}{T_{1/2}} \tag{4.35}$$

The number of radioactive particles at any time t can be determined by solving the first order differential equation given in *Eqn. 4.33*. This can be done by several methods as described in the following sub-sections.

4.8.1.1 Built-in Scilab function

The differential equation can be defined in Scilab in the following manner.

```
function Ndot = f(t,N);
Ndot = -lamda*N;
endfunction
```

The built-in 'ode' function can be called inside a loop to find the solution of the differential equation at different values of time.

```
j = N(1);
while(j < 40000);
    ode_result = ode(N(1),t(1),j-N(1),f);
    disp('ode : Number at t = ' +string(j-N(1))+ ' is :
'        +string(ode_result));
    j = j+h;
end
```

The input parameters used for this example are

```
t(1) = 0;                              //Initial time
N(1) = 100;                  //Initial number of particles
half_life = 10000;                   //Half-life in years
lamda = 0.693/half_life;
h = 5000;                                  //Step size
```

The result of the Scilab program from the built-in function is as follows.

```
ode : Number at t = 0 is : 100
ode : Number at t = 5000 is : 70.715885
ode : Number at t = 10000 is : 50.007364
ode : Number at t = 15000 is : 35.363152
ode : Number at t = 20000 is : 25.007364
ode : Number at t = 25000 is : 17.684178
ode : Number at t = 30000 is : 12.505522
ode : Number at t = 35000 is : 8.8433896
```

This implies that after consecutive half-lives, the number of particles reduces to 50% of the initial number.

4.8.1.2 Euler's method

The solution of *Eqn. 4.33* can be determined from Euler's numerical technique by defining a function for Euler's formula (as explained in *Section 4.2*) and a function for the differential equation (as done in *Section 4.8.1.1*). The initial values can be user-defined, as shown in the following.

```
t(1) = input ('Enter initial time : ');
N(1) = input('Enter initial number of particles :');
half_life = input('Enter half-life in years : ');
lamda = 0.693/half_life;                //Decay constant
final = 4*half_life;                        //Final time
h = half_life/2;                           //Step size
```

The function for Euler's formula is then invoked so as to get the number of particles (N) at any time t.

```
[t,N] = euler(t(1),N(1),h,final);
```

The input parameters used for this example are

```
t(1) = 0;                                    //Initial time
N(1) = 100;                      //Initial number of particles
half_life = 10000;                     //Half-life in years
```

The result of the Scilab program from Euler's method is as follows.

```
Euler_Method : Number at t = 0 is : 100
Euler_Method : Number at t = 5000 is : 65.35
Euler_Method : Number at t = 10000 is : 42.706225
Euler_Method : Number at t = 15000 is : 27.908518
Euler_Method : Number at t = 20000 is : 18.238217
Euler_Method : Number at t = 25000 is : 11.918675
Euler_Method : Number at t = 30000 is : 7.7888538
Euler_Method : Number at t = 35000 is : 5.090016
```

In this case, a large step size of 5000 was taken, and the result is obviously not accurate. The method was repeated for a step size of 10 and it gave a reasonably accurate result as follows.

```
Euler_Method : Number at t = 0 is : 100
Euler_Method : Number at t = 5000 is : 70.707388
Euler_Method : Number at t = 10000 is : 49.995347
Euler_Method : Number at t = 15000 is : 35.350404
Euler_Method : Number at t = 20000 is : 24.995348
Euler_Method : Number at t = 25000 is : 17.673558
Euler_Method : Number at t = 30000 is : 12.496511
Euler_Method : Number at t = 35000 is : 8.8359565
```

4.8.1.3 Modified Euler's method

The solution of *Eqn. 4.33* can be determined from the modified Euler's numerical technique by defining its function (as explained in *Section 4.3*) and a function for the differential equation (as done in *Section 4.8.1.1*). The initial values can be user-defined, as shown below.

```
t(1) = input ('Enter initial time : ');
N(1) = input('Enter initial number of particles :');
half_life = input('Enter half-life in years : ');
lamda = 0.693/half_life;                    //Decay constant
final = 4*half_life;                        //Final time
h = half_life/2;                            //Step size
```

The function for the modified Euler's formula is then invoked so as to get the number of particles (*N*) at any time *t*.

```
[t,N] = modeuler(t(1),N(1),h,final);
```

The input parameters used for this example are

```
t(1) = 0;                            //Initial time
N(1) = 100;                 //Initial number of particles
half_life = 10000;                   //Half-life in years
```

The result of the Scilab program from the modified Euler's method is as follows.

```
Modified Euler_Method : Number at x = 5000 is : 71.353
Modified Euler_Method : Number at x = 10000 is : 46.629
Modified Euler_Method : Number at x = 15000 is : 30.472
Modified Euler_Method : Number at x = 20000 is : 19.914
Modified Euler_Method : Number at x = 25000 is : 13.014
Modified Euler_Method : Number at x = 30000 is : 8.5043
Modified Euler_Method : Number at x = 35000 is : 5.5576
Modified Euler_Method : Number at x = 40000 is : 3.6319
```

Notice that for a given step size, the modified Euler's method gives a more accurate result than Euler's method.

4.8.1.4 Second order Runge–Kutta Method

The solution of *Eqn. 4.33* can be determined from the second order Runge–Kutta method by defining its function (as explained in *Section 4.4*) and a function for the differential equation (as done in *Section 4.8.1.1*). The initial values can be user-defined, as shown in the following.

```
t(1) = input('Enter initial time : ');
N(1) = input('Enter initial number of particles :');
half_life = input('Enter half-life in years : ');
lamda = 0.693/half_life;              //Decay constant
final = 4*half_life;                     //Final time
h = half_life/2;                          //Step size
```

The function for the second order Runge–Kutta method is then invoked so as to get the number of particles (*N*) at any time *t*.

```
[t,N] = rk2(t(1),N(1),h,final);
```

The input parameters used for this example are

```
t(1) = 0;                                 //Initial time
N(1) = 100;                //Initial number of particles
half_life = 10000;                    //Half-life in years
```

Result of the Scilab program from the second order Runge–Kutta method is as follows.

```
Second_Order_Runge-Kutta : Number at x = 0 is : 100
Second_Order_Runge-Kutta : Number at x = 5000 is : 71.353
Second_Order_Runge-Kutta : Number at x = 10000 is : 50.913
Second_Order_Runge-Kutta : Number at x = 15000 is : 36.328
Second_Order_Runge-Kutta : Number at x = 20000 is : 25.921
Second_Order_Runge-Kutta : Number at x = 25000 is : 18.495
Second_Order_Runge-Kutta : Number at x = 30000 is : 13.197
Second_Order_Runge-Kutta : Number at x = 35000 is : 9.4165
```

4.8.1.5 Fourth order Runge–Kutta method

The solution of *Eqn. 4.33* can be determined from the fourth order Runge–Kutta method by defining its function (as explained in *Section 4.5*) and a function for the differential equation (as done in *Section 4.8.1.1*). The initial values can be user-defined, as shown in the following.

```
t(1) = input('Enter initial time : ');
N(1) = input('Enter initial number of particles :');
half_life = input('Enter half-life in years : ');
lamda = 0.693/half_life;              //Decay constant
final = 4*half_life;                     //Final time
h = half_life/2;                          //Step size
```

The function for the fourth order Runge–Kutta method is then invoked so as to get the number of particles (N) at any time t.

```
[t,N] = rk4(t(1),N(1),h,final);
```

The input parameters used for this example are

```
t(1) = 0;                         //Initial time
N(1) = 100;             //Initial number of particles
half_life = 10000;                //Half-life in years
```

Result of the Scilab program is as follows.

```
Fourth_Order_Runge-Kutta : Number at x = 0 is : 100
Fourth_Order_Runge-Kutta : Number at x = 5000 is : 70.72
Fourth_Order_Runge-Kutta : Number at x = 10000 is : 50.013
Fourth_Order_Runge-Kutta : Number at x = 15000 is : 35.369
Fourth_Order_Runge-Kutta : Number at x = 20000 is : 25.013
Fourth_Order_Runge-Kutta : Number at x = 25000 is : 17.689
Fourth_Order_Runge-Kutta : Number at x = 30000 is : 12.51
Fourth_Order_Runge-Kutta : Number at x = 35000 is : 8.8468
```

4.8.1.6 Graphical representation of the solution

The Scilab program for solving the differential equation consists of the following steps,

- Define a function for the differential equation

```
function Ndot = f(t,N)
Ndot = -lamda*N;
endfunction
```

- Mention the initial value (starting value) of time (t_0) and the initial number of radioactive atoms/nuclei (N_0). For example,

```
t0 = 0;
N0 = input("Enter initial number of atoms : ");
```

- Mention the half-life of the substance (preferably in years).

```
T_half = input("Enter half-life in years : ");
```

- Calculate the decay constant

```
lamda = log(2)/T_half;
```

- Give a range for the -axis and invoke the 'ode' command.

```
t = 0:0.1:4*T_half;
N = ode(N0,t0,t,f);
```

- Finally, plot the solution.

```
plot2d(t/T_half,N)
```

The graph in *Figure 4.6* has been generated using the aforementioned Scilab program. The readers are advised to use their plotting skills developed in *Chapter 2* and reproduce this graph. The input parameters for this graph were

- $t_0 = 0$
- $N_0 = 100$
- $T_{1\backslash 2} = 10000$

4.8.2 Orthogonal trajectory

Orthogonal trajectories are a set of two curves that are orthogonal to each other. These two curves always intersect at right angles with each other. Ordinary first order differential equations are a useful tool to determine these orthogonal set of curves. The mechanism is as follows.

Suppose the equation of the first family of curves is

$$f(x, y, \alpha) = 0 \qquad (4.36)$$

In *Eqn. 4.36*, α is a constant parameter. The equation for the second family of curves is

$$g(x, y, \beta) = 0 \qquad (4.37)$$

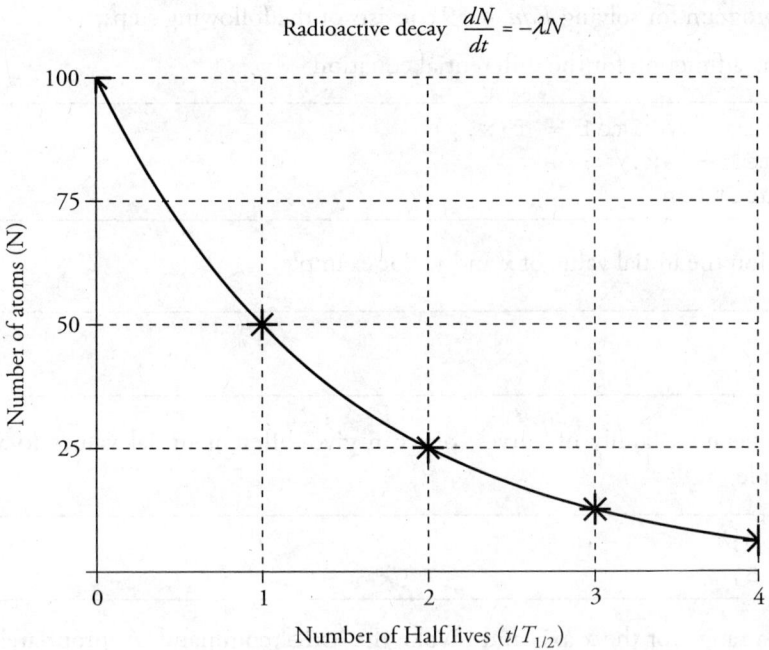

Figure 4.6   Graph showing radioactive decay

In *Eqn. 4.37*, β is a constant parameter. These two curves will be orthogonal if *Eqn. 4.38* holds true.

$$g'(x, y) = \frac{-1}{f'(x, y)} \tag{4.38}$$

It is interesting to plot the orthogonal trajectories in Scilab. Assume that the differential equation for the first curve is given by *Eqn. 4.39*.

$$\frac{dy}{dx} = -\frac{x}{y} \tag{4.39}$$

The analytical solution of this differential equation is a circle (*Eqn. 4.40*).

$$x^2 + y^2 = \alpha \tag{4.40}$$

The Scilab program for solving *Eqn. 4.39* consists of the following steps,

- Define a function for the differential equation

```
function yprime1 = f(x,y)
yprime1 = -x./y;
endfunction
```

- Mention the initial value of *x* and *y*. For example,

```
x0 = 0;
y0 = 3;
```

- To generate a family of curves, one can give different initial values to *x* and *y*. For example,

```
x0 = 0;
y0 = 5;
```

- Give a range for the *x*-axis and invoke the 'ode' command. Appropriately chose step size and the final value for *x* variable

```
x = 0 : Step Size : Final Value;
y = ode(y0,x0,x,f);
```

- Finally, plot the solution in the four quadrants.

```
plot2d(x,y)
plot2d(x,-y)
plot2d(-x,-y)
plot2d(-x,y)
```

The differential equation of the orthogonal curve is given in *Eqn. 4.41*.

$$\frac{dy}{dx} = \frac{y}{x} \tag{4.41}$$

The analytical solution of *Eqn. 4.41* is a straight line (*Eqn. 4.42*).

$$y = \beta x \tag{4.42}$$

The Scilab program for solving *Eqn. 4.41* consists of the following steps,

- Define a function for the differential equation

```
function yprime2 = f(x,y)
yprime2 = y/x;
endfunction
```

- Mention the initial value of x and y. For example,

```
x0 = 0.0001;
y0 = 0;
```

- To generate a family of curves, one can give different initial values of x and y. For example,

```
x0 = 0;
y0 = 0.0001;
```

- The last step is to give a range for the x-axis and invoke the 'ode' command. Appropriately chose step size and the final value for x variable

```
x = x0 : Step Size : Final Value;
y = ode(y0,x0,x,f);
```

- Finally, plot the solution in the four quadrants.

```
plot2d(x,y)
plot2d(-x,y)
plot2d(-x,-y)
plot2d(x,-y)
```

Figure 4.7 shows the orthogonal trajectories for this example.

4.8.3 Square wave ↔ Triangular wave

Consider the function given in *Eqn. 4.43*.

$$\frac{dy}{dt} = \begin{cases} 1, \ 0 \leq t \leq \pi \\ -1, \ \pi \leq t \leq 2\pi \end{cases} \tag{4.43}$$

Eqn. 4.43 represents a square wave having a period 2π and 50% duty cycle.

The analytical solution of *Eqn. 4.43* is given in *Eqns. 4.44–4.45*.

$$y = \int\limits_0^t \frac{dy}{dt} dt = \int\limits_0^\pi dt + \int\limits_\pi^{2\pi} -dt \qquad (4.44)$$

$$y = \begin{cases} t, \, 0 \le t \le \pi \\ -t, \, \pi \le t \le 2\pi \end{cases} \qquad (4.45)$$

Orthogonal trajectory

$$Y' = -\frac{X}{Y} \,\, (\text{Circle}) \quad ; \quad Y' = \frac{Y}{X} \,\, (\text{Line})$$

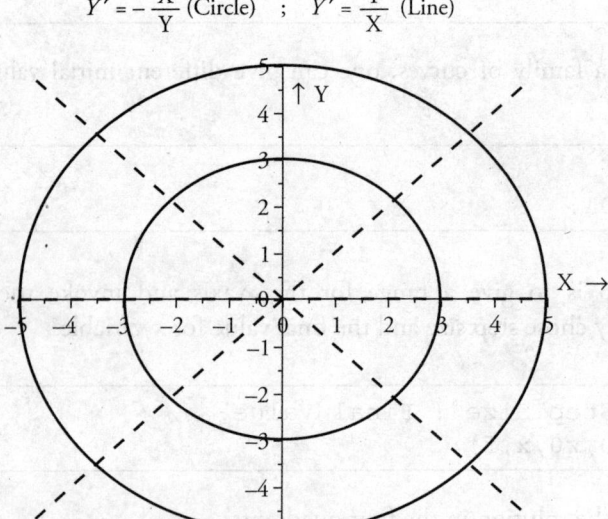

Figure 4.7   Example of orthogonal trajectory

4.8.3.1 Built-in function and graphical representation

The following Scilab program uses the built-in command 'ode' to solve the differential equation representing a square wave. The corresponding graph is shown in *Figure 4.8*.

```
function ydot = f(t,N)
ydot = squarewave(t,50);
endfunction

t = 0:0.1:4*%pi;
fplot2d(t,f)

y1 = ode(0,0,t,f);
plot2d(t,y1)
```

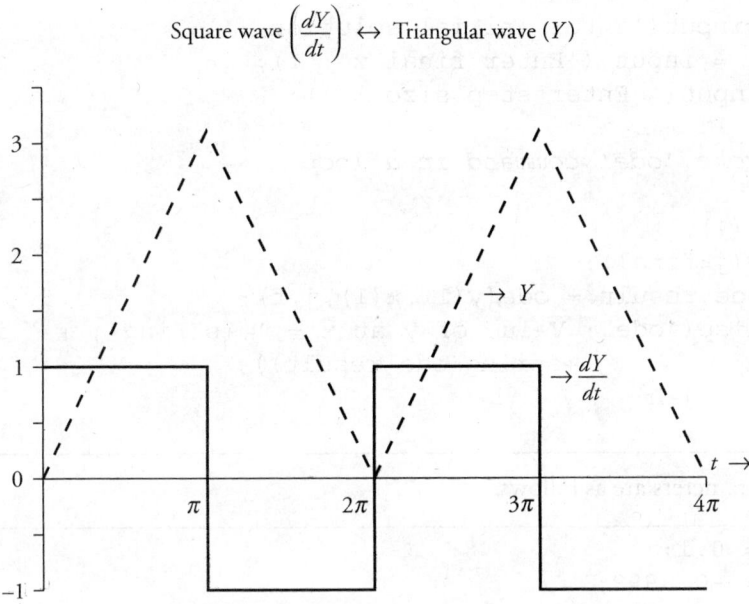

Figure 4.8   Solution of the differential equation of a square wave

4.8.4 Sinusoidal wave

Consider the function given in *Eqn. 4.46*. The analytical solution of *Eqn. 4.46* is given in *Eqn. 4.47*.

$$\frac{dy}{dx} = \sin x \tag{4.46}$$

$$y = -\cos x \tag{4.47}$$

4.8.4.1 Built-in function

The following Scilab program shows the solution of *Eqn. 4.46* using the built-in 'ode' function.

```
//Define the differential equation
function yprime = f(x,y);
yprime = sin(x);
endfunction

//Enter the initial values
x(1) = input ('Enter initial x : ');
```

```
y(1)=input('Enter initial solution :');
final = input ('Enter final x : ');
h = input(' Enter step size : ');

//Invoke 'ode' command in a loop

j = x(1);
while(j<final);
    ode_result = ode(y(1),x(1),j,f);
    disp('ode : Value of y at x = ' +string(j)+ ' is : '
                +string(ode_result));
    j = j+h;
end
```

The input parameters are as follows.

```
x(1) = 0.1;
y(1) = -0.9999;
final = 1;
h = 0.1;
```

The output is as follows.

```
ode : Solution at x = 0.1 is : -0.9999
ode : Solution at x = 0.2 is : -0.9850
ode : Solution at x = 0.3 is : -0.9602
ode : Solution at x = 0.4 is : -0.9260
ode : Solution at x = 0.5 is : -0.8825
ode : Solution at x = 0.6 is : -0.8302
ode : Solution at x = 0.7 is : -0.7697
ode : Solution at x = 0.8 is : -0.7016
ode : Solution at x = 0.9 is : -0.6265
ode : Solution at x = 1 is : -0.5452
```

4.8.4.2 Euler's method

The solution of *Eqn. 4.46* can be determined from Euler's method by defining its function (as explained in *Section 4.2*) and a function for the differential equation (as done in *Section 4.8.4.1*). The initial values can be user-defined, as shown in the following.

```
x(1) = input ('Enter initial x : ');
y(1) = input ('Enter initial solution :');
final = input ('Enter final x : ');
h = input ('Enter step size : ');
```

The function for Euler's method is then invoked so as to get the solution.

```
[x,y]  =  euler(x(1),y(1),h,final);
```

The input parameters used for this example are

```
x(1)  =  0.1;
y(1)  =  -0.9999;
final  =  1;
h  =  0.1;
```

Result of the Scilab program from Euler's method is as follows.

```
Euler_Method  :  Solution at x = 0.1 is : -0.9999
Euler_Method  :  Solution at x = 0.2 is : -0.9899
Euler_Method  :  Solution at x = 0.3 is : -0.9700
Euler_Method  :  Solution at x = 0.4 is : -0.9405
Euler_Method  :  Solution at x = 0.5 is : -0.9016
Euler_Method  :  Solution at x = 0.6 is : -0.8536
Euler_Method  :  Solution at x = 0.7 is : -0.7971
Euler_Method  :  Solution at x = 0.8 is : -0.7327
Euler_Method  :  Solution at x = 0.9 is : -0.6610
Euler_Method  :  Solution at x = 1 is : -0.5827
```

4.8.4.3 Modified Euler's method

The solution of *Eqn. 4.46* can be determined from the modified Euler's method by defining its function (as explained in *Section 4.3*) and a function for the differential equation (as done in *Section 4.8.4.1*). The initial values can be user-defined, as shown in the following.

```
x(1)  =  input ('Enter initial x : ');
final  =  input ('Enter final x : ');
y(1) =  input('Enter initial solution :');
h  =  input('Enter step size : ');
```

The function for the modified Euler's method is then invoked so as to get the solution.

```
[x,y]  =  modeuler(x(1),y(1),h,final);
```

The input parameters used for this example are

```
x(1)  =  0.1;
y(1)  =  -0.9999;
final  =  1;
h  =  0.1;
```

The result of the Scilab code from the modified Euler's method is as follows.

```
Modified Euler_Method : Solution at x = 0.2 is : -0.9850
Modified Euler_Method : Solution at x = 0.3 is : -0.9652
Modified Euler_Method : Solution at x = 0.4 is : -0.9358
Modified Euler_Method : Solution at x = 0.5 is : -0.8971
Modified Euler_Method : Solution at x = 0.6 is : -0.8494
Modified Euler_Method : Solution at x = 0.7 is : -0.7932
Modified Euler_Method : Solution at x = 0.8 is : -0.7291
Modified Euler_Method : Solution at x = 0.9 is : -0.6577
Modified Euler_Method : Solution at x = 1 is : -0.5798
Modified Euler_Method : Solution at x = 1.1 is : -0.4960
```

4.8.4.4 Second order Runge–Kutta method

Solution of *Eqn. 4.46* can be determined from the second order Runge–Kutta method by defining its function (as explained in *Section 4.4*) and a function for the differential equation (as done in *Section 4.8.4.1*). The initial values can be user-defined, as shown in the following.

```
x(1) = input('Enter initial x : ');
final = input('Enter final x : ');
y(1)=input('Enter initial solution :');
h = input('Enter step size : ');
```

The function for the second order Runge–Kutta method is then invoked so as to get the solution.

```
[x,y] = rk2(x(1),y(1),h,final);
```

The input parameters used for this example are

```
x(1) = 0.1;
y(1) = -0.9999;
final = 1;
h = 0.1;
```

Result of the Scilab code from the second order Runge–Kutta method is as follows.

```
Second_Order_Runge-Kutta : Solution at x = 0.1 is : -0.9999
Second_Order_Runge-Kutta : Solution at x = 0.2 is : -0.9850
Second_Order_Runge-Kutta : Solution at x = 0.3 is : -0.9603
Second_Order_Runge-Kutta : Solution at x = 0.4 is : -0.9260
```

```
Second_Order_Runge-Kutta : Solution at x = 0.5 is : -0.8826
Second_Order_Runge-Kutta : Solution at x = 0.6 is : -0.8304
Second_Order_Runge-Kutta : Solution at x = 0.7 is : -0.7699
Second_Order_Runge-Kutta : Solution at x = 0.8 is : -0.7019
Second_Order_Runge-Kutta : Solution at x = 0.9 is : -0.6268
Second_Order_Runge-Kutta : Solution at x = 1 is : -0.5456
```

4.8.4.5 Fourth order Runge–Kutta method

Solution of *Eqn. 4.46* can be determined from the fourth order Runge–Kutta method by defining its function (as explained in *Section 4.5*) and a function for the differential equation (as done in *Section 4.8.4.1*). The initial values can be user-defined, as shown in the following.

```
x(1) = input('Enter initial x : ');
final = input('Enter final x : ');
y(1)=input('Enter initial solution :');
h = input('Enter step size : ');
```

The function for the fourth order Runge–Kutta method is then invoked so as to get the solution.

```
[x,y] = rk4(x(1),y(1),h,final);
```

The input parameters used for this example are

```
x(1) = 0.1;
y(1) = -0.9999;
final = 1;
h = 0.1;
```

Result of the Scilab program from the fourth order Runge–Kutta method is as follows.

```
Fourth_Order_Runge-Kutta : Solution at x = 0.1 is : -0.9999
Fourth_Order_Runge-Kutta : Solution at x = 0.2 is : -0.9850
Fourth_Order_Runge-Kutta : Solution at x = 0.3 is : -0.9602
Fourth_Order_Runge-Kutta : Solution at x = 0.4 is : -0.9260
Fourth_Order_Runge-Kutta : Solution at x = 0.5 is : -0.8825
Fourth_Order_Runge-Kutta : Solution at x = 0.6 is : -0.8302
Fourth_Order_Runge-Kutta : Solution at x = 0.7 is : -0.7697
Fourth_Order_Runge-Kutta : Solution at x = 0.8 is : -0.7016
Fourth_Order_Runge-Kutta : Solution at x = 0.9 is : -0.6265
Fourth_Order_Runge-Kutta : Solution at x = 1 is : -0.5452
```

4.8.4.6 Graphical representation

The following Scilab program uses the built-in command 'ode' to solve the differential equation for a sine wave. The graph is shown in *Figure 4.9*.

```
//Define the function for differential equation
function yprime = f(x,y)
yprime = sin(x);
endfunction


x = 0:0.1:4*%pi;                    //Range of x-axis
fplot2d(x,f)                        //Plot the function
y = ode(-1,0,x,f);          //Solution at every value of x
plot2d(x,y)                         //Plot the solution
```

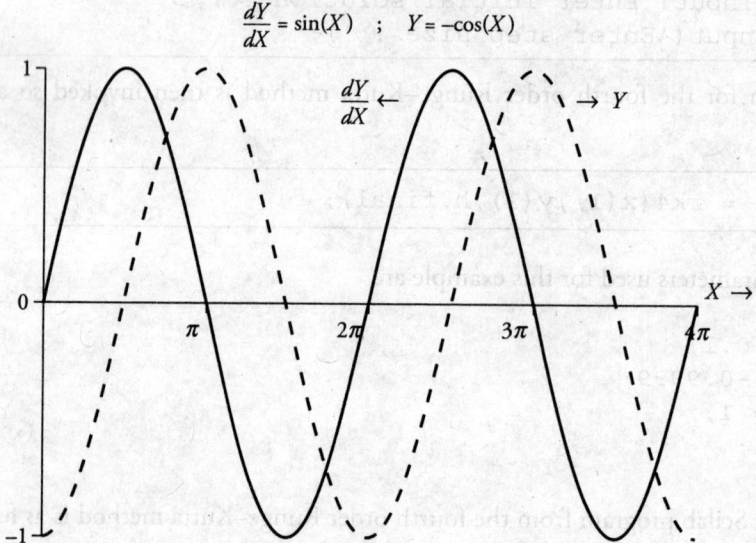

Sine wave $\leftrightarrow$ Cosine wave

$$\frac{dY}{dX} = \sin(X) \quad ; \quad Y = -\cos(X)$$

Figure 4.9   Solution of the differential equation of a sinusoidal wave

4.8.5 Freely falling object

Suppose an object is placed at a height h_0. It is released with an initial velocity v_0. Assuming that there is no air resistance, the object falls under the influence of a uniform gravitational field with a velocity $v(t)$ and covers a distance x. The acceleration of the freely falling object is given by *Eqn. 4.48*.

$$\frac{d^2 x}{dt^2} = \frac{dv}{dt} = -g \tag{4.48}$$

The aim is to solve *Eqn. 4.48* and determine the position and velocity of the object as a function of time. The following Scilab program has been written to trace the position of the object when it is released with different initial velocities. The position–time and the velocity–time graphs are shown in *Figure 4.10*.

1. **Runge–Kutta's fourth order approximation**

   For using this technique, the second order differential equation is first written in terms of two first order differential equations (*Eqns. 4.49–4.50*), which are then solved simultaneously. These equations are

$$\frac{dx}{dt} = v \tag{4.49}$$

$$\frac{dv}{dt} = -g \tag{4.50}$$

```
//Load the *.sci file which contains function for Runge-
Kutta method
exec('differentiation.sci',-1)

//Define functions for the two first order differential
equations corresponding
function x_dot = f1(t,x,v)
x_dot = v;
endfunction

function v_dot = f2(t,x,v)
v_dot = -g;
endfunction

g = 9.82;                        //Acceleration due to gravity
height_initial = 100;                   //Initial height
v_initial = 0;                          //Initial velocity
t_initial = 0;                          //Initial time
h = 0.1                                 //Step size

final = ((v_initial)+(sqrt((v_initial*v_
initial)+2*g*height_initial)))/g

//Call the function for rk4 method
[t,x,v] = rk42(t_initial, height_initial, v_
initial,h,final);

plot2d(t,x)                             //Plot x-t graph
plot2d(t,v)                             //Plot v-t graph
```

2. Built-in Scilab 'ode' function

For using this technique, the second order differential equation is converted into two first order differential equations by constructing matrices as shown in *Eqns. 4.51–4.52*.

$$X = \begin{pmatrix} x \\ \dfrac{dx}{dt} \end{pmatrix} \tag{4.51}$$

$$\frac{dX}{dt} = \begin{pmatrix} \dfrac{dx}{dt} \\ \dfrac{d^2 x}{dt^2} \end{pmatrix} \tag{4.52}$$

Eqns. 4.51–4.52 imply that

$$\frac{dX}{dt}(1) = X(2) \tag{4.53}$$

$$\frac{dX}{dt}(2) = -g \tag{4.54}$$

Therefore, based on *Eqn. 4.53* and *Eqn. 4.54*, the initial value of displacement and velocity of the object are given in the form of a matrix as shown in *Eqn. 4.55*.

$$\begin{pmatrix} h_0 \\ v_0 \end{pmatrix} \tag{4.55}$$

```
//Define the system of equations
function xdash = f(t,x)
xdash(1) = x(2);
xdash(2) = -g;
endfunction

g = 9.82;                        //Acceleration due to gravity
height_initial = 100;                     //Initial height
v_initial = 0;                         //Initial velocity
t_initial = 0;                           //Initial time
t = t_initial:0.3:((v_initial) + (sqrt((v_initial*v_
initial)+2*g*height_initial))))/g;

//Call the built-in function
x = ode([height_initial;v_initial],t_initial,t,f);
```

```
subplot(211)
plot2d(t,x(1,:));                        //Plot x-t graph
v_initial = 10;                          //Initial velocity
t = t_initial:0.3:((v_initial) + (sqrt((v_initial*v_
initial)+2*g*height_initial)))/g;

// Call the built-in function
x = ode([height_initial;v_initial],t_initial,t,f);
plot2d(t,x(1,:));                         //Plot x-t graph

subplot(212)
v_initial = 0;
t = t_initial:0.3:((v_initial) + (sqrt((v_initial*v_
initial)+2*g*height_initial)))/g;

//Call the built-in function
x = ode([height_initial;v_initial],t_initial,t,f);
plot2d(t,x(2,:));                         //Plot v-t graph

v_initial = 10;
t = t_initial:0.3:((v_initial) + (sqrt((v_initial*v_
initial)+2*g*height_initial)))/g;

// Call the built-in function
x = ode([height_initial;v_initial],t_initial,t,f);
plot2d(t,x(2,:));                         //Plot v-t graph
```

Some interesting points that should be noticed in these graphs (*Figure 4.10*) are as follows:

- The position–time graph is a curved line.
- It implies an accelerated motion.
- The object starts moving with a slow velocity and ends with a larger velocity.
- The velocity–time graph is a straight line with a negative slope.
- This implies a constant and negative accelerated motion.
- The velocity increases in the downward direction. The slope shows the acceleration of the object.

4.8.6 Atwood's machine

The diagram of Atwood's machine is given in *Figure 4.11*. It consists of two objects of different mass, hung vertically from a pulley. Suppose $m_1 > m_2$, length of the string is l and radius of the pulley is r.

It is assumed that the string is massless and inextensible; and the pulley is ideal such that its moment of inertia is very small. If the system is released from rest, then the acceleration of the objects will be given by *Eqn. 4.56*. The equation of motion is (*Eqn. 4.57*).

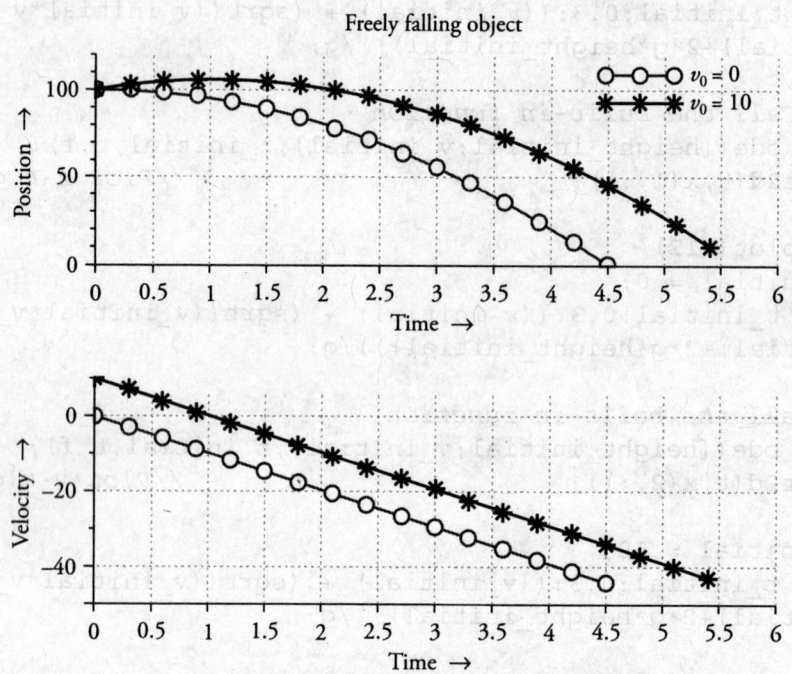

Figure 4.10 Trajectory of a freely falling object

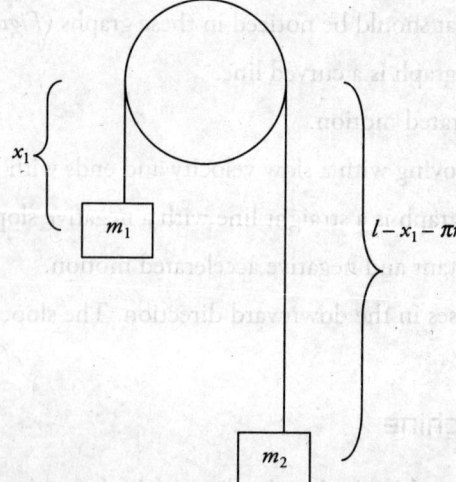

Figure 4.11 Diagram of Atwood's machine

$$a = g\left(\frac{m_1 - m_2}{m_1 + m_2}\right) \tag{4.56}$$

$$\frac{d^2x}{dt^2} = \frac{dv}{dt} = a \tag{4.57}$$

The following Scilab program determines the position–time graph of the two objects. Two methods have been used for solving the equation of motion. The position–time and the velocity–time graphs are shown in *Figure 4.12*.

1. **Runge–Kutta's fourth order approximation**

   For using this technique, the second order differential equation is first written in terms of two first order differential equations which are then solved simultaneously. These equations are given in *Eqn. 4.58* and *Eqn. 4.59*.

   $$\frac{dx}{dt} = v \tag{4.58}$$

   $$\frac{dv}{dt} = a \tag{4.59}$$

```
//Load the *.sci file which contains the function for
the Runge-Kutta method
exec('differentiation.sci',-1)

//Define functions for the two first order differential
equations corresponding
function x_dot = f1(t,x,v)
x_dot = v;
endfunction

function v_dot = f2(t,x,v)
v_dot = a;
endfunction

g = 9.82;                 //Acceleration due to gravity
m1 = 100;                      //Mass of first object
m2 = 30;                     //Mass of second object
x1 = 5;                 //Initial position of first mass
l = 100;                       //Length of the string
t0 = 0;                            //Initial time
v0 = 0;                          //Initial velocity
radius = 2;                     //Radius of the pulley
a = g*(m1-m2)/(m1+m2);      //Acceleration of the system
```

```
h = 0.1                                        //Step size
final = sqrt(2*a*(1-x1-(radius*%pi)))/a

//Call the function for rk4 method
[t,x,v] = rk42(t0, x1, v0,h,final);

//Plot x-t graph of the first object
plot2d(t,x)

//Plot x-t graph of the second object
plot2d(t,1-x-(radius*%pi));
```

2. **Built-in Scilab 'ode' function**

   For using this technique, the second order differential equation is converted into two first order differential equations by constructing the matrices shown in *Eqns. 4.60–4.61*.

$$X = \begin{pmatrix} x \\ \dfrac{dx}{dt} \end{pmatrix} \tag{4.60}$$

$$\frac{dX}{dt} = \begin{pmatrix} \dfrac{dx}{dt} \\ \dfrac{d^2x}{dt^2} \end{pmatrix} \tag{4.61}$$

Eqns. 4.60–4.61 imply that

$$\frac{dX}{dt}(1) = X(2) \tag{4.62}$$

$$\frac{dX}{dt}(2) = a \tag{4.63}$$

Therefore, based on *Eqns. 4.62–4.63*, the initial value of displacement and velocity of the object are given in the form of a matrix as shown in *Eqn. 4.64*.

$$\begin{pmatrix} x_1 \\ v_0 \end{pmatrix} \tag{4.64}$$

```
g = 9.82;                      //Acceleration due to gravity
m1 = 100;                        //Mass of first object
m2 = 30;                        //Mass of second object
x1 = 5;                  //Initial position of first mass
l = 100;                          //Length of the string
t0 = 0;                               //Initial time
v0 = 0;                           //Initial velocity
radius = 2;                      //Radius of the pulley
a = g*(m1-m2)/(m1+m2);      //Acceleration of the system

//Define the system of equations
function xdot = f(t,x)
xdot(1) = x(2);
xdot(2) = a;
endfunction
t = t0:0.3:sqrt(2*a*(l-x1-(radius*%pi)))/a;

//Call the built-in function
x = ode([x1;v0],t0,t,f);

//Plot position-time graph of first object
plot2d(t,x(1,:));

//Plot position-time graph of second object
plot2d(t,l-x(1,:)-(radius*%pi));
```

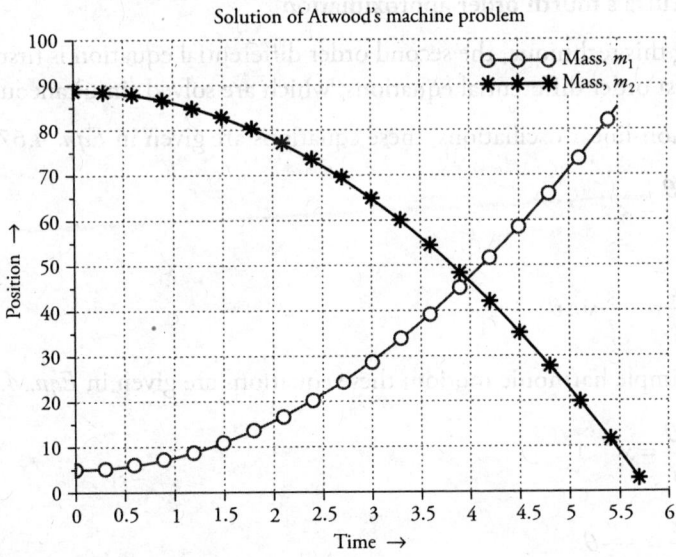

Figure 4.12 Position–time graph of the objects in Atwood's machine

4.8.7 Simple pendulum

It consists of a small mass (m) hung from an almost mass less inextensible string of length l. The mass executes oscillatory motion when it is displaced from the equilibrium position by an angle θ_0 and then released. The equation of motion of the vibrating pendulum is given by *Eqn. 4.65*.

$$\frac{d^2\theta}{dt^2} + \frac{g}{L}\sin\theta = 0 \qquad (4.65)$$

In *Eqn. 4.65*,

- θ is the angular displacement of mass from equilibrium position at any time t

- g is acceleration due to gravity

An important concept involved in the motion of a simple pendulum is the 'small angle approximation', according to which, motion of the pendulum can be assumed to be simple harmonic if the initial displacement (θ_0) is small. In such a case,

$$\sin\theta \cong \theta$$

$$\frac{d^2\theta}{dt^2} + \frac{g}{L}\theta = 0 \qquad (4.66)$$

The following Scilab program has been written to show the behaviour of a simple pendulum for different initial displacements. Two methods have been used for solving the equation of motion.

1. **Runge–Kutta's fourth order approximation**

   For using this technique, the second order differential equation is first written in terms of two first order differential equations, which are solved simultaneously.

   - For non-linear oscillations, these equations are given in *Eqn. 4.67*.

   $$\frac{d\theta}{dt} = x$$

   $$\frac{dx}{dt} = -\frac{g}{L}\sin\theta \qquad (4.67)$$

   - For simple harmonic motion, these equations are given in *Eqn. 4.68*.

   $$\frac{d\theta}{dt} = x$$

   $$\frac{dx}{dt} = -\frac{g}{L}\theta \qquad (4.68)$$

```
//Load the *.sci file which contains function for Runge-
Kutta method
exec('differentiation.sci',-1)

//Define functions for both the first order differential
equations (non-linear oscillation of the pendulum)
function theta_dot = f1(t,theta,x)
theta_dot = x;
endfunction
function x_dot = f2(t,theta,x)
x_dot = -(g/l)*sin(theta);
endfunction

t0 = 0;                              //Initial time
theta0 = 5*%pi/180;      //Initial angle (in radians)
x0 = 0;                       //Initial velocity(dθ/dt)

final = 10;                          //Final time
h = 0.1;                             //Step size
l = 1;

g = 9.8;

//Call the function for rk4 method
[t,theta,x] = rk42(t0,theta0,x0,h,final);

plot2d(t,theta)                    //Plot θ vs t graph

//Define functions for two first order differential
equations (small angle approximation)
function theta_dot = f1(t,theta,x)
theta_dot = x;
endfunction
function x_dot = f2(t,theta,x)
x_dot = -(g/l)*(theta);
endfunction

//Call the function for rk4 method
[t,theta,x] = rk42(t0,theta0,x0,h,final);

plot2d(t,theta)                    //Plot θ vs t graph
```

2. Built-in Scilab 'ode' function

For using this technique, the second order differential equation is converted into two first order differential equations by constructing the matrices shown in *Eqns. 4.69–4.70*.

$$\varphi = \begin{pmatrix} \theta \\ \dfrac{d\theta}{dt} \end{pmatrix} \tag{4.69}$$

$$\frac{d\varphi}{dt} = \begin{pmatrix} \dfrac{d\theta}{dt} \\ \dfrac{d^2\theta}{dt^2} \end{pmatrix} \tag{4.70}$$

Eqns. 4.69–4.70 imply that

$$\frac{d\varphi}{dt}(1) = \varphi(2) \tag{4.71}$$

$$\frac{d\varphi}{dt}(2) = -\frac{g}{L}\sin\theta = -\frac{g}{L}\sin\varphi(1) \tag{4.72}$$

Based on *Eqns. 4.71–4.72*, the initial value of the angular displacement and velocity of the pendulum are given in the form of a matrix (*Eqn. 4.73*).

$$\begin{pmatrix} \theta_0 \\ \dfrac{d\theta}{dt}\Big|_0 \end{pmatrix} \tag{4.73}$$

```
t = 0:0.1:10;                          //Time range
t0 = 0;                                //Initial time
phi0 = [5*%pi/180; 0];//Initial displacement, velocity
g = 9.8;
l = 1;

//Define the system of equations
function phi_dot = f(t,phi)
phi_dot(1) = phi(2);
phi_dot(2)=-(g/l)*phi(1);          //Small angle approx.
phi_dot(2) = -(g/l)*sin(phi(1));   //Non-linear approx.
endfunction

phi = ode(phi0,t0,t,f);    //Call the built-in function
plot2d(t,phi(1,:));                  //Plot the result
```

Figure 4.13 graphically shows the transformation of oscillations of a simple pendulum from non-linearity ($\theta_0 = 60°$) to being simple harmonic ($\theta_0 = 5°$). For the case when $\theta_0 = 60°$ (*Figure 13a*), the exact solution is different from the simple harmonic (small angle approximation) solution. The two solutions overlap when the initial displacement is 5° (*Figure 4.13b*).

4.8.8 Mass–spring system

A mass–spring system consists of a mass (m) connected to a rigid support with the help of a spring. A restoring force acts on the system whenever the mass is displaced from the equilibrium position.

The second order differential equation for this mass–spring system is given by

$$m\frac{d^2x}{dt^2} + c\frac{dx}{dt} + kx = 0 \qquad (4.74)$$

In *Eqn. 4.74*,

- x is the extension produced in the stretched/compressed spring at time t

- c is the damping constant and k is the spring constant

- $\dfrac{c^2}{4mk}$ is the damping ratio

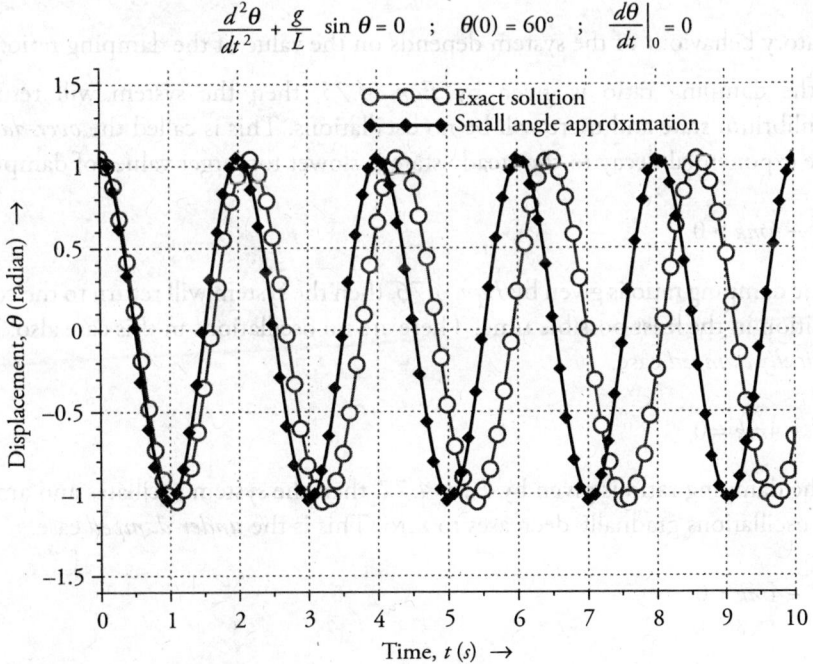

$$\frac{d^2\theta}{dt^2} + \frac{g}{L}\sin\theta = 0 \quad ; \quad \theta(0) = 60° \quad ; \quad \frac{d\theta}{dt}\bigg|_0 = 0$$

Figure 4.13a   Graph showing non-linear behaviour of a simple pendulum

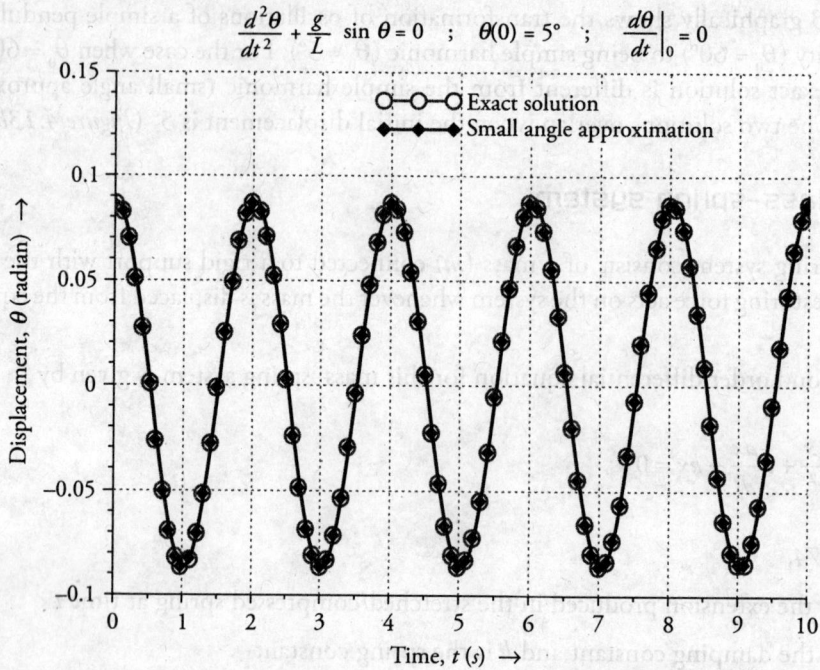

$$\frac{d^2\theta}{dt^2} + \frac{g}{L}\sin\theta = 0 \quad ; \quad \theta(0) = 5° \quad ; \quad \frac{d\theta}{dt}\Big|_0 = 0$$

Figure 4.13b Graph showing the small angle approximation

The oscillatory behaviour of the system depends on the value of the damping ratio.

- If the damping ratio is given by *Eqn. 4.75*, then the system will return to the equilibrium state and there will be no oscillations. This is called the *over-damped* case. The exponential decay to the steady state is slower for larger values of damping ratio.

$$c^2 - 4mk > 0 \tag{4.75}$$

- If the damping ratio is given by *Eqn. 4.76*, then the system will return to the equilibrium position in the least possible time. There are no oscillations in this case also. This is the *critically damped* case.

$$c^2 - 4mk = 0 \tag{4.76}$$

- If the damping ratio is given by *Eqn. 4.77*, then the system oscillates and amplitude of the oscillations gradually decreases to zero. This is the *under-damped* case.

$$c^2 - 4mk < 0 \tag{4.77}$$

The following Scilab program has been written to show the oscillations of a mass–spring system for the under-damped case. Value of the damping ratio can be changed to understand the other cases.

Two methods have been used for solving the equation of motion.

1. **Runge–Kutta's fourth order approximation**

   For using this technique, the second order differential equation is first written in terms of two first order differential equations, which are then solved simultaneously. These equations are given in *Eqn. 4.78* and *Eqn. 4.79*.

$$\frac{dx}{dt} = y \tag{4.78}$$

$$\frac{dy}{dt} = -\frac{(cy + kx)}{m} \tag{4.79}$$

```
//Load the *.sci file which contains function for Runge-
Kutta method
exec('differentiation.sci',-1)

//Define function for first order differential equations
function x_dot = f1(t,x,y)
x_dot = y;
endfunction

function y_dot = f2(t,x,y)
y_dot = -(c*y + k*x)/m;
endfunction
t0 = 0;                          //Initial time
x0 = 2;                          //Initial displacement
xdot_0 = 0;                      //Initial velocity
final = 20;                      //Final time
h = 0.1;                         //Step size
m = 50;                          //Mass
c = 20;                          //Damping constant
k = 128;                         //Spring constant

//Call the function for fourth order Runge-Kutta method
[t,x,y] = rk42(t0,x0,xdot_0,h,final);

//Plot the displacement vs. time graph
plot2d(t,x);
```

```
//Plot the envelope of the decaying amplitude
t = 0:0.1:20;
plot2d(t,x0*exp(-(c*t)/(2*m)));
plot2d(t,-x0*exp(-(c*t)/(2*m)));
```

2. Built-in Scilab 'ode' function

For using this technique, the second order differential equation is converted into two first order differential equations by constructing the matrices given in *Eqns. 4.80–4.81*.

$$X = \begin{pmatrix} x \\ \dfrac{dx}{dt} \end{pmatrix} \tag{4.80}$$

$$\frac{dX}{dt} = \begin{pmatrix} \dfrac{dx}{dt} \\ \dfrac{d^2x}{dt^2} \end{pmatrix} \tag{4.81}$$

Eqns. 4.80–4.81 imply that

$$\frac{dX}{dt}(1) = X(2) \tag{4.82}$$

$$\frac{dX}{dt}(2) = -\frac{(cy + kx)}{m} = -\frac{(cy + k\,X(1))}{m} \tag{4.83}$$

Therefore, based on *Eqns. 4.82–4.83*, the initial value of displacement and velocity of the mass–spring system are given in the form of a matrix, as shown in *Eqn. 4.84*.

$$X_0 = \begin{pmatrix} x_0 \\ \dfrac{dx}{dt}\bigg|_0 \end{pmatrix} \tag{4.84}$$

```
t = 0:0.1:20;                        //Time range
t0 = 0;                              //Initial time
x0 = [2; 0];            //Initial displacement, velocity
m = 50;                                      //Mass
c = 20;                              //Damping constant
k = 128;                             //Spring constant
```

```
//Define a system of equations
function x_dot = f(t,x)
x_dot(1) = x(2);
x_dot(2) = -(c*x(2) + k*x(1))/m;
endfunction

x = ode(x0,t0,t,f);          //Call the built-in function
plot2d(t,x(1,:));                        //Plot the result
```

Figure 4.14 graphically shows the damped oscillations of the mass–spring constant.

$$m\frac{d^2x}{dt^2} + c\frac{dx}{dt} + kx = 0 \quad ; \quad x(0) = 2 \quad ; \quad \frac{dx}{dt}\Big|_0 = 0$$

Underdamped motion ($c^2 - 4mk < 0$)

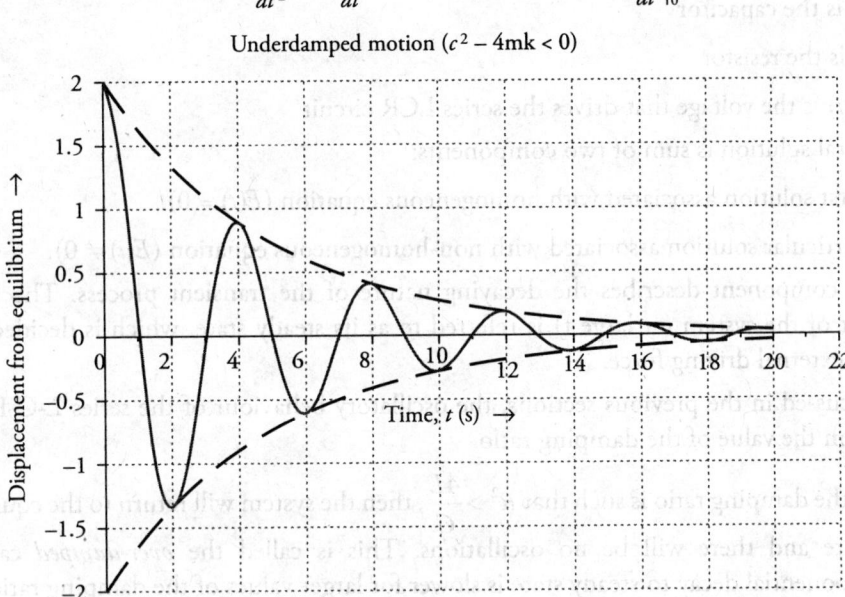

Figure 4.14   Under-damped motion of a mass–spring system

4.8.9 Series L-C-R circuit

The diagrammatical representation of a series L-C-R circuit is shown in *Figure 4.15*.

The second order differential equation for this system is given by *Eqn. 4.85*.

$$L\frac{d^2Q}{dt^2} + R\frac{dQ}{dt} + \frac{Q}{C} = E(t) \tag{4.85}$$

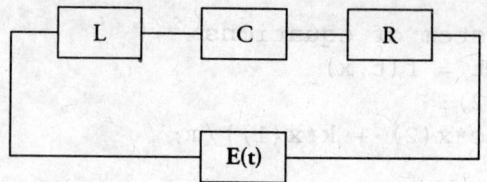

Figure 4.15   Series L-C-R circuit

In *Eqn. 4.85*,

- *Q* is the amount of charge in the circuit at time *t*
- *L* is the inductor
- *C* is the capacitor
- *R* is the resistor
- *E(t)* is the voltage that drives the series LCR circuit

The general solution is sum of two components:

- First solution associated with homogeneous equation ($E(t) = 0$).
- Particular solution associated with non-homogeneous equation ($E(t) \neq 0$).

The first component describes the decaying nature of the transient process. The ultimate behaviour of the system (at large t) is referred to as its steady state, which is decided by the non-zero external driving force.

As discussed in the previous sections, the oscillatory behaviour of the series L-C-R circuit depends on the value of the damping ratio.

- If the damping ratio is such that $R^2 > \dfrac{4L}{C}$, then the system will return to the equilibrium state and there will be no oscillations. This is called the *over-damped* case. The exponential decay to steady state is slower for larger values of the damping ratio.

- If the damping ratio is such that $R^2 = \dfrac{4L}{C}$, then the system will return to the equilibrium position in the least possible time. There will be no oscillations in this case also. This is the *critically damped* case.

- If the damping ratio is such that $R^2 < \dfrac{4L}{C}$, then the system oscillates and the amplitude of the oscillations gradually decreases to zero. This is the *under-damped* case.

The following Scilab program shows the steady state behaviour of the system for a critically damped case. It is assumed that the driving force is (*Eqn. 4.86*),

$$E(t) = 5\cos(2t)$$

$$(4.86)$$

For comparison, the homogeneous solution of the differential equation has also been calculated. The corresponding graph is shown in *Figure 4.16*. The reader is advised to change the value of the damping ratio and see its effect on the behaviour of the system.

Two methods have been used for solving the second order differential equation for series L-C-R circuits.

1. **Runge–Kutta's fourth order approximation**

   For using this technique, the second order differential equation is first written in terms of two first order differential equations, which are then solved simultaneously. These equations are given in *Eqns. 4.87–4.88*.

   $$\frac{dQ}{dt} = y \tag{4.87}$$

   $$\frac{dy}{dt} = \frac{5\cos(2t)}{L} - \frac{Q}{LC} - \frac{Ry}{L} \tag{4.88}$$

```
//Load the *.sci file which contains function for Runge-
Kutta method
exec('differentiation.sci',-1)

//Define function for first order differential equations
function charge_dot = f1(t,charge,y)
charge_dot = y
endfunction

function y_dot = f2(t,charge,y)
y_dot = (5*cos(w*t)/L) - (charge/(L*C)) - (R*y/L)
endfunction

L = 2;                                    //Inductor
C = 0.5;                                  //Capacitor
R = 4;                                    //Resistor
w = 2;                        //Frequency of driving force
t0 = 0;
charge0 = 2;
y0 = 0;
final = 12;
h = 0.1;

//Call the function for fourth order Runge-Kutta method
[t,charge,charge_dot] = rk42(t0,charge0,y0,h,final);

//Plot the charge vs. time graph
plot2d(t,charge);
```

2. Built-in Scilab 'ode' function

For using this technique, the second order differential equation is converted into two first order differential equations by constructing matrices, as shown in *Eqns. 4.89–4.90*.

$$\text{Charge} = \begin{pmatrix} Q \\ \dfrac{dQ}{dt} \end{pmatrix} \tag{4.89}$$

$$\text{Charge\_dot} = \begin{pmatrix} \dfrac{dQ}{dt} \\ \dfrac{d^2Q}{dt^2} \end{pmatrix} \tag{4.90}$$

Eqns. 4.89–4.90 imply that

$$\text{Charge\_dot}(1) = \text{Charge}(2) \tag{4.91}$$

$$\begin{aligned}
\text{Charge\_dot}(2) &= \frac{5\cos(2t)}{L} - \frac{Q}{LC} - \frac{R\dfrac{dQ}{dt}}{L} \\
&= \frac{5\cos(2t)}{L} - \frac{\text{Charge}(1)}{LC} - \frac{R(\text{Charge}(2))}{L}
\end{aligned} \tag{4.92}$$

Based on *Eqns. 4.91–4.92*, the initial values of charge and current in the circuit are given in the form of a matrix, as shown in *Eqn. 4.93*.

$$\text{Charge\_0} = \begin{pmatrix} Q_0 \\ \dfrac{dQ}{dt} \end{pmatrix}\Bigg|_0 \tag{4.93}$$

```
L = 2;                                        //Inductor
C = 0.5;                                       //Capacitor
R = 4;                                          //Resistor
w = 2;                            //Frequency of driving signal
charge_0 = [2;0]                    //Initial charge, current
t = 0.1:0.1:12;                                //Time range

//Define a system of equations.
//Put E(t)= 0 for homogeneous solution
//Put E(t)= 5cos(2t) for the full solution
function charge_dot = f(t,charge)
charge_dot(1)= charge(2);
charge_dot(2)=((E(t))-(R*charge(2)) - (charge(1)/C))/L;
endfunction
```

```
charge = ode(charge_0,0,t,f); //Call built-in function
plot2d(t,charge(1,:));                //Plot the result
```

Figure 4.16 shows the transient and steady state response of the circuit to the driving signal.

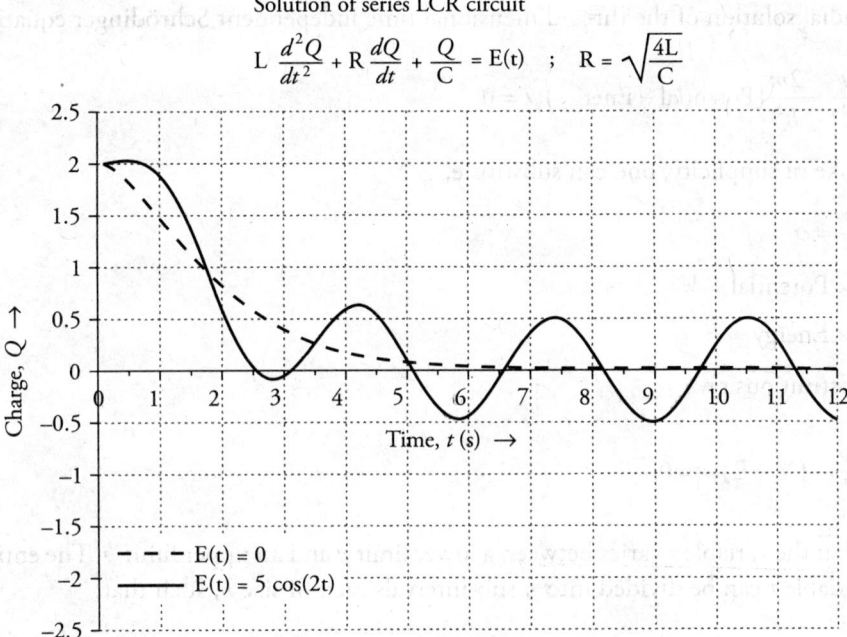

Solution of series LCR circuit

$$L \frac{d^2Q}{dt^2} + R \frac{dQ}{dt} + \frac{Q}{C} = E(t) \quad ; \quad R = \sqrt{\frac{4L}{C}}$$

Figure 4.16   Transient and steady state response of series L-C-R circuit

In *Figure 4.16*,

- The values of L, C and R have been taken such that they satisfy the condition for a critically damped system.

- If there is no driving signal, then the charge in the system decays to zero in the shortest possible time. After this time, the system achieves its steady state. The amplitude of charge in the steady state is given by

$$\frac{E_0}{\omega \sqrt{R^2 + \left(\omega L - \frac{1}{\omega C} \right)^2}}$$

4.8.10 Schrödinger equation

The Schrödinger equation explains the behaviour of quantum systems (such as atoms and molecules) and is used to determine their allowed energy states (energy eigen values). It also gives the probability (through associated wave functions) of finding particles at a certain position in space and time. This section briefly discusses the concept of finding the radial solution of the time independent Schrödinger equation.

The radial solution of the three-dimensional time independent Schrödinger equation is

$$\frac{d^2\psi}{dr^2} - \frac{2m}{\hbar^2}\left(\text{Potential} - \text{Energy}\right)\psi = 0 \tag{4.94}$$

For the sake of simplicity, one can substitute,

- $\dfrac{2m}{\hbar^2} = \alpha$
- $\alpha \times \text{Potential} = V$
- $\alpha \times \text{Energy} = E$

These substitutions give

$$\frac{d^2\psi}{dr^2} - \left(V - E\right)\psi = 0 \tag{4.95}$$

Assume that the variable r varies between a lower limit a and an upper limit b. The entire range for the variable r can be divided into n sub-intervals each of size h, such that,

$$a < r_2 < r_3 < \ldots < r_n < b$$

$$r_{i+1} - r_i = h$$

From the finite-difference method,

$$\frac{d^2\psi_n}{dr^2} = \frac{\psi_{n-1} - 2\psi_n + \psi_{n+1}}{h^2} \tag{4.96}$$

A comparison of *Eqn. 4.95* with the *Eqn. 4.96* gives

$$\frac{\psi_{n-1} - 2\psi_n + \psi_{n+1}}{h^2} - V_n\psi_n = -E\psi_n \tag{4.97}$$

Eqn. 4.97 implies that

$$2\psi_n - \psi_{n-1} - \psi_{n+1} + h^2 V_n\psi_n = h^2 E\psi_n \tag{4.98}$$

At the boundary,

- $\psi_{n-1} = \psi_1 = 0$

- $\psi_{n+1} = 0$

Therefore *Eqn. 4.98* becomes

$$\left(2 + h^2 V_n\right)\psi_n = h^2 E \psi_n \tag{4.99}$$

For other values in-between the boundary, it is imperative to write *Eqn. 4.98* in the form of a matrix, such that

$$H \begin{pmatrix} \psi_2 \\ \vdots \\ \psi_n \end{pmatrix} = h^2 E \begin{pmatrix} \psi_2 \\ \vdots \\ \psi_n \end{pmatrix} \tag{4.100}$$

In *Eqn. 4.100*,

- H is called the Hamiltonian operator that acts on wave function ψ. The result is proportional to ψ, which is called the stationary state. The proportionality constant, $h^2 E$, is called the energy eigen value of the eigen state ψ.

- The diagonal elements of the Hamiltonian matrix H are given by $(2 + h^2 V_n)$, where n starts from 2 and goes till the number of sub-intervals.

- The elements of matrix H adjacent to the diagonal elements are equal to -1.

- The matrix H is therefore written as

$$H = \begin{pmatrix} 2 + h^2 V_2 & -1 & & & & \\ -1 & 2 + h^2 V_3 & -1 & & & \\ & -1 & 2 + h^2 V_4 & -1 & & \\ & & & & \ddots & -1 \\ & & & & -1 & 2 + h^2 V_n \end{pmatrix}$$

In general, the characteristic equation of a matrix (say, A) is given by *Eqn. 4.101*.

$$AX = \lambda X \tag{4.101}$$

In *Eqn. 4.101*, X is called the eigen vector corresponding to the eigen value λ of the square matrix A. Comparison of this equation with *Eqn. 4.100* implies that

- The energy eigen values of matrix H are given by $h^2 E$.

- The eigen vectors corresponding to these eigen values are given by $\begin{pmatrix} \psi_2 \\ \vdots \\ \psi_n \end{pmatrix}$.

The Scilab programs written in the subsequent sections perform the following tasks.

- Define the functional form of potential energy.

- Generate the Hamiltonian matrix H.

- Determine the eigen values (λ) of the matrix H.

- Determine the energy of the particle using the following conversion,

$$\lambda = h^2 E$$

$$E = \frac{\lambda}{h^2}$$

$$Energy = \frac{E}{\alpha}$$

- Compare this energy with the analytical solution of the specific case.

- Plot the wave function for different energy values

4.8.10.1 Infinite potential well

The functional form of potential energy for an infinite potential well is (*Eqn. 4.102*),

$$V = \begin{cases} 0 & \text{for } a \leq r \leq b \\ \infty & \text{otherwise} \end{cases} \tag{4.102}$$

The Scilab program for determining the energy eigen values and energy eigen vectors for an electron confined within an infinite potential well is as follows.

```
a = 0;                              //Lower boundary
b = 8;                              //Upper boundary
h = 0.01;                             //Step size
n = (b-a)/h;                      //Number of intervals
m = 0.511d6;                  //Mass of electron (eV/c²)

hbar = 1973;                         // hc/2π (in eV Å)

e = 3.795;              //Electron charge in (eV Å)^{1/2}
alpha = 2*m/(hbar*hbar);
V = 0;                            //Potential is zero
```

```
//Construction of the Hamiltonian matrix
A = zeros(n,n);
r = zeros(1,n);

r(1) = r(1) + h;
A(1,1) = 2 + (V*h*h/r(1));
A(1,2) = -1;

for i = 2:n-1;
    r(i) = r(i-1) + h;
    A(i,i-1) = -1;
    A(i,i) = 2 + (V*h*h/r(i));
    A(i,i+1) = -1;
end

r(n) = r(n-1) + h;
A(n,n-1) = -1;
A(n,n) = 2 + (V*h*h/r(n));

//Determine eigen vector and eigen value for matrix A
[eigenvector,eigenvalue] = spec(A);
E = diag(eigenvalue)/(alpha*h*h);

//Columns of the matrix [eigenvector] correspond to
different eigen vectors for different eigen values,
starting from the ground state. Plot the eigen vectors
corresponding to particular eigen value.
plot2d(r, eigenvector(:,1))
xgrid(13)

//Display the ground state energy
mprintf("\n Energy of ground state = %f",E(1));
mprintf("\n Energy of ground state (from formula) =
%f",%pi*%pi*hbar*hbar/(2*m*b*b));
```

The output is given in *Table 4.2*. The corresponding graphs are shown in *Figure 4.17*.

Table 4.2   Value of energy eigen values for the infinite potential well

State	Energy (eV) (Finite difference method)	Energy (eV) (Direct expression)
Ground ($n = 1$)	0.585919	0.587385
First excited ($n = 2$)	2.343666	2.349541
Second excited ($n = 3$)	5.273215	5.286467
Third excited ($n = 4$)	9.374520	9.398164

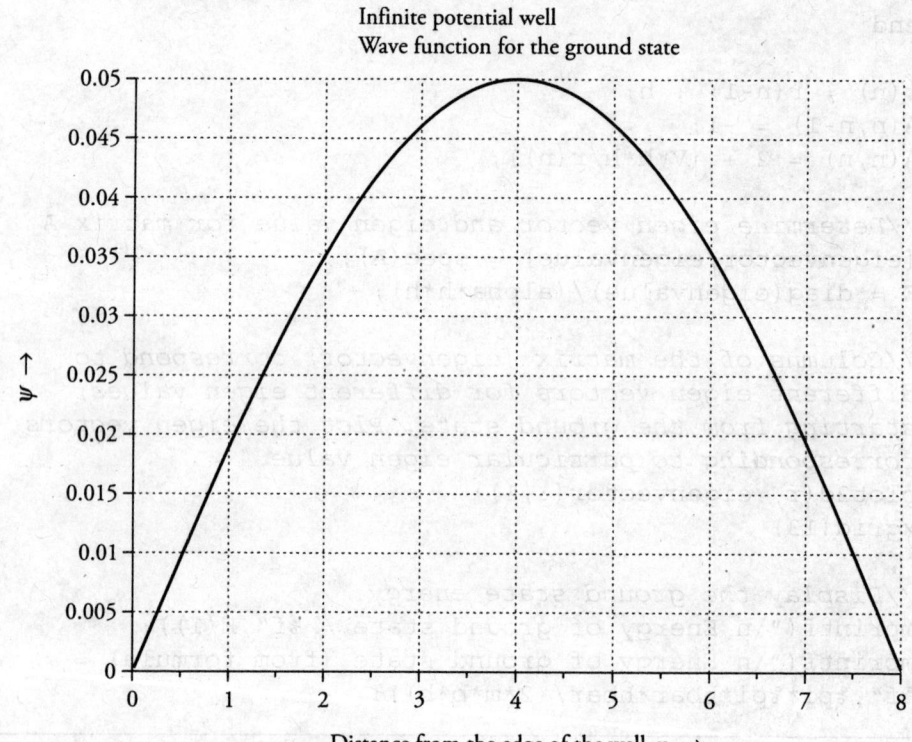

Infinite potential well
Wave function for the ground state

Distance from the edge of the well, x →

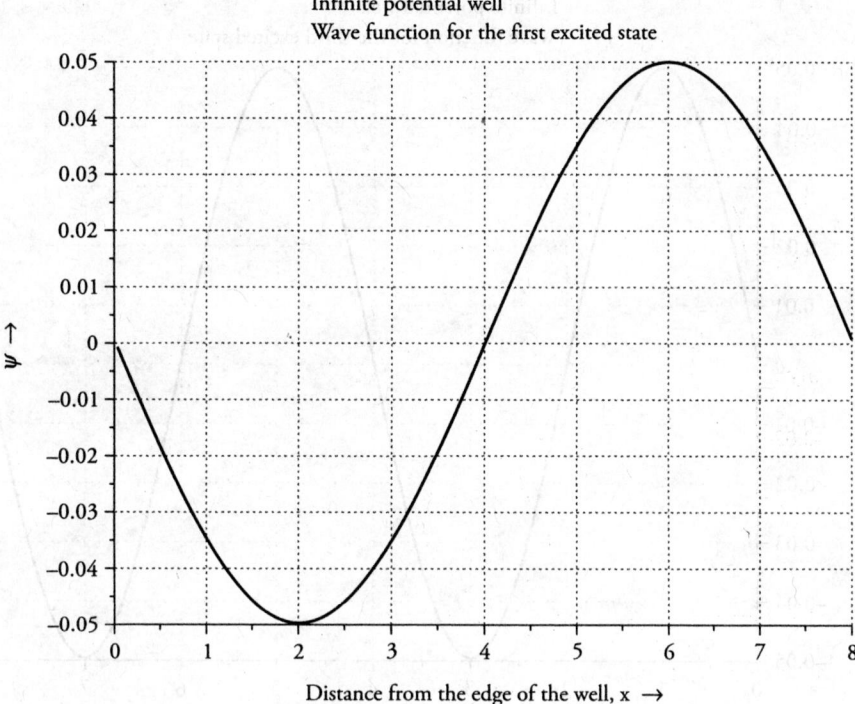

Infinite potential well
Wave function for the first excited state

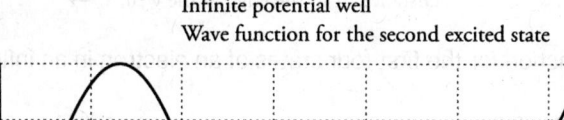

Distance from the edge of the well, x →

Infinite potential well
Wave function for the second excited state

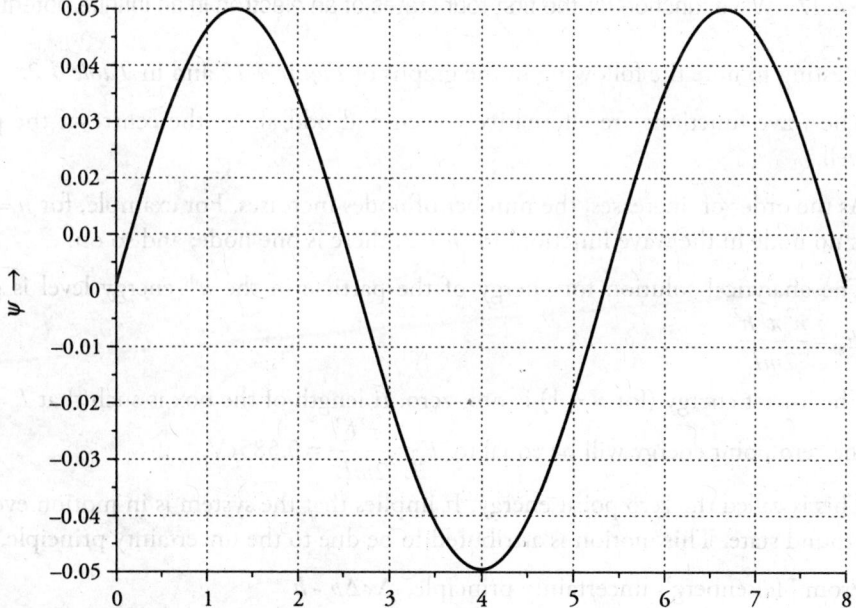

Distance from the edge of the well, x →

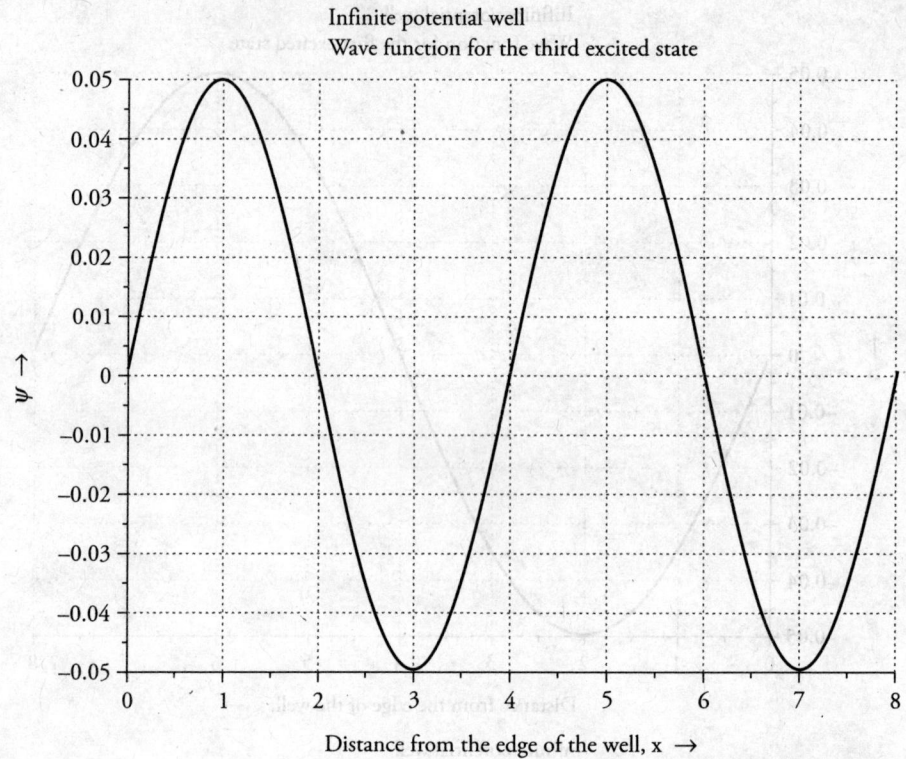

Figure 4.17 Wave function for the first four states of an electron in an infinite potential well

It is interesting to note the following in the graphs of *Figure 4.17* and in *Table 4.2*.

- The wave functions are alternatively even and odd about the centre of the potential well.

- As the order of increases, the number of nodes increases. For example, for $n = 1$, there is no node in the wave function; for $n = 2$, there is one node; and so on.

- The analytical solution for energy of the particle in the nth energy level is given by

$$E_n = \frac{n^2 \pi^2 \hbar^2}{2mL^2}$$

- The lowest energy (for $n = 1$) is non-zero. If length of the box is such that $L = 8$, then the zero point energy will be equal to $E_1 = \frac{\pi^2 \hbar^2}{2mL^2} = 0.585\,\text{eV}$

- This is called the zero point energy. It implies that the system is in motion even in the ground state. This motion is attributed to be due to the uncertainty principle.

- From Heisenberg's uncertainty principle, $\Delta x \, \Delta p \sim \hbar$

- If the length of the box is L, then uncertainty in the position of the particle is given by

$$\Delta x = L$$

This implies that the minimum uncertainty in the momentum of the particle will be

$$\Delta p = \frac{\hbar}{L}$$

This implies that the momentum is at least of the order of the uncertainty in the momentum. Therefore, minimum energy will be given by

$$E = \frac{p^2}{2m} = \frac{\hbar^2}{2mL^2}$$

4.8.10.2 Hydrogen atom: Coulomb potential

The Coulomb potential describing the interaction of a positively charged nucleus with a negatively charged electron is given by *Eqn. 4.103*.

$$V = -\frac{e^2}{r} \tag{4.103}$$

The Scilab program to determine the radial wave functions for different orbitals of the hydrogen atom is as follows. The scheme is similar to that discussed in *Section 4.8.10.1*.

```
a = 0;                              //Lower boundary
b = 20;                             //Upper boundary
h = 0.02;                           //Step size
n = (b-a)/h;                        //Number of intervals
m = 0.511d6;                        //Mass of electron (eV/c²)

hbar = 1973;                        //ℏc (in eV Å)

e = 3.795;          //Electron charge(in(eV Å)^(1/2))
alpha = 2*m/(hbar*hbar);
V = -alpha*e*e;                     //Potential

//Construction of the Hamiltonian matrix
A = zeros(n,n);
r = zeros(1,n);


r(1) = r(1) + h;
A(1,1) = 2 + (V*h*h/r(1));
A(1,2) = -1;

for i = 2:n-1;
    r(i) = r(i-1) + h;
```

```
      A(i,i-1) = -1;
      A(i,i) = 2 + (V*h*h/r(i));
      A(i,i+1) = -1;
   end

   r(n) = r(n-1) + h;
   A(n,n-1) = -1;
   A(n,n) = 2 + (V*h*h/r(n));

   //Determine eigen vector and eigen value for matrix A
   [eigenvector,eigenvalue] = spec(A);
   E = diag(eigenvalue)/(alpha*h*h);

   E(1)
   E(2)
   E(3)

   //Plot ground state wave function
   plot2d(r, eigenvector(:,1))

   //Plot first excited state wave function
   plot2d(r, eigenvector(:,2))

   //Plot second excited state wave function
   plot2d(r, eigenvector(:,3))

   //Plot ground state electron probability
   plot2d(r, eigenvector(:,1).* eigenvector(:,1))

   //Plot first excited state electron probability
   plot2d(r, eigenvector(:,2).* eigenvector(:,2))

   //Plot second excited state electron probability
   plot2d(r, eigenvector(:,3).* eigenvector(:,3))
   xgrid(13)
```

The output for energy eigen values (in eV) is as follows.

```
E(1) = - 13.609078    (ground state)
E(2) = - 3.4031811    (first excited state)
E(3) = - 1.5124975    (second excited state)
```

The analytical value of energy $\left(=\dfrac{-13.6}{n^2}\,eV\right)$ in these energy states are as follows.

Ground state (1s orbital): –13.6 eV

First excited state (2s orbital): –3.4 eV

Second excited state (3s orbital): –1.51 eV

The radial wave functions and the electron probability densities for 1s, 2s and 3s orbitals have been plotted in *Figure 4.18* and *Figure 4.19*, respectively.

It is interesting to note the following in *Figures 4.18* and *4.19*:

- The most probable radius (distance from the nucleus) increases as one goes from '1s' to '2s' to '3s' orbital.

- The higher orbitals have a certain probability at lower distances also.

- For the '1s' orbital, the peak in the electron probability density graph occurs at a distance of 0.52 Å. This value corresponds to the Bohr radius (r_0). It is a physical constant whose value is approximately equal to the most probable distance between the electron and nucleus of the hydrogen atom in its ground state.

$$r_0 = \frac{4\pi\varepsilon_0\hbar^2}{m_e e^2} \equiv \frac{(\hbar c)^2}{me^2} = 528944 \times 10^{-6} = 0.529\,\text{Å}$$

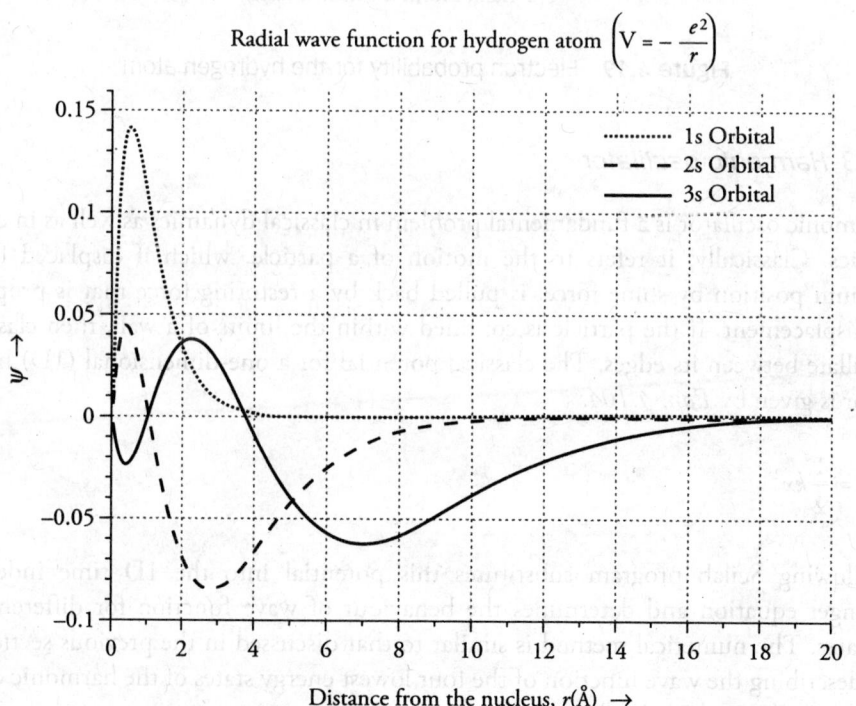

Figure 4.18 Radial wave function for the hydrogen atom

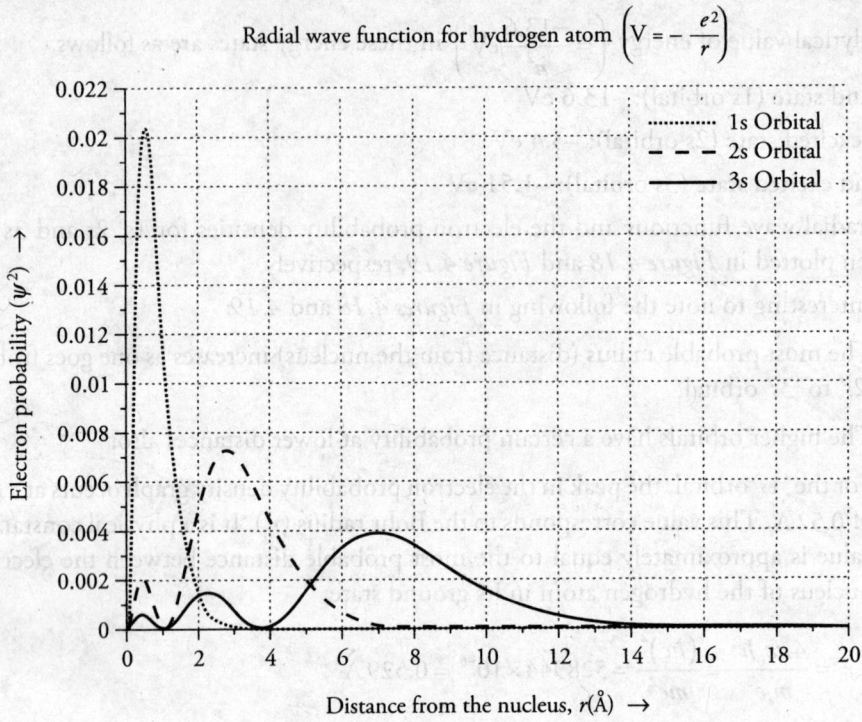

Figure 4.19   Electron probability for the hydrogen atom

4.8.10.3 Harmonic oscillator

The harmonic oscillator is a fundamental problem in classical dynamics as well as in quantum mechanics. Classically, it refers to the motion of a particle, which if displaced from the equilibrium position by some force, is pulled back by a restoring force that is proportional to the displacement. If the particle is confined within the limits of a wall, then classically it will oscillate between its edges. The classical potential for a one-dimensional (1D) harmonic oscillator is given by *Eqn. 4.104*.

$$V = \frac{1}{2}kx^2 \tag{4.104}$$

The following Scilab program substitutes this potential into the 1D time independent Schrödinger equation and determines the behaviour of wave function for different energy eigen states. The numerical method is similar to that discussed in the previous sections. The graphs describing the wave function of the four lowest energy states of the harmonic oscillator are shown in *Figure 4.20 (a–d)*.

```
a = 0;                              //Lower boundary
b = 16;                             //Upper boundary
h = 0.01;                              //Step size
n = (b-a)/h;                    //Number of intervals
m = 0.511d6;                //Mass of electron (eV/c²)

hbar = 1973;                        //ℏc (in eV Å)

e = 3.795;          //Electron charge ( in( eV Å )^1/2 )
alpha = 2*m/(hbar*hbar);
k = 1;                          //Positive constant
A = zeros(n,n);
r = zeros(1,n)-8;         //Boundary shifted to [-8,8]

r(1) = r(1) + h;
A(1,1) = 2 + (h*h*alpha*0.5*k*r(1)^2);
A(1,2) = -1;

for i = 2:n-1;
    r(i) = r(i-1) + h;
    A(i,i-1) = -1;
    A(i,i) = 2 + (h*h*alpha*0.5*k*r(i)^2);
    A(i,i+1) = -1;
end

r(n) = r(n-1) + h;
A(n,n-1) = -1;
A(n,n) = 2 + (h*h*alpha*0.5*k*r(n)^2);

[c,d] = spec(A);
E = diag(d)/(alpha*h*h);
eigenvector = c(:,1);  //Wave function of ground state
plot2d(r,eigenvector)        //Plot the wave function

x = [sqrt(2*E(1)/k), sqrt(2*E(1)/k)];
y = [min(eigenvector) max(eigenvector)];
plot2d(x,y)                 //Plot the classical limit
plot2d(-x,y)
xgrid(13)
```

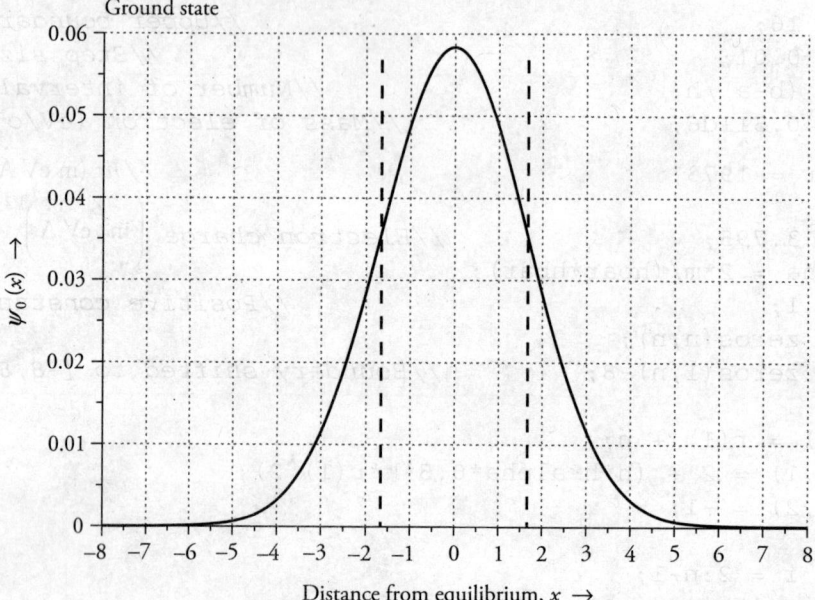

Figure 4.20(a) Wave function described in *Section 4.8.10.3*

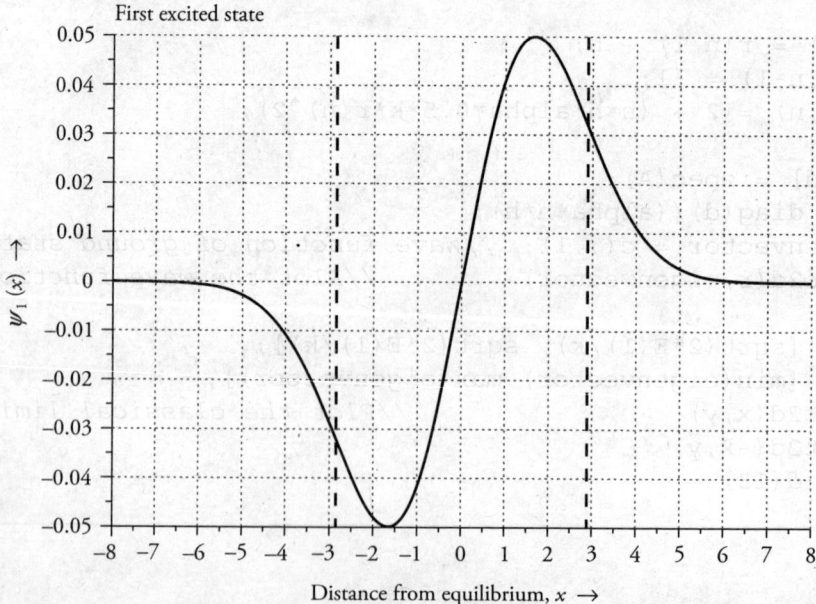

Figure 4.20(b) Wave function described in *Section 4.8.10.3*

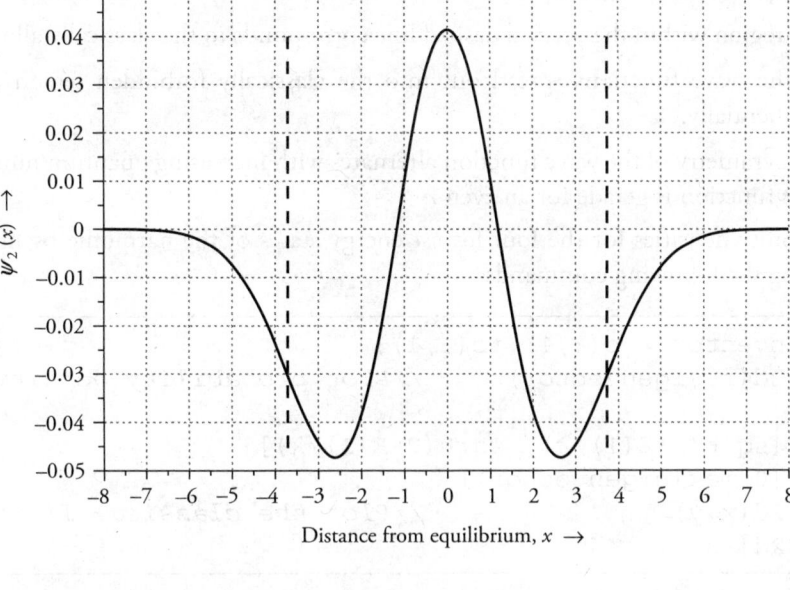

Figure 4.20(c)   Wave function described in *Section 4.8.10.3*

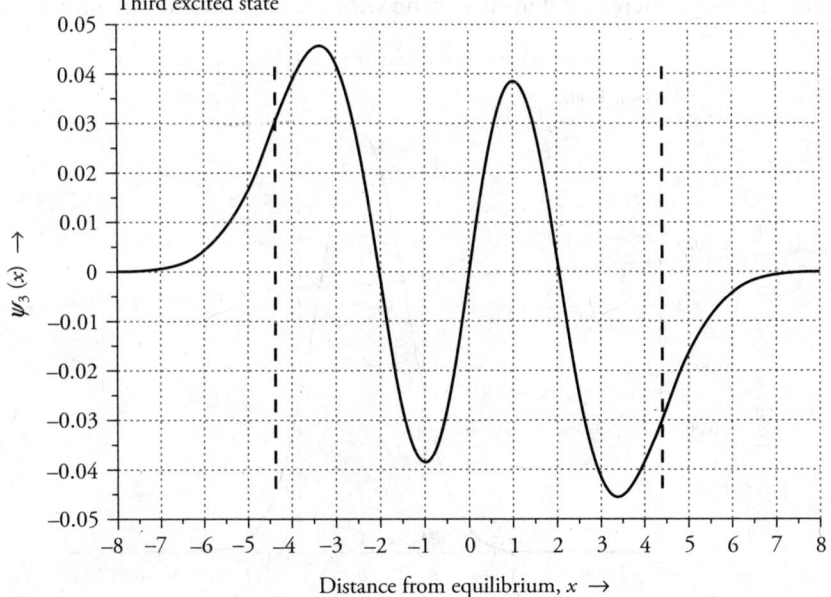

Figure 4.20(d)   Wave function described in *Section 4.8.10.3*

The following are clear from *Figure 4.20*.

- These graphs show the variation of the wave function with distance of the particle from the equilibrium position.

- The region within the vertical dashed line corresponds to the classically allowed region.

- All the wave functions spread out into the classically forbidden region and fall off exponentially.

- The symmetry of the wave function alternates with increasing quantum number *n*. The wave function is gerade for an even *n*.

The probability densities for the four lowest energy states of the harmonic oscillator can be plotted using the following commands.

```
eigenvector = c(:,1).*c(:,1);
plot2d(r,eigenvector)        //Plot probability density

x = [sqrt(2*E(1)/k), sqrt(2*E(1)/k)];
y = [0 max(eigenvector)]
plot2d(x,y)                  //Plot the classical limit
plot2d(-x,y)
```

These are shown in *Figure 4.21(a–d)*. It is clear that as far as the quantum oscillator is concerned, there is a certain significant probability of the particle being present in the classically forbidden region. The quantum probability gets smaller and smaller and approaches the classical limit as the quantum number *n* increases. This is in sync with the correspondence principle.

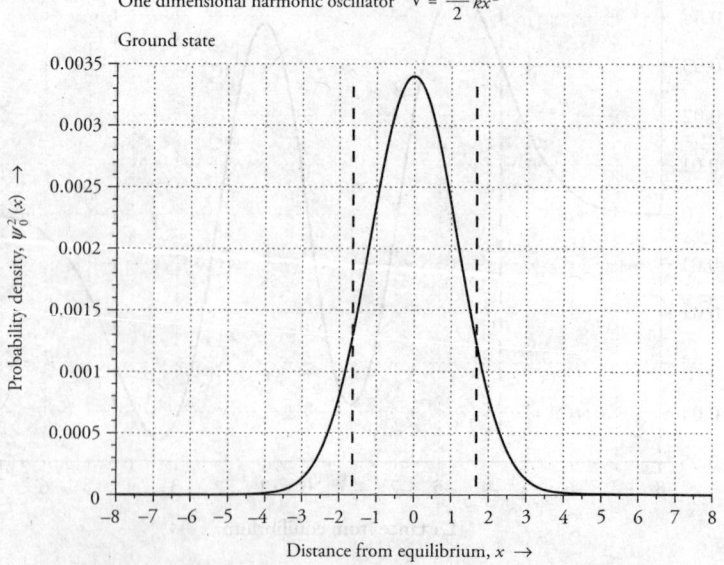

Figure 4.21(a)   Probability density for *Section 4.8.10.3*

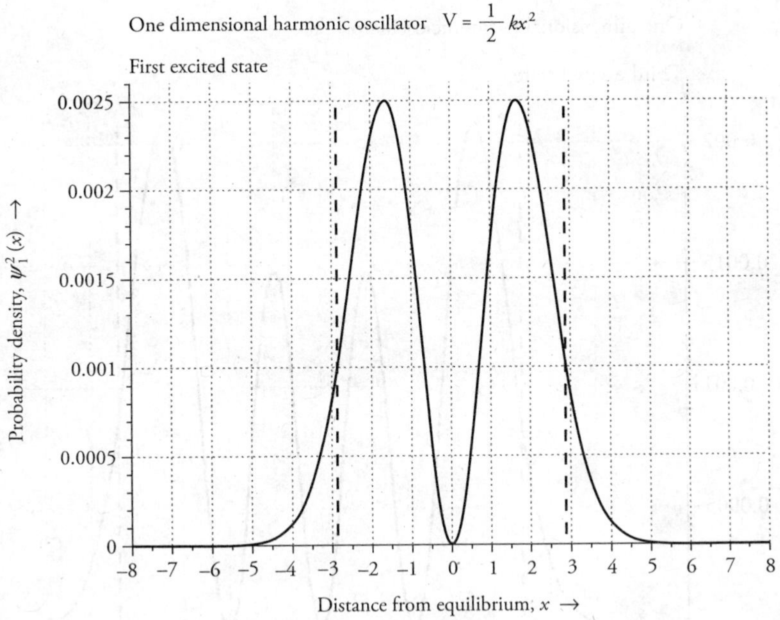

One dimensional harmonic oscillator $V = \dfrac{1}{2} kx^2$

First excited state

Figure 4.21(b)   Probability density for *Section 4.8.10.3*

One dimensional harmonic oscillator $V = \dfrac{1}{2} kx^2$

Second excited state

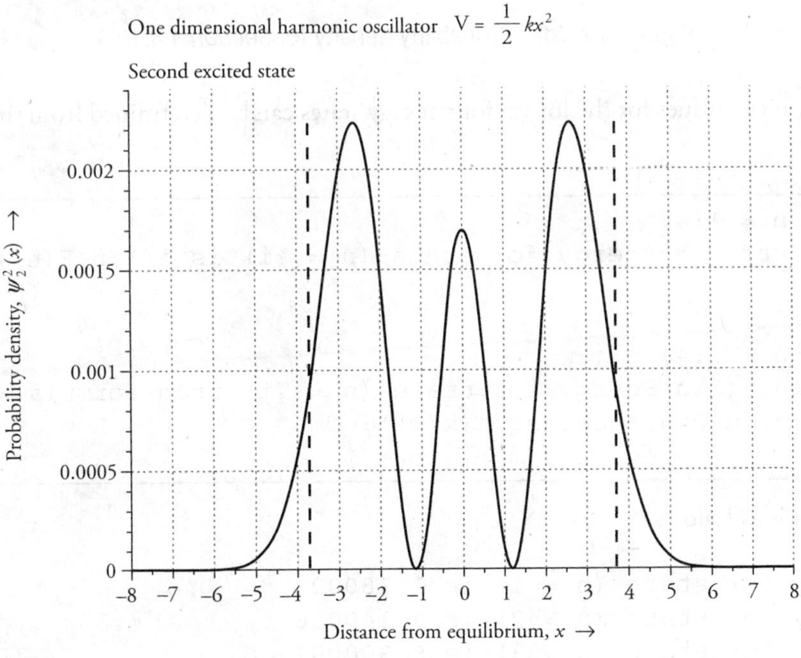

Figure 4.21(c)   Probability density for *Section 4.8.10.3*

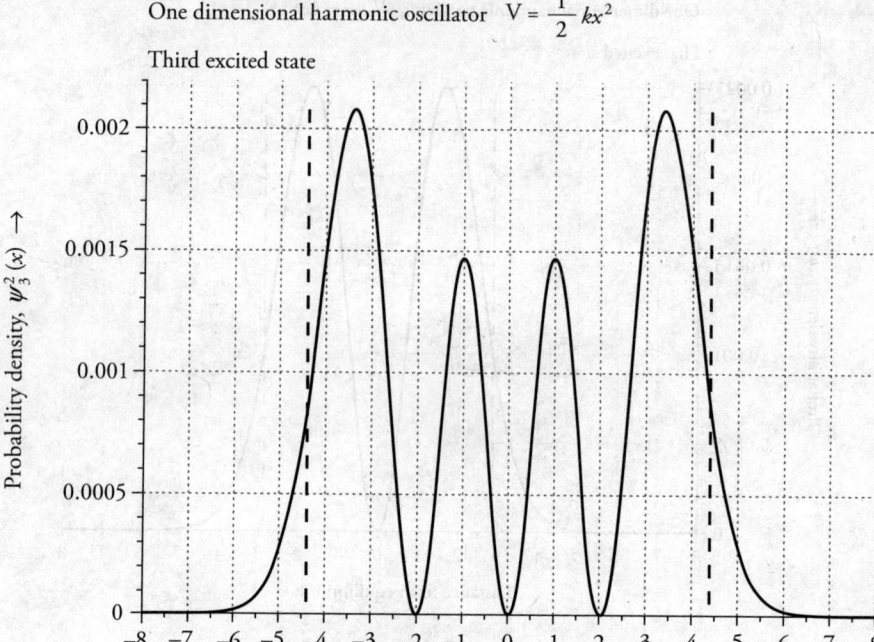

One dimensional harmonic oscillator $V = \dfrac{1}{2}kx^2$

Third excited state

Figure 4.21(d)   Probability density for *Section 4.8.10.3*

The energy eigen values for the lowest four energy states can be determined from the following program.

```
for n = 1:4;
mprintf("\n Energy for state (n = %i) is %f",n,E(n));
end

for n = 1:4;
mprintf("\n Energy for state (n = %i) from formula =
%f",n,(n-0.5)*hbar*sqrt(k/m));
end
```

The output is as follows.

```
Energy for state (n = 1) is 1.380021
Energy for state (n = 2) is 4.140056
Energy for state (n = 3) is 6.900081
Energy for state (n = 4) is 9.660105
```

```
Energy for state (n = 1) from formula = 1.380024
Energy for state (n = 2) from formula = 4.140072
Energy for state (n = 3) from formula = 6.900120
Energy for state (n = 4) from formula = 9.660168
```

4.8.11 Lagrangian dynamics

Lagrangian mechanics is widely used in solving mechanical problems in physics. The Lagrangian is used to summarize the dynamics of the system. It is a mathematical function of time and the generalized coordinates along with their derivatives. The Lagrangian (L) for a system of particles is defined by

$$L = K - V \tag{4.105}$$

In *Eqn. 4.105*,

- K is the total kinetic energy

- V is the potential energy

Lagrange's equation of motion (Euler–Lagrange equation) is given by

$$\frac{d}{dt}\left(\frac{\partial L}{\partial \dot{q}_j}\right) - \frac{\partial L}{\partial q_j} = 0 \tag{4.106}$$

In *Eqn. 4.106*,

- The position vector of the particle depends on n-generalized coordinates. Therefore q_j is the jth generalized coordinate and $\dot{q}_j$ is the velocity component of the jth generalized coordinate.

- The Lagrangian is often written in the form, $L(q, \dot{q}, t)$

It should be noticed that as opposed to the Newtonian mechanics, the Lagrangian mechanics deals with the dynamics of systems that involve conservative forces. In case of non-conservative forces, a set of modified Euler–Lagrange equations are used to account for driven and dissipative forces.

In the following, the Lagrangian formalism of classical mechanics has been applied for a coupled pendulum system. The coupled pendulum consists of two pendulums coupled at their midpoints with a spring having spring constant k. The length of both the pendulum is equal to l and mass of both the bobs is m.

The kinetic energy is given by

$$K = \frac{1}{2}ml^2\left(\dot{\theta}_1^2 + \dot{\theta}_2^2\right) \tag{4.107}$$

The potential energy is approximately equal to

$$V = -mgl\left(\cos\theta_1 + \cos\theta_2\right) + \frac{1}{2}k\left(\frac{l}{2}\right)^2\left(\theta_1 - \theta_2\right)^2 \tag{4.108}$$

From *Eqns. 4.107–4.108*, the Lagrangian of the coupled pendulum system will be given by

$$L = K - V = \frac{1}{2}ml^2\left(\dot{\theta}_1^2 + \dot{\theta}_2^2\right) + mgl\left(\cos\theta_1 + \cos\theta_2\right) - \frac{1}{2}k\left(\frac{l}{2}\right)^2\left(\theta_1 - \theta_2\right)^2 \tag{4.109}$$

The Lagrangian in *Eqn. 4.109* can be simplified by replacing the cosine terms by their small angle approximation. Thus,

$$L = \frac{1}{2}ml^2\left(\dot{\theta}_1^2 + \dot{\theta}_2^2\right) - \frac{mgl}{2}\left(\theta_1^2 + \theta_2^2\right) - \frac{1}{2}k\left(\frac{l}{2}\right)^2\left(\theta_1 - \theta_2\right)^2 \tag{4.110}$$

Lagrange's equations of motion for this system are given by *Eqns. 4.111–4.112*.

$$\frac{d}{dt}\left(\frac{\partial L}{\partial\dot{\theta}_1}\right) - \frac{\partial L}{\partial\theta_1} = 0 \tag{4.111}$$

$$\frac{d}{dt}\left(\frac{\partial L}{\partial\dot{\theta}_2}\right) - \frac{\partial L}{\partial\theta_2} = 0 \tag{4.112}$$

Substitution of L from *Eqn. 4.110* in *Eqns. 4.111–4.112* will give

$$\ddot{\theta}_1 + \frac{g}{l}\theta_1 + \frac{k}{4m}\left(\theta_1 - \theta_2\right) = 0 \tag{4.113}$$

$$\ddot{\theta}_2 + \frac{g}{l}\theta_2 + \frac{k}{4m}\left(\theta_2 - \theta_1\right) = 0 \tag{4.114}$$

A re-arrangement of the terms of *Eqns. 4.113–4.114* will give

$$\ddot{X} + \frac{g}{l}X = 0 \tag{4.115}$$

$$\ddot{Y} + \left(\frac{g}{l} + \frac{k}{2m}\right)Y = 0 \tag{4.116}$$

In *Eqns. 4.115–4.116*,

- $X = \theta_1 + \theta_2$
- $Y = \theta_1 - \theta_2$

Case (I) In-phase motion

In the in-phase mode, both the pendulums oscillate with the same phase and amplitude. Therefore,

- $\theta_1 = \theta_2$
- $X \neq 0$
- $Y = 0$
- Frequency of oscillation of the system is same as that of the individual pendulum.
- The spring has no effect on the motion of the pendulums.

The algorithm for writing the Scilab program for in-phase motion of the coupled pendulum is as follows.

- Write the user-defined function for *Eqn. 4.115*, such that,

$$Z = \begin{pmatrix} X \\ \dot{X} \end{pmatrix}$$

$$\dot{Z} = \begin{pmatrix} \dot{X} \\ \ddot{X} \end{pmatrix} = \begin{pmatrix} Z(2) \\ -\dfrac{g}{l} Z(1) \end{pmatrix}$$

- Write the user-defined function for *Eqn. 4.116*, such that,

$$Z = \begin{pmatrix} Y \\ \dot{Y} \end{pmatrix}$$

$$\dot{Z} = \begin{pmatrix} \dot{Y} \\ \ddot{Y} \end{pmatrix} = \begin{pmatrix} Z(2) \\ -\left(\dfrac{g}{l} + \dfrac{k}{2m}\right) Z(1) \end{pmatrix}$$

- The initial conditions are chosen such that $\theta_1 = \theta_2 = 1$. Therefore,

$$X \big|_{t=0} = 2$$

$$\dot{X} \big|_{t=0} = 0$$

$$Y \big|_{t=0} = 0$$

$$\dot{Y}\,|_{t=0} = 0$$

$$\frac{g}{l} = \pi^2$$

$$\frac{g}{l} + \frac{k}{2m} = 4\pi^2$$

- Determine the value of X for a range of time values.
- Determine the value of Y for a range of time values.
- Plot the variation of θ_1 and θ_2 as a function of time.

The Scilab program is written as follows. The time–angular displacement graph is shown in *Figure 4.22*.

```
function zdot = f1_X(t,z)
zdot(1) = z(2);
zdot(2) = -alpha_0*z(1);
endfunction

function zdot = f2_Y(t,z)
zdot(1) = z(2);
zdot(2) = -alpha*z(1);
endfunction

X_0 = 2;
V_0 = 0;
alpha_0 = %pi^2;
t = 0:0.1:10;
X = ode([X_0;V_0],0,t,f1_X);

Y_0 = 0;
V_0 = 0;
alpha = (2*%pi)^2;
t = 0:0.1:10;
Y = ode([Y_0;V_0],0,t,f2_Y);

subplot(211)
plot2d(t,0.5*(X(1,:)-Y(1,:)));

subplot(212)
plot2d(t,0.5*(X(1,:)+Y(1,:)));
```

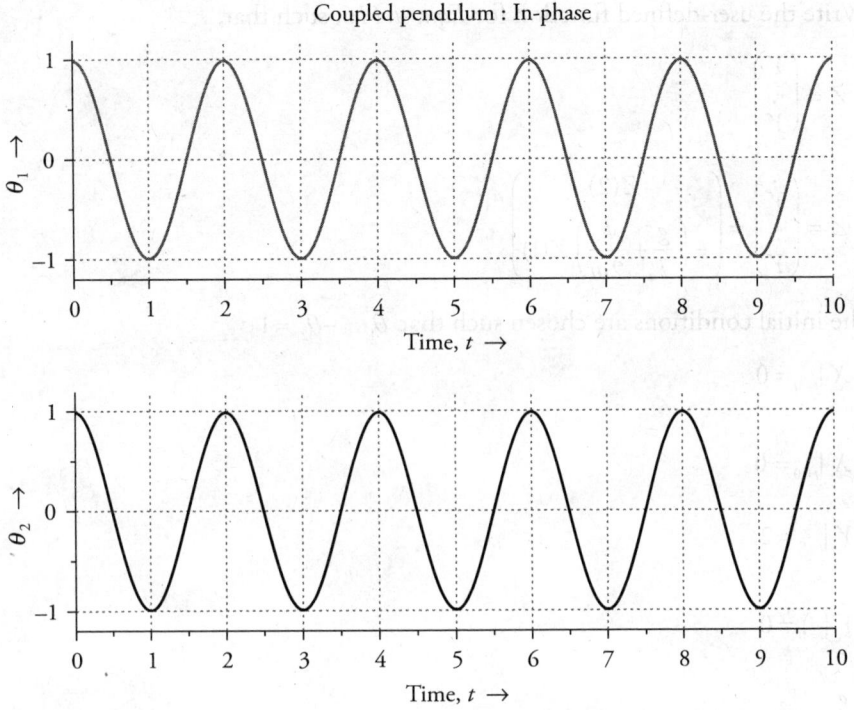

Figure 4.22 In-phase motion of a coupled pendulum

Case (II) Out-of-phase motion

In the out-of-phase mode, the two pendulums oscillate with opposite phase. Therefore,

- $\theta_1 = -\theta_2$
- $X = 0$
- Effectively, only *Eqn. 4.116* is valid.
- Frequency of oscillation of the system is greater than that of an individual pendulum.

Algorithm for writing the Scilab program for out-of-phase motion of the coupled pendulum is as follows.

- Write the user-defined function for *Eqn. 4.115*, such that,

$$Z = \begin{pmatrix} X \\ \dot{X} \end{pmatrix}$$

$$\dot{Z} = \begin{pmatrix} \dot{X} \\ \ddot{X} \end{pmatrix} = \begin{pmatrix} Z(2) \\ -\dfrac{g}{l} Z(1) \end{pmatrix}$$

- Write the user-defined function for *Eqn. 4.116*, such that,

$$Z = \begin{pmatrix} Y \\ \dot{Y} \end{pmatrix}$$

$$\dot{Z} = \begin{pmatrix} \dot{Y} \\ \ddot{Y} \end{pmatrix} = \begin{pmatrix} Z(2) \\ -\left(\dfrac{g}{l} + \dfrac{k}{2m}\right) Z(1) \end{pmatrix}$$

- The initial conditions are chosen such that $\theta_1 = -\theta_2 = 1$

$$X\big|_{t=0} = 0$$

$$\dot{X}\big|_{t=0} = 0$$

$$Y\big|_{t=0} = 2$$

$$\dot{Y}\big|_{t=0} = 0$$

$$\frac{g}{l} = \pi^2$$

$$\frac{g}{l} + \frac{k}{2m} = 4\pi^2$$

- Determine the value of X for a range of time values.
- Determine the value of Y for a range of time values.
- Plot the variation of θ_1 and θ_2 as a function of time.

The Scilab program for this case is written as follows. The time-angular displacement graph is shown in *Figure 4.23*.

```
function zdot = f1_X(t,z)
zdot(1) = z(2);
zdot(2) = -alpha_0*z(1);
endfunction

function zdot = f2_Y(t,z)
zdot(1) = z(2);
zdot(2) = -alpha*z(1);
endfunction

X_0 = 0;
```

```
V_0 = 0;
alpha_0 = %pi^2;
t = 0:0.1:10;
X = ode([X_0;V_0],0,t,f1_X);

Y_0 = 2;
V_0 = 0;
alpha = (2*%pi)^2;
t = 0:0.1:10;
Y = ode([Y_0;V_0],0,t,f2_Y);
subplot(211)
plot2d(t,0.5*(X(1,:)-Y(1,:)));

subplot(212)
plot2d(t,0.5*(X(1,:)+Y(1,:)));
```

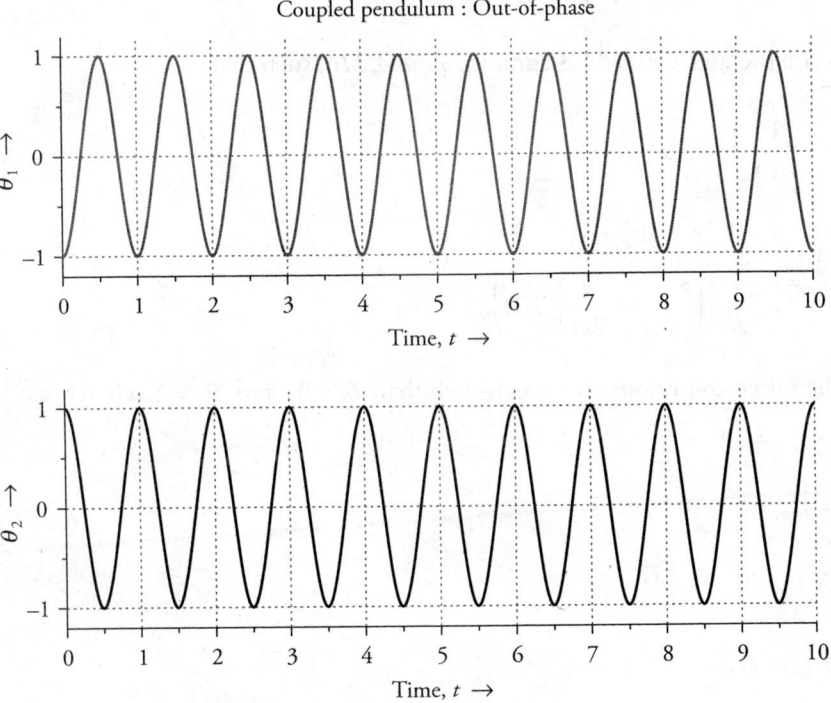

Figure 4.23 Out-of-phase motion of a coupled pendulum

Case (III) Resonant motion

In the resonant mode, only one pendulum is displaced from its equilibrium position. The second pendulum is kept stationary. Therefore,

- $\theta_1 = 0$
- $\theta_2 \neq 0$

The algorithm for writing the Scilab program for the resonant mode of the coupled pendulum is as follows.

- Write the user-defined function for *Eqn. 4.115*, such that,

$$Z = \begin{pmatrix} X \\ \dot{X} \end{pmatrix}$$

$$\dot{Z} = \begin{pmatrix} \dot{X} \\ \ddot{X} \end{pmatrix} = \begin{pmatrix} Z(2) \\ -\dfrac{g}{l} Z(1) \end{pmatrix}$$

- Write the user-defined function for *Eqn. 4.116*, such that,

$$Z = \begin{pmatrix} Y \\ \dot{Y} \end{pmatrix}$$

$$\dot{Z} = \begin{pmatrix} \dot{Y} \\ \ddot{Y} \end{pmatrix} = \begin{pmatrix} Z(2) \\ -\left(\dfrac{g}{l} + \dfrac{k}{2m}\right) Z(1) \end{pmatrix}$$

- The initial conditions are chosen such that, $\theta_1 = 0$ and $\theta_2 = 2$. Therefore,

$$X\big|_{t=0} = 2$$

$$Y\big|_{t=0} = 2$$

$$\dot{X}\big|_{t=0} = 0$$

$$\dot{Y}\big|_{t=0} = 0$$

$$\frac{g}{l} = \pi^2$$

$$\frac{g}{l} + \frac{k}{2m} = 1.21\pi^2$$

- Determine the value of X for a range of time values.

- Determine the value of Y for a range of time values.

- Plot the variation of θ_1 and θ_2 as a function of time.

The Scilab program for this case is written as follows. The time–angular displacement graph is shown in *Figure 4.24*.

```
function zdot = f1_X(t,z)
zdot(1) = z(2);
zdot(2) = -alpha_0*z(1);
endfunction

function zdot = f2_Y(t,z)
zdot(1) = z(2);
zdot(2) = -alpha*z(1);
endfunction


X_0 = 2;
V_0 = 0;
alpha_0 = %pi^2;
t = 0:0.1:40;
X = ode([X_0;V_0],0,t,f1_X);


Y_0 = 2;
V_0 = 0;
alpha = (1.1*%pi)^2;
t = 0:0.1:40;
Y = ode([Y_0;V_0],0,t,f2_Y);

subplot(211)
plot2d(t,0.5*(X(1,:)-Y(1,:)));

subplot(212)
plot2d(t,0.5*(X(1,:)+Y(1,:)));
```

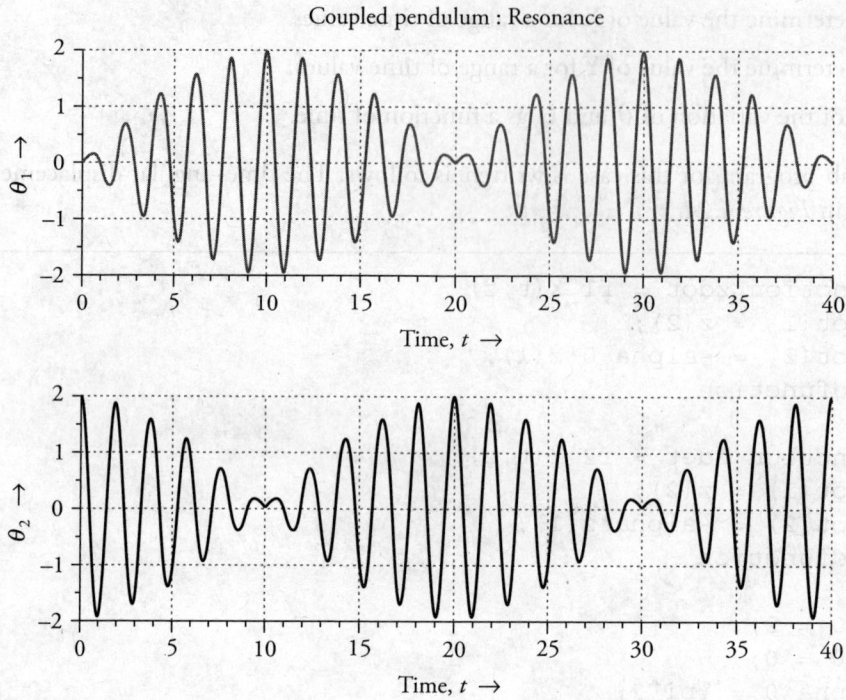

Figure 4.24   Resonant motion of a coupled pendulum

4.9 Exercises

1.   Consider the following differential equation.

$$\frac{dY}{dX} = -Y$$

The initial condition is

$Y(\text{at } X = 0) = 1$

Graphically compare the solution of this differential equation obtained from the built-in 'ode' function of Scilab with the estimated result from Euler's and Runge–Kutta methods.

2.   Write a Scilab program to graphically show the significance of step size in Euler's method on estimation of solution for the equation,

$$\frac{dy}{dt} = -2y$$

The initial condition is

$y \,(\text{at } t = 0) = 5$

3. Write a Scilab program to graphically show the significance of a large step size in Euler's method on estimation of solution for the equation

$$\frac{dy}{dx} = \frac{e^x}{\left(1+e^x\right)^2}$$

The initial condition is

y (at $x = -4$) = 0.018

4. Using an appropriate example, write a Scilab program to prove that Euler's method leads to an overestimation when the solution curve is concave downwards; and it leads to an underestimation when the solution curve is concave upwards.

5. Write a Scilab program for solving the following equation and draw the family of curves representing the solution. Moreover, draw the family of orthogonal curves on the same graph.

$$\frac{dy}{dx} = \frac{y^2 - x^2}{2xy}$$

6. Write a Scilab program for solving the following coupled equations by making use of Euler's method (step size 0.1 and 0.2) and the Runge–Kutta fourth order method (step size 0.2). Compare the result with the analytical solution.

$$\frac{dx}{dt} = y$$

$$\frac{dy}{dt} = -x$$

The initial conditions are
* $x_0 = 1$
* $y_0 = 0$

7. An object is heated to a temperature T_0 and is then kept in a room having a temperature T_s. From Newton's law, the cooling rate $\left(\frac{dT}{dt}\right)$ of a heated object is proportional to the difference between T and T_s,

$$\frac{dT}{dt} = \alpha(T_s - T)$$

Write a Scilab program for solving this differential equation by making use of the Runge–Kutta method and draw the solution curves for two different values of α. Take the initial temperature of the object and the surrounding environment to be equal to 80 °C and 20 °C respectively.

8. Write a Scilab program to solve the logistic equation of the following population growth model.

    $$\frac{dN}{dt} = \alpha N(\beta - N)$$

    Take the initial values to be equal to
    - $\alpha = 0.02, 0.025$
    - $\beta = 100$
    - $N(\text{at } t = 0) = 1$

9. Write a Scilab program for solving the following equation using Euler's method, Runge–Kutta fourth order method and the built-in Scilab function. Compare all the estimated results with the analytical solution.

    $$\frac{d^2y}{dx^2} + y = 0$$

    The initial conditions are
    - $y(0) = 3$
    - $\left.\dfrac{dy}{dx}\right|_0 = 5$

10. For the simple pendulum discussed in *Section 4.8.7*, write a Scilab program to draw the phase-space plot (graph of $\dot{\theta}$ vs. θ) for angular amplitudes 30°, 120° and 175°.

11. A 50 g mass in the mass–spring system is displaced from the equilibrium position by 1 cm. It is released with zero velocity. The spring constant is 128 dyne/cm. Draw the resultant motion for over-damped, critically damped and under-damped cases.

12. Solve the following second order differential equations for the boundary conditions mentioned. Plot the result in each case.

    a) $\dfrac{d^2y}{dx^2} + y = 0$, $y(0) = 1$, $y\left(\dfrac{\pi}{2}\right) = 1$

    b) $x^2\dfrac{d^2y}{dx^2} + 3xy = 8$, $y(1) = 1$, $y(5) = 1$

    c) $x^2\dfrac{d^2y}{dx^2} - 2x\dfrac{dy}{dx} + 5xy = 10$, $y(1) = 1$, $y(5) = 1$

    d) $\dfrac{d^2y}{dx^2} - \dfrac{dy}{dx} + 5y = -3$, $y(1) = 1$, $y(5) = 1$

13. An object is kept at a height of 100 m from the ground. It is released from rest at time $t = 0$. It falls under the effect of air drag and uniform gravitational field. Write a Scilab program to compare the velocity–time graph for the motion of the object under free fall and in the presence of air resistance.

14. Write a Scilab program to show the effect of resistance (damping factor) on the amount of charge present in a series L-C-R circuit at any time t. It can be assumed that there is no external driving signal, the initial charge in the circuit is $2C$ and there is no initial current. Take the parameter values as
    - $L = 2H$
    - $C = 0.5\ F$
    - $R = 1\Omega,\ 4\Omega,\ 7\Omega.$

15. With reference to the previous question, write a Scilab program to draw the charge and current profile of the series L-C-R circuit under the critically damped condition.

16. With reference to the previous question, write a Scilab program to draw the homogeneous and steady state behaviour of charge present in the series L-C-R circuit for the under-damped condition. Take the driving signal to be equal to $3 \sin t$.

17. Write a Scilab program to determine the solution of the following equation. Plot the solution curve also.

$$\frac{d^3y}{dx^3} + 3\frac{dy}{dx} - 5\sin 2x = 0$$

The initial conditions are

- $y|_{x=0} = 0$

- $\dfrac{dy}{dx}\big|_{x=0} = 0$

- $\dfrac{d^2y}{dx^2}\big|_{x=0} = 0$

18. The following questions are based on the Schrödinger equation for hydrogen atom (*Section 4.8.10.2*) where an electron is moving under the Coulomb potential. Write a Scilab program to
    a) Show that the wave functions of hydrogen-like atoms are orthogonal.
    b) Determine the value of the Bohr radius.
    c) Calculate the one-dimensional probability that the electron in 1s orbital lies in the range,
       - $0 \le r \le r_{Bohr}$
       - $r_{Bohr} \le r \le 2r_{Bohr}$
       - $0 \le r \le 10r_{Bohr}$
       - $r_{Bohr} \le r \le 10r_{Bohr}$
    d) Calculate the one-dimensional probability that the electron in 2s orbital lies in the range,
       - $0 \le r \le r_{Bohr}$
       - $4r_{Bohr} \le r \le 6r_{Bohr}$
    e) Calculate the one-dimensional probability that the electron in 3s orbital lies in the range,
       - $0 \le r \le r_{Bohr}$.
       - $4r_{Bohr} \le r \le 6r_{Bohr}$.
       - $12r_{Bohr} \le r \le 14r_{Bohr}$.

19. Write a Scilab program to plot the ground state radial wave function for an atom which is subject to a screened Coulomb potential. The potential describing the behaviour of an atom at a distance r from the centre is given by

$$V = -\frac{e^2}{r} \exp\left(-\frac{r}{a}\right)$$

Take
- Mass of electron $= m = 0.511 \times 10^6 \dfrac{eV}{c^2}$
- $\hbar c = 1973 eV\,\text{Å}$
- Charge $= e = 3.795\left(eV\text{Å}\right)^{1/2}$
- Screening constant $= a = 3\text{Å},5\text{Å},7\text{Å}$

20. This question is in context to the harmonic oscillator problem discussed in *Section 4.8.10.3*. Write a Scilab program to determine the energy eigen values of the three lowest energy states, if the boundary is taken between 0 to 16 instead of –8 to 8. Explain the result.

21. Write a Scilab program to determine the energy eigen value of the three lowest energy states of the s-wave radial Schrödinger equation for a harmonic oscillator.

    Take
    - Harmonic potential $= V = 50x^2$
    - Mass of neutron $= m = 940\dfrac{MeV}{c^2}$
    - $\hbar c = 197.3 MeV\,fm$
    - Take the radial distance between –L to L

    How will the energy eigen values change if the radial distance is taken between 0 to 2L? Give reasons.

22. Using a Scilab program, determine the ground state energy (in MeV) and plot the corresponding wave function for a neutron having mass 940 MeV/c² and moving under the influence of anharmonic potential given by

$$V(r) = \frac{1}{2}kr^2 + \frac{1}{3}br^3$$

The input conditions are as follows.
- $k = 100$ MeV fm^{-2}
- $b = 0,10,30$ MeV fm^{-3}
- $\hbar c = 197.3 MeV - fm$

23. Determine the lowest vibrational energy of a hydrogen molecule that is subject to the following Morse potential. Moreover, plot the corresponding wave function as a function of radial distance.

$$V(r) = D\left(e^{-2\alpha r'} - e^{-\alpha r'}\right)$$

Here,

- $r' = \dfrac{r - r_0}{r}$
- $m = 940 \times 10^6 \ eV / c^2$
- Dissociation energy $= D = 0.755501 \ eV$
- Factor controlling the width of the potential $= \alpha = 1.44$
- Equilibrium bond distance $= r_0 = 0.1313 \mathring{A}$

24. Write a Scilab program for solving Lagrange's equation of motion for under-damped simple pendulum. Plot the position–time $(\theta - t)$ and the phase-plane $(\dot\theta - \theta)$ graphs.

    Take the initial values (in appropriate units) as follows.
    - Length of the pendulum $= l = 1$
    - Acceleration due to gravity $= g = 9.82$
    - Initial angular displacement (in radian) $= \theta|_{t=0} = 0.5$
    - Initial angular velocity $= \dot\theta|_{t=0} = 1$
    - Damping coefficient $= 0.2$

25. Write a Scilab program for solving Lagrange's equations of motion for a spring pendulum. Plot the spring phase–plane and pendulum phase–plane graphs. Take the initial values (in appropriate units) as follows.
    - Equilibrium length of the pendulum $= l = 3$
    - Acceleration due to gravity $= g = 9.82$
    - Spring constant $= k = 5$
    - Mass of the bob $- m = 3$
    - Extension in the length of pendulum $= r|_{t=0} = 4$
    - Initial velocity $= \dot r|_{t=0} = 0$
    - Initial angular displacement $= \theta|_{t=0} = 0.2$
    - Initial angular velocity $= \dot\theta|_{t=0} = 0$

26. Write a Scilab program for solving Lagrange's equation of motion for a double pendulum. Plot the position graph of both the masses. Take the initial values (in appropriate units) as follows.
    - Acceleration due to gravity $= g = 9.82$
    - Length of the first pendulum $= l_1 = 1$
    - Length of the second (lower) pendulum $= l_2 = 2$
    - Mass of the first pendulum $= m_1 = 2$
    - Mass of the second pendulum $= m_2 = 1$
    - Initial angular displacement of first pendulum $= \theta_1|_{t=0} = \pi$
    - Initial angular velocity of first pendulum $= \dot\theta_1|_{t=0} = 0$

- Initial angular displacement of the second pendulum = $\theta_2 \mid_{t=0} = \dfrac{\pi}{2}$
- Initial angular velocity of second pendulum = $\dot{\theta}_2 \mid_{t=0} = 0$

27. Write a Scilab program for solving Lagrange's equation of motion for a simple pendulum that is attached to a rotating pivot. Take the initial values (in appropriate units) as follows.

- Acceleration due to gravity = $g = 9.82$
- Radius of the rotating pivot = 0.2
- Angular frequency of the rotating pivot = 20
- Length of the pendulum = 1
- Initial angular displacement of the bob = $\theta \mid_{t=0} = \pi/6$
- Initial angular velocity = $\dot{\theta} \mid_{t=0} = 0$

28. Write s Scilab program for solving Lagrange's equation of motion for a simple pendulum that is attached to a pivot that is moving in a horizontal plane. Take the initial values (in appropriate units) as follows.

- Acceleration due to gravity = $g = 9.82$
- Length of pendulum = $l = 1$
- Amplitude of oscillating pivot = $a = 0.2$
- Oscillating frequency of the pivot = $w = 100$
- Initial displacement of the bob= $\theta \mid_{t=0} = 0.1$
- Initial angular velocity = $\dot{\theta} \mid_{t=0} = 0$

29. Write a Scilab program for solving Lagrange's equation of motion for a simple pendulum that is attached to a pivot that is moving in a vertical plane. Take the same initial conditions as in the previous question.

30. A box of mass M is sliding down a frictionless inclined ramp. The inclined plane has a mass m and is moving towards the right. Write a Scilab program for solving Lagrange's equation of motion for this system and plot the position–time graph for this box. Take the initial conditions as

- Acceleration due to gravity = $g = 9.82$
- $M = 2$
- $m = 3$
- Initial position of the inclined ramp = $x_1 \mid_{t=0} = 0$
- Velocity of the moving ramp = $\dot{x}_1 \mid_{t=0} = 2$
- Initial position of the box = $x_2 \mid_{t=0} = 5$
- Velocity of the sliding box = $\dot{x}_2 \mid_{t=0} = 0.2$

Integration and Differentiation

After studying this chapter, the reader should be able to

◊ Perform numerical integration using Scilab

◊ Use the trapezoidal rule for solving definite integrals

◊ Use Simpson's 1/3 rule for solving definite integrals

◊ Use Simpson's 3/8 rule for solving definite integrals

◊ Apply these integration rules to solve common physics problems

◊ Perform differential calculus using Scilab

5.1 Introduction

The fundamental theorem of calculus closely relates integration to differentiation; it implies that integration can be considered as an inverse operation of differentiation, in the sense that, when a continuous function is integrated and then differentiated, it gives back the original function.

Integration refers to the area under the curve defined by a function f in a given interval $[a, b]$. If x is variable of integration, then the definite integral (y) is given by *Eqn. 5.1*.

$$y = \int_a^b f(x)\,dx \tag{5.1}$$

There are innumerable examples in physics that require integral calculus. For instance, simulations of classical mechanics problems may require evaluation of velocity from the acceleration of a body; and displacement of an object from its velocity profile. However, sometimes direct calculations become formidable or intractable, and direct integration rules have to be replaced by approximate numerical methods.

Differentiation refers to finding the rate of change of a dependent quantity (y) with respect to a change in independent quantity (x), i.e. $\dfrac{dy}{dx}$. There are countless applications of differential calculus in physics, such as determination of slope and tangent of geometric curves, especially when the rate of change is not constant.

This chapter starts with a discussion on various numerical techniques used for estimating the definite integral of a function. Improper integrals will be discussed in *Chapter 6* (on Special Functions). This is followed by a quick overview of the methods of differential calculus which are often used in Scilab.

The layout of this chapter is as follows. In *Section 5.2*, the built-in Scilab functions dedicated to computation of definite integrals are discussed. User-defined customized Scilab functions based on *trapezoidal* and *Simpson's* methods are discussed in *Sections 5.3* to *5.5*. The method of differentiation is discussed in *Section 5.6*. The knowledge acquired in all these sections is applied to various advanced physics problems in *Section 5.7*.

5.2 Built-in Scilab Functions for Integration

Scilab has several built-in functions to calculate definite integrals. In this book, two built-in functions have been used for integration, namely,

- `intg`
- `integrate`

5.2.1 intg

Consider the definite integral given in *Eqn. 5.2*.

$$\int_0^1 \frac{1}{1+x}\,dx \tag{5.2}$$

In order to use the built-in Scilab function, the first step is to define the function that has to be integrated. This is shown in the following.

```
function alpha = f(x)
alpha = 1 ./(1+x);
endfunction
```

The next step is to define the lower and upper limits of the definite integral.

```
a = 0;
b = 1;
```

The last step is to call the built-in function. `intg(a,b,f)` evaluates the definite integral of the continuous function *f*, from the lower limit *a* to the upper limit *b*.

```
intg(a,b,f)
```

5.2.2 integrate

In this case also, the integrand has to be defined first. This is followed by giving the lower and upper limits of the integral. The built-in function of Scilab is then called to determine the integral.

According to the syntax, `integrate(expr,v,a,b)` evaluates the definite integral of the continuous function `expr` for the variable `v` from the lower limit `a` to the upper limit `b`.

```
integrate('f(x)','x',a,b)
```

5.3 Trapezoidal Rule

The trapezoidal rule evaluates a definite integral by approximating the area under the curve by that of a trapezoid with the abscissae and ordinates of the integration limits as its vertices. For example, as shown in *Figure 5.1*, the integral of a function *y* within the limits *a* and *b* is approximated by calculating the area of the trapezoid PQRS. Integral of the function is given by *Eqn. 5.3*. Here, y_a (corresponding to PS) is the value of *y* at *x* = *a* and y_b (corresponding to QR) is the value of *y* at *x* = *b*

$$\int_a^b y \, dx \approx (b-a)\left\{\frac{y_b + y_a}{2}\right\}$$

(5.3)

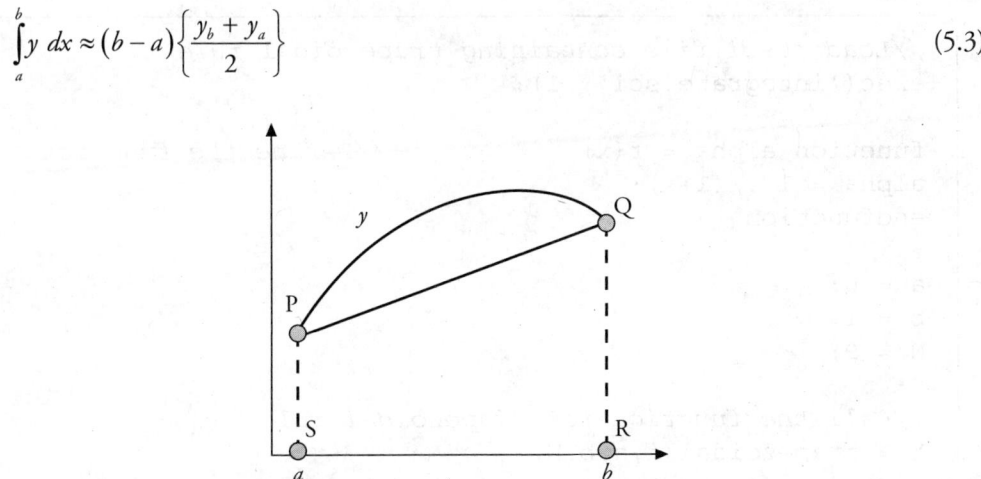

Figure 5.1   Trapezoidal rule

It is clear from *Figure 5.1* that the trapezoidal method will generally either overestimate or underestimate the area under the curve. Therefore, for improving accuracy, it is advisable to partition the domain of integration ($[a,b]$) into multiple segments ($[x_1 \equiv x_a < x_2 < x_3 < \ldots < x_{N-1} < x_b \equiv x_N]$), each of length h; and add the area of all the trapezoids therein.

If the domain of integration is divided into N nodes (equivalent to N-1 segments), then according to the trapezoidal rule of integration,

$$\int_a^b y \, dx = \frac{h}{2}\left[y_a + 2\left(y_2 + y_3 + \ldots + y_{N-1} \right) + y_b \right]$$

(5.4)

It can be noted that for consecutive segments, the function is evaluated twice at every interior point $(x_2, x_2, \ldots, x_{N-1})$. Therefore, $y_2, y_3, \ldots, y_{N-1}$ are counted twice in the aforementioned expression. The following Scilab function evaluates the sum given on the right side of *Eqn. 5.4*. This function can be saved to a file in '*.sci*' format and can be called whenever required.

```
function Y = trapezoidal(f,a,b,N)
h = (b - a)/(N-1);
x = linspace(a,b,N);
y = feval(x,f);
Y = (y(1) + 2.*sum(y(2:N-1)) + y(N)).*(h/2);
endfunction
```

Once the function for the trapezoidal rule is defined, it becomes straightforward to integrate any function. For example, consider the integral given in *Eqn. 5.2*. The following Scilab program shows the evaluation using the trapezoidal rule.

```
//Load *.sci file containing trapezoidal rule
exec('integrate.sci',-1);

function alpha = f(x)            //Define the function
alpha = 1 ./(1+x);
endfunction


a = 0;
b = 1;
N = 9;


//Call the function for trapezoidal rule
Y = trapezoidal(f,a,b,N)
```

The trapezoidal method is one of the simplest methods of numerical integration. For an interval of size h, it gives an error estimate of the order h^2. A higher accuracy can be achieved by applying the Romberg integration, which is based on the Richardson extrapolation to the trapezoid rule. This method uses a weighted estimation from two trapezoids (within two different sub-intervals of different widths) to obtain a more accurate third estimate. It therefore gives a better approximation of the integral by reducing the true error.

5.4 Simpson's 1/3 – Rule

Simpson's 1/3 rule is an extension of the trapezoidal rule. For the same number of intervals, it gives a more accurate result as compared to the trapezoidal rule. The reason lies in the fact that the trapezoidal rule approximates the integrand locally with a first order polynomial (straight line), whereas Simpson's 1/3 rule uses second order polynomial (parabola) as an approximation for the integrand.

Suppose, the interval a to b is divided into even number of sub-intervals, each of size h. According to Simpson's 1/3 rule,

$$\int_{a}^{b} y\, dx = \frac{h}{3}\left[y_a + 2\left(y_3 + y_5 + \ldots + y_{n-2}\right) + 4\left(y_2 + y_4 + \ldots + y_{n-1}\right) + y_b \right] \tag{5.5}$$

The following Scilab function evaluates the sum given on the right side of *Eqn. 5.5*.

```
function Y = simpson_1_3(f,a,b,N)
h = (b - a)/(N-1);
x = linspace(a,b,N);
y = feval(x,f);
Y = (y(1) + 2.*sum(y(3:2:N-2)) + 4.*sum(y(2:2:N-1)) +
y(N)).*(h/3.0);
endfunction
```

Consider the same integral taken in *Section 5.2.1* (*Eqn. 5.2*). The following program shows the evaluation of this integral by making use of Simpson's 1/3 - rule.

```
//Load *.sci file containing Simpson's 1/3 - rule
exec('integrate.sci',-1);

function alpha = f(x)                    //Define the function
alpha = 1 ./(1+x);
endfunction

a = 0;
b = 1;
```

```
N = 9;              //9 nodes correspond to 8 sub-intervals

//Call the function of Simpson's 1/3 - rule
Y = simpson_1_3(f,a,b,N)
```

5.5 Simpson's 3/8 – Rule

This rule approximates the integrand segment-wise by a cubic polynomial, as opposed to the linear and quadratic estimation in the previous two methods. According to Simpson's 3/8 rule,

$$\int_a^b y\,dx = \frac{3h}{8}\left[y_a + 3y_2 + 3y_3 + 2y_4 + 3y_5 + 3y_6 + 2y_7 + \ldots + 2y_{n-3} + 3y_{n-2} + 3y_{n-1} + y_b\right] \quad (5.6)$$

The following Scilab function evaluates the series given on the right side of *Eqn. 5.6*. It is important to remember that the step size should be such that the interval *a* to *b* is divided into sub-intervals in multiples of 3, and each is of size *h*.

```
function Y = simpson_3_8(f,a,b,N)
h = (b-a)/(N-1);
x = linspace(a,b,N);
y = feval(x,f)
Y = (y(1) + 3.*sum(y(2:3:N-2)) + 3.*sum(y(3:3:N-1)) +
2.*sum(y(4:3:N-3)) + y(N)).*(3.*h/8.0);
endfunction
```

Consider the same integral as taken in the previous sections (*Eqn. 5.2*). The following Scilab program shows the evaluation of this integral using Simpson's 3/8 - rule.

```
//Load *.sci file containing Simpson's 3/8 - rule
exec('integrate.sci',-1);

function alpha = f(x)              //Define the function
alpha = 1 ./(1+x);
endfunction

a = 0;
b = 1;
N = 34;          //34 nodes correspond to 33 sub-intervals

//Call the function for the Simpson's 3/8 - rule
Y = simpson_3_8(f,a,b,N)
```

Table 5.1 gives the result obtained from all the methods discussed earlier for the integral given in *Eqn. 5.2*. The effect of step size is mentioned for all the methods. The error with respect to the analytical solution of the integral (0.6931472) is also mentioned. It should be noted that the built-in integration functions give a more accurate estimate of the integral. The trapezoidal and Simpsons' rules follow the Newton–Cotes quadrature method, which divides the interval into evenly spaced interpolation points. The built-in functions (`intg` and `integrate`) are based on the Gaussian quadrature, which evaluates the integral with a higher degree of precision for a smaller number of interpolation points. It uses different step size depending on slow/ rapid variation of the function in the given interval.

Table 5.1 Results from different integration methods

Method	Step size	Value of integral	Error w.r.t. analytical solution
Built-in function – intg	--	0.6931472	0
Built-in function - integrate	--	0.6931472	0
Trapezoidal rule	0.125	0.6941219	0.0009747
	0.25	0.6970238	0.0038766
	0.5	0.7083333	0.0151861
Simpson's 1/3 rule	0.125	0.6931545	0.0000073
	0.25	0.6932540	0.0001068
	0.5	0.6944444	0.0012972
Simpson's 3/8 rule	0.03	0.6931472	0
	0.333	0.69375	0.0006028

5.6 Differentiation

Differentiation of a function *f* with respect to a variable *x* is defined as (*Eqn. 5.7*),

$$\frac{df}{dx} = \lim_{x_2 - x_1 \to 0} \frac{f(x_2) - f(x_1)}{x_2 - x_1} \tag{5.7}$$

Discrete derivatives can be easily calculated using the built-in Scilab function, `diff`. This function computes the difference of the value of the function at two consecutive points. If two consecutive points are x_1 and x_2, then the command `diff` will calculate,

$$f(x_2) - f(x_1)$$

Dividing this output by an appropriate step size results in a discrete derivative of the function. This is explained with the help of the following example. The function to be differentiated over the interval [0,2] is given by *Eqn. 5.8*.

$$f(x) = x^2 \tag{5.8}$$

The following Scilab program determines the first and second derivative of this function in the given range and plots the result, which is shown in *Figure 5.2*.

```
x = [0:0.01:2];                    //Range of 'x' variable
y = x.^2;

dy = diff(y)/0.01;                 //First derivative
d2y = diff(dy)/0.01;               //Second derivative
x1 = x(1:$-1);
x2 = x1(1:$-1);
plot2d(x,y)                        //Plot the original curve
plot2d(x1,dy)                      //Plot first derivative
plot2d(x2,d2y)                     //Plot second derivative
```

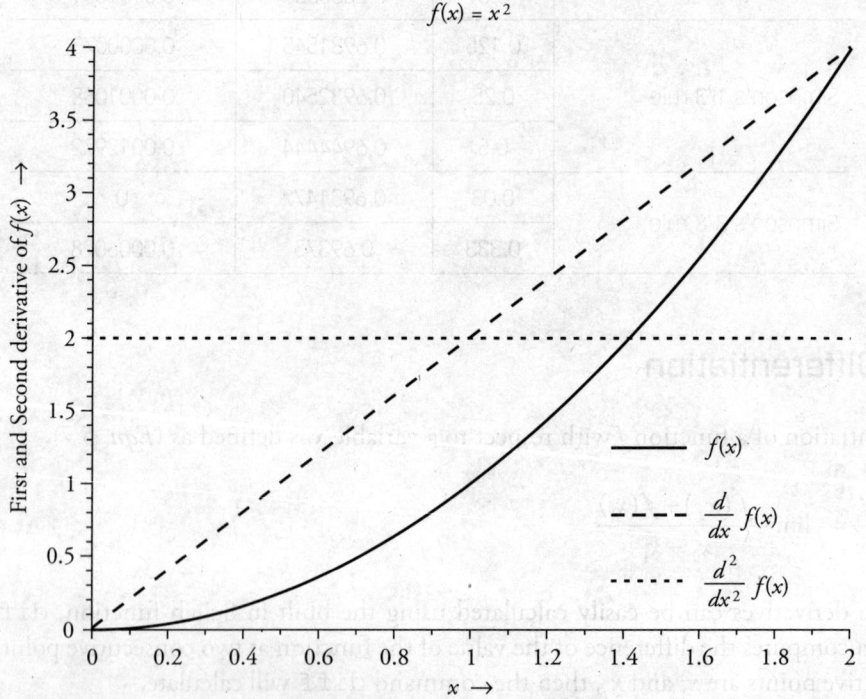

Figure 5.2   Example of discrete derivative

5.7 Applications

5.7.1 Integration in cylindrical coordinates

5.7.1.1 Line integral

The coordinates of a point in the cylindrical coordinate system are specified by (r, θ, z). The differential length element $(\vec{dl})$ in the cylindrical coordinate system is given by *Eqn. 5.9.*

$$\vec{dl} = dr\,\hat{a}_r + r\,d\theta\,\hat{a}_\theta + dz\,\hat{a}_z \qquad (5.9)$$

Therefore, the line integral along a curve will have radial (change in coordinates along r), angular (change in coordinates along θ), and azimuthal (change in coordinates along z) components.

The line integral in cylindrical coordinates can be defined through a Scilab function in the following manner.

```
function line_int =
line_integral(radius_1,radius_2,theta_1,theta_2,height_1
,height_2)
if (radius_1 == radius_2) & (height_1 == height_2) then
line_int =
radius_1.*integrate('1','phi',theta_1,theta_2)

elseif (radius_1 == radius_2) & (theta_1 == theta_2)
then
line_int = integrate('1','z',height_1,height_2)

elseif (theta_1 == theta_2) & (height_1 == height_2)
then
line_int = integrate('1','r',radius_1,radius_2);
end
endfunction
```

It is now simple to calculate the line integral for any configuration in cylindrical coordinates.

For example, consider the diagram in *Figure 5.3*, where the radius of the cylinder is 3 units and its height is 5 units. The Cartesian coordinates have also been marked for some corners of the cylinder.

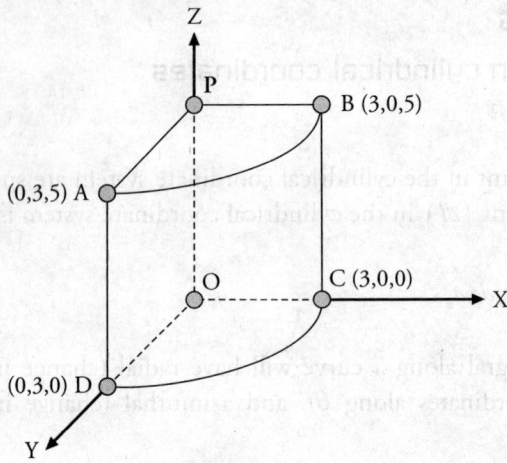

Figure 5.3   Diagram for explaining integration in cylindrical coordinates

The first step in calculating the line integral along different paths is to transform the coordinates of the marked points (x, y, z) to cylindrical system (r, θ, z). This can be done using the following Scilab command (for example for point A),

```
A = [0 3 5]                          //Cartesian coordinates
cartesian_to_cylindrical(A)
```

Repeating the aforementioned command for all the marked points will give,

$$A(0,3,5) \rightarrow A\left(3, \frac{\pi}{2}, 5\right)$$

$$B(3,0,5) \rightarrow B(3,0,5)$$

$$C(3,0,0) \rightarrow C(3,0,0)$$

$$D(0,3,0) \rightarrow D\left(3, \frac{\pi}{2}, 0\right)$$

The Scilab programs to calculate line integral along the different paths are given as follows.

• Along DA (This will be the same as that along CB)

Along this path, the radial and angular components do not change. Therefore the line integral is given by *Eqn. 5.10*.

$$\int_0^5 dz = z\big|_0^5 = 5 \tag{5.10}$$

```
//Load the *.sci file which contains the function for
line integral
exec(' integrate.sci ',-1);
line_integral(3,3, %pi/2, %pi/2,0,5)
```

The answer will be equal to 5.

- Along BA (This will be the same as that along CD)

    Along this path, the radial and azimuthal components do not change. Therefore the line integral is given by *Eqn. 5.11.*

$$\int_0^{\pi/2} r\, d\theta = 3\theta\big|_0^{\pi/2} = 1.5\pi \tag{5.11}$$

```
//Load the *.sci file which contains the function for
line integral
exec('integrate.sci',-1);
line_integral(3,3,0,%pi/2,5,5)
```

The answer will be equal to 4.712389

5.7.1.2 Surface integral

With reference to *Figure 5.3*, the differential normal surface area element $\left(\vec{dS}\right)$ in cylindrical coordinate system is given by *Eqn. 5.12.*

$$\vec{dS} = r\, d\theta\, dz\, \hat{a}_r + dr\, dz\, \hat{a}_\theta + r\, dr\, d\theta\, \hat{a}_z \tag{5.12}$$

The surface integral of any area element will have radial, angular, and azimuthal components. The surface integral in cylindrical coordinates can be defined through a Scilab function in the following manner.

```
function surface_int =
surface_integral(radius_1,radius_2,theta_1,theta_2,heigh
t_1,height_2)

if (radius_1 == radius_2) then
alpha_z = integrate('1','z',height_1,height_2)
surface int =
```

```
radius_1.*alpha_z.*integrate('1','phi',theta_1,theta_2)

elseif (theta_1 == theta_2) then
alpha_z = integrate('1','z',height_1,height_2)
surface_int = alpha_z.*integrate('1','r',radius_1,radi
us_2)
elseif (height_1 == height_2) then
alpha_phi = integrate('1','phi',theta_1,theta_2);
surface_int = alpha_phi.*integrate('r','r',radius_1,radi
us_2);
end
endfunction
```

It is now easy to calculate the surface integral for any configuration in cylindrical coordinates. For example, consider the diagram in *Figure 5.3*, where the radius of the cylinder is 3 units and its height is 5 units. The Scilab programs to calculate surface integral for the different surfaces are as follows.

- The surface ABCD

  For this surface, the normal surface area element is the radial component. Therefore, the surface integral is given by *Eqn. 5.13*.

$$\int_{\theta=0}^{\pi/2}\int_{z=0}^{5} r\,d\theta\,dz = 15\frac{\pi}{2} \tag{5.13}$$

```
//Load the *.sci file which contains the function for
surface integral
exec('integrate.sci',-1);
surface_integral(3,3,0,%pi/2,0,5)
```

The answer will be equal to 23.561945

- The surface CDO (This will be the same as that of ABP)

  For this surface, the normal surface area element is the azimuthal component. Therefore, the surface integral is given by *Eqn. 5.14*.

$$\int_{r=0}^{3}\int_{\theta=0}^{\pi/2} r\,dr\,d\theta = 9\frac{\pi}{4} \tag{5.14}$$

```
//Load the *.sci file which contains the function for
surface integral
exec('integrate.sci',-1);
surface_integral(0,3,0,%pi/2,0,0)
```

The answer will be equal to 7.0685835

- The surface APOD (This will be the same as that of BPOC)

    For this surface, the normal surface area element is the angular component. Therefore, the surface integral is given by *Eqn. 5.15*.

$$\int_{r=0}^{3}\int_{z=0}^{5} dr\, dz = 15 \tag{5.15}$$

```
//Load the *.sci file which contains the function for
surface integral
exec('integrate.sci',-1);
surface_integral(0,3,0,0,0,5)
```

The answer will be equal to 15

5.7.1.3 Volume Integral

The differential volume element in cylindrical coordinates is given by *Eqn. 5.16*.

$$dV = r\, dr\, d\theta\, dz \tag{5.16}$$

The volume integral can be defined through a Scilab function in the following manner.

```
function volume_int = volume_integral(radius_1,radius_2,
theta_1,theta_2,height_1,height_2)
radial = integrate('r','r',radius_1,radius_2)
angular = integrate('1','phi',theta_1,theta_2)
z_direction = integrate('1','z',height_1,height_2);
volume_int = radial.*angular.*z_direction
endfunction
```

The volume of the cylindrical portion ABCDOP is given by *Eqn. 5.17*.

$$V = \int_{r=0}^{3}\int_{\theta=0}^{\pi/2}\int_{z=0}^{5} r\, dr\, d\theta\, dz = \frac{45\pi}{4} \tag{5.17}$$

```
//Load the *.sci file which contains the function for
volume integral
exec('integrate.sci',-1);
volume_integral(0,3,0,%pi/2,0,5)
```

The answer will be equal to 35.342917.

5.7.2 Total charge

A course on 'Electricity and Magnetism' invariably uses integral calculus for calculating the charge enclosed within a given configuration. For example, the integral calculus is useful for calculating the electric field due to charge distributed over a region in space.

A simple application can be the calculation of total charge on a sheet bounded by $-1 < x < 1$ and $-1 < y < 1$. Assuming that the charge density on the $z = 0$ plane is equal to $3|y|^3$, the total charge on the sheet will be equal to (*Eqn. 5.18*),

$$\int_{-1}^{1} dx \int_{-1}^{1} 3\,|\,y\,|^3 \, dy \qquad\qquad (5.18)$$

The following Scilab program calculates this integral.

```
integrate('1','x',-1,1)*integrate('3*(abs(y)^3)',
'y',-1,1)
```

5.7.3 Electric flux density

Another interesting application of definite integrals is the calculation of flux due to electric field density $\left(\vec{D}\right)$. According to Gauss's law, the total electric flux through a closed surface is equal to the total charge enclosed by that surface.

Consider a charged sphere having radius a and charge density ρ. If charge is uniformly distributed within a three-dimensional volume, then the definite integral should be calculated over the volume element.

Total charge enclosed = $Q_{enc} = \int_V \rho\, dv = \rho \int_V dv$

Total electric flux through the closed surface = $\oint_S \vec{D} \circ \vec{dS} = D\,4\pi r^2$

Therefore, the radial distribution of $\vec{D}$ is given by *Eqn. 5.19*.

$$\vec{D} = \begin{cases} \dfrac{1}{4\pi r^2}\,\rho \int dV = \dfrac{1}{4\pi r^2}\,\rho \int_{r=0}^{r} r^2\,dr \int_{\theta=0}^{\pi} \sin\theta\, d\theta \int_{\varphi=0}^{2\pi} d\varphi & \text{For } r \le a \\[4ex] \dfrac{1}{4\pi r^2}\,\rho \int dV = \dfrac{1}{4\pi r^2}\,\rho \int_{r=0}^{a} r^2\,dr \int_{\theta=0}^{\pi} \sin\theta\, d\theta \int_{\varphi=0}^{2\pi} d\varphi & \text{For } r \ge a \end{cases} \qquad (5.19)$$

The following Scilab program calculates this integral and also plots the radial profile of electric flux density, which is shown in *Figure 5.4*.

```
a = 3;    //Radius of the sphere
rho = 4;  //Charge density
i = 1;
for j = 0.01:0.1:10;
    distance(i) = j;        //Distance from the center
    if j > a then
        value(i) = (1/(4*%pi*j.*j))*rho*integrate('r*r','r',
0,a)*integrate('sin(theta)','theta',0,%pi)*integrate('1','ph
i',0,2*%pi);
    elseif j <= a then
        value(i) = (1/(4*%pi*j.*j))*rho*integrate('r*r','r',
0,j)*integrate('sin(theta)','theta',0,%pi)*integrate('1','ph
i',0,2*%pi);
    end
i = i+1;
end

plot2d(distance,value)        //Plot radial profile of |D|
```

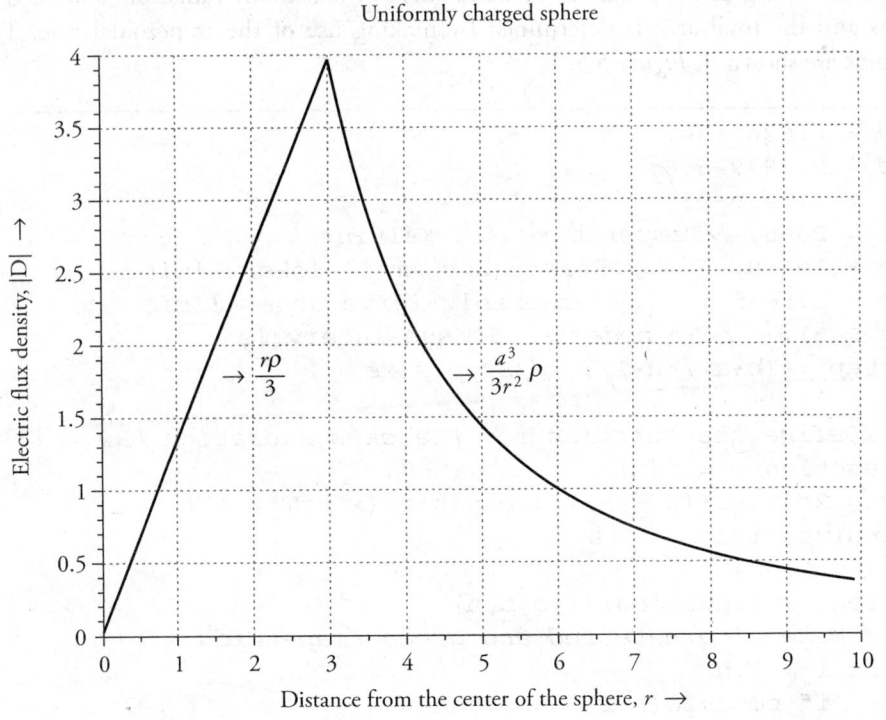

Figure 5.4   Radial profile of electric flux density

5.7.4 Planck's law for blackbody radiation

Planck's law for blackbody radiation states that the spectral radiance (amount of energy emitted at temperature T per unit surface area of the blackbody per unit time, per unit solid angle over which the radiation is measured, per unit wavelength) is given by *Eqn. 5.20*.

$$f(\lambda) = \frac{2hc^2}{\lambda^5} \frac{1}{\exp\left(\dfrac{hc}{\lambda kT}\right) - 1} \tag{5.20}$$

When the spectral radiance given by Planck's law is integrated over all possible values of wavelength (from zero to infinity), the resultant quantity is called the irradiance (power per unit area) (*Eqn. 5.21*).

$$E = \int_0^\infty E_\lambda \, d\lambda \tag{5.21}$$

The trapezoidal rule can be used to find the area under the intensity–wavelength curve of Planck's law for blackbody radiation.

In the following Scilab program, Planck's curve of blackbody radiation is divided into 50 sections and the total area is determined by making use of the trapezoidal rule. These 50 trapezoids are shown in *Figure 5.5*.

```
h = 6.626e-34;
c = 2.997925d8;
k = 1.381e-23;
T = 2000; //Temperature (in Kelvin)
a = 1e-10;      //Reasonably small lower limit
b = 1.5e-5;     //Reasonably large upper limit
n = 51;   //51 nodes -- 50 sub-intervals
step = (b-a)/(n-1);   //Step size

//Define the function for Planck's Radiation Law
function y = f(x)
y = 2*h*c*c*(x^(-5))/((exp(h*c/(x*k*T))-1))
endfunction

area = trapezoidal(f,a,b,n)
// Make alternate red and green trapezoids
for i = 1:n
    if pmodulo(i,2) == 0 then
        j = 3;
    elseif pmodulo(i,2) == 1 then
```

```
    j = 5;
    end
    x1 = a+(step*(i-1));
    x2 = a+(step*i);
    y1 = 2*h*c*c*(x1^(-5))/((exp(h*c/(x1*k*T))-1));
    y2 = 2*h*c*c*(x2^(-5))/((exp(h*c/(x2*k*T))-1));
    xpts = [x1*1e6, x2*1e6, x2*1e6, x1*1e6];
    ypts = [y1*1e-10, y2*1e-10, 0, 0];
    scf(0);
    plot2d(x1,y1*1e-10);
    xfpoly(xpts,ypts,j);
end
```

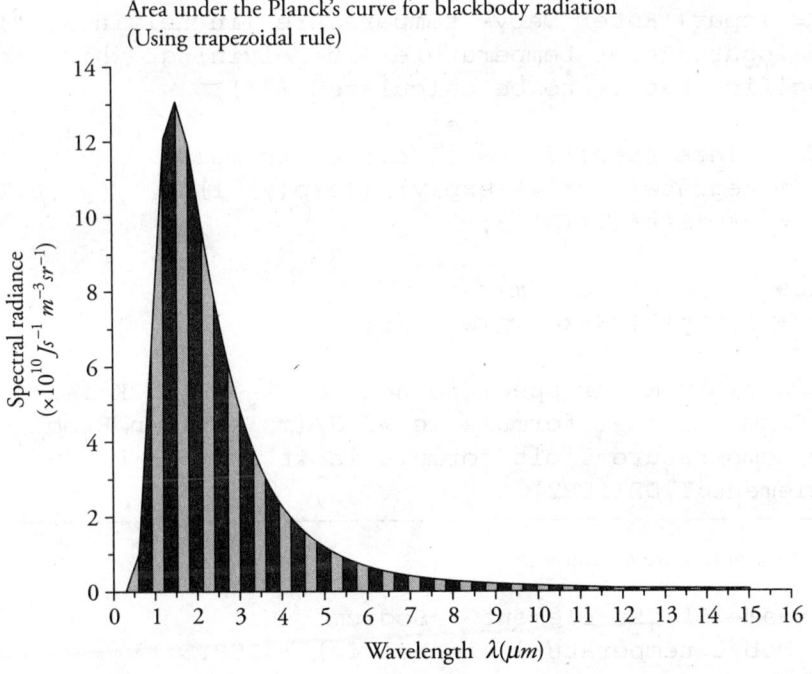

Figure 5.5 Planck's law for blackbody radiation

5.7.5 Specific heat of solids

According to Debye's model of specific heat for metals, if T_D is Debye's temperature, then the molar specific heat at different temperatures (T) can be determined from *Eqn. 5.22*.

$$C_v = 9Nk\left(\frac{T}{T_D}\right)^3 \int_0^{T_D/T} \frac{x^4 e^x}{\left(e^x - 1\right)^2}\,dx \qquad (5.22)$$

In the low temperature limit $(T \ll T_D)$, *Eqn. 5.22* approximates to *Eqn. 5.23*.

$$C_v = \frac{12\pi^4 Nk}{5}\left(\frac{T}{T_D}\right)^3 \tag{5.23}$$

The following Scilab program compares the molar specific heat of different metals using *Eqns. 5.22–5.23*.

```
k = 1.381e-23;
N = 6.022e23;
h = 6.626e-34;

element = input("Enter name of the element : ","string");
TD = input("Enter Debye temperature (in Kelvin) : ");
T = input("Enter temperature (in Kelvin) at which molar
specific heat is to be calculated : ");

//Calculate specific heat (direct formula)
m = integrate('(y**4)*exp(y)/((exp(y)-1)^2)','y',0,TD/T);
DB1 = 9*m*N*k*(T/TD)^3;

//Low temperature limit
DB2 = 12*%pi^4*N*k*(T/TD)^3/5;

mprintf("\n Molar specific heat of %s at %i K is :
\n From original formula is %f J/(mole-K) \n From
low temperature limit formula is %f J/(mole-K) \n
",element,T,DB1,DB2)
```

The input parameters are as follows.

```
Enter name of the element : Sodium
Enter Debye temperature (in Kelvin) : 158.5
Enter temperature (in Kelvin) at which molar specific heat
is to be calculated : 5
```

The answer is.

```
Molar specific heat of Sodium at 5 K is :
From original formula is 0.061033 J/(mole-K)
From low temperature limit formula is 0.061033 J/(mole-K)
```

The results for some other elements are given in *Table 5.2*. The readers are advised to check all these results.

Table 5.2 Molar specific heat of some elements

Element	Debye's Temperature	Molar Specific Heat ($J\ mole^{-1}K^{-1}$)				
			5 K	10 K	20 K	30 K
Copper	340 K	Eqn. 5.22	0.0062	0.0494	0.3956	1.32
		Eqn. 5.23	0.0062	0.0494	0.3957	1.33
Calcium	229 K	Eqn. 5.22	0.020	0.162	1.28	3.87
		Eqn. 5.23	0.020	0.162	1.29	4.37
Gold	162 K	Eqn. 5.22	0.057	0.457	3.34	8.07
		Eqn. 5.23	0.057	0.457	3.64	12.35
Sodium	158.5 K	Eqn. 5.22	0.061	0.488	3.53	8.39
		Eqn. 5.23	0.061	0.488	3.90	13.18
Hydrogen	122 K	Eqn. 5.22	0.133	1.064	6.41	12.3
		Eqn. 5.23	0.133	1.070	8.56	28.9

5.7.6 Dirac delta function (Shifting property)

According to the shifting property of Dirac delta function (*Eqn. 5.24*),

$$\int_{-\infty}^{\infty} \delta(x-a) f(x)\, dx = f(a) \tag{5.24}$$

In the following Scilab program,

- The functional representation for the Dirac delta function is given by *Eqn. 5.25*.

$$\delta(x) = \frac{1}{\sqrt{2\pi\sigma^2}} \exp\left(-\frac{(x-4)^2}{2\sigma^2}\right) \tag{5.25}$$

- The function $f(x)$ is given by *Eqn. 5.26*.

$$f(x) = x + 3 \tag{5.26}$$

- Therefore, the expected result over a reasonably large range of integration is given by *Eqn. 5.27*.

$$\int_{a}^{b} \delta(x-4) f(x)\, dx = f(4) = 7 \tag{5.27}$$

The following program calculates this integral using Simpson's methods and the built-in Scilab function.

```
//Load all the functions of integration rules
exec('integrate.sci',-1);

//Define the function which has to be integrated
function alpha = dirac(x)
alpha = (exp((-(x-4).^(2))/(2.*sigma.*sigma)).*(x+3))/
sqrt(2*%pi.*sigma.*sigma)
endfunction

sigma = 0.2;     //Standard Deviation
a = 0;      //Lower limit
b = 8;      //Upper limit
n = 41;                   //n nodes - n-1 sub-intervals
simpson_1_3(dirac,a,b,n)
n = 61;
simpson_3_8(dirac,a,b,n)

intg(a,b,dirac)
```

5.7.7 Cornu's spiral and Fresnel's diffraction pattern

Cornu's spiral is a graphical method that computes and predicts the Fresnel diffraction pattern due to various obstacles. A quick recapitulation with the help of *Figure 5.6* is given next.

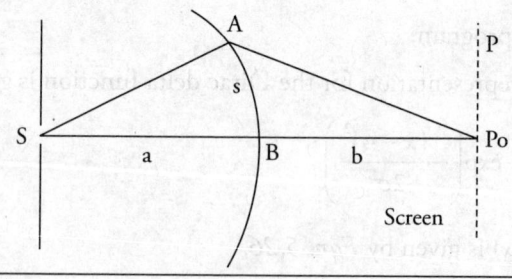

Figure 5.6   Geometry of cylindrical wave front in Fresnel's diffraction

In *Figure 5.6*, S is the source of monochromatic light of wavelength λ. Fresnel's diffraction pattern is observed on the screen placed at a perpendicular distance of $a + b$ from the source. A dimensionless variable is generally introduced to reduce the number of variables and make calculation easier. It is given in *Eqn. 5.28*.

$$v = \sqrt{\frac{2(a+b)}{ab\lambda}}s \qquad (5.28)$$

The intensity (I) of radiation received on the screen, at any point P_o from the portion AB of the wave front is proportional to the sum of square of Fresnel's integrals.

$$I \propto C^2 + S^2 \qquad (5.29)$$

In *Eqn. 5.29*,

- C and S are Fresnel's integrals given in *Eqns. 5.30–5.31*.

$$C = \int_0^v \cos\frac{1}{2}\pi u^2 \, du \qquad (5.30)$$

$$S = \int_0^v \sin\frac{1}{2}\pi u^2 \, du \qquad (5.31)$$

Cornu's spiral is a graph between C (plotted on the x-axis) and S (plotted on the y-axis). The following Scilab program draws Cornu's spiral using the built-in function for integration. The graph is shown in *Figure 5.7*.

```
function alpha = f1(x)              //Fresnel's Integral
alpha = cos(%pi*x*x/2);
endfunction
function alpha = f2(x)              //Fresnel's Integral
alpha = sin(%pi*x*x/2);
endfunction


i = 1;
for v = -5:0.01:5;                            //Range of v
    x(i) = integrate('f1(x)','x',0,v);
    y(i) = integrate('f2(x)','x',0,v);
    i = i+1;
end

plot2d(x,y)                        //Plot Cornu's spiral

x1 = 0.5;
y1 = 0.5;
plot(x1,y1)                        //Mark the eye of spiral
plot(-x1,-y1)                      //Mark the eye of spiral
```

In *Figure 5.7,*

- The dimensionless variable v is measured along the curve.

- As $v \to \pm\infty$, the value of Fresnel's integrals approaches ± 0.5. This is shown with a plus marker in the figure.

- The intensity at point P_o is due to the contribution from the upper and lower wave fronts such that points A and B approach infinity. Therefore, amplitude of radiation at P_o due to the upper wave front will be proportional to

$$\sqrt{0.5^2 + 0.5^2} = \frac{1}{\sqrt{2}}$$

- The amplitude of radiation due to the lower wave front will be proportional to

$$\sqrt{(-0.5)^2 + (-0.5)^2} = \frac{1}{\sqrt{2}}$$

- Total intensity at P_o is equal to 2.

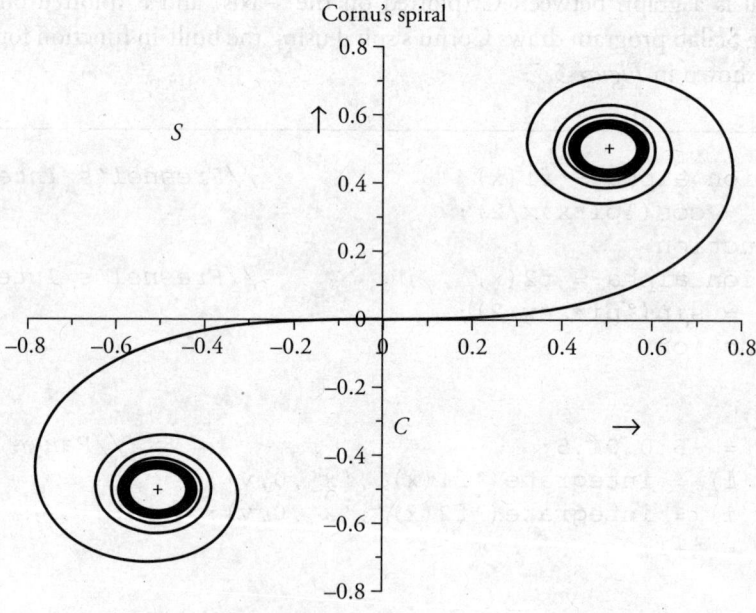

Figure 5.7   Cornu's spiral

Suppose an obstacle in the form of a straight edge is placed in front of the wave front as shown in *Figure 5.8.* The following Scilab program determines Fresnel's diffraction pattern on the screen for various positions of the observation point P. *Figure 5.9* shows the diffraction pattern.

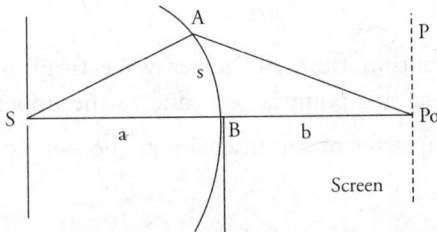

Figure 5.8   Fresnel's diffraction due to a straight edge

```
function alpha = f1(x)                    //Fresnel's Integral
alpha = cos(%pi*x*x/2);
endfunction

function alpha = f2(x)                    //Fresnel's Integral
alpha = sin(%pi*x*x/2);
endfunction


i = 1;
for v = -0.99:0.01:7;
    v1(i) = v;
    x(i) = integrate('f1(x)','x',0,v);
    y(i) = integrate('f2(x)','x',0,v);
    intensity(i) = ((0.5+x(i))^2 + (0.5+y(i))^2);
    i = i+1;
end
plot2d(v1,intensity);        //Plot the diffraction pattern
```

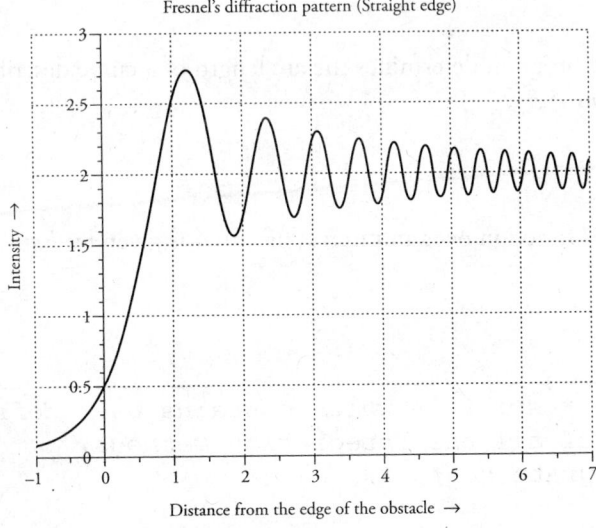

Figure 5.9   Fresnel's diffraction pattern due to straight edge

From *Figures 5.8* and *5.9*,

- At point P_0, contribution from the lower wave front is blocked by the obstacle. Therefore, intensity at this point is only due to the upper wave front. This is equal to $\dfrac{1}{2}$, which is one-quarter of the intensity at the same point in the absence of any obstacle.

- As the observer moves upwards from P_0, the lower wave front also starts contributing to the radiation received at point P, in addition to the whole of the upper front.

- Depending on the location of point P, the intensity will be proportional to

$$(0.5+C)^2 +(0.5+S)^2$$

- The region below the point P_o is called the geometrical shadow region. There is little penetration of light in this region too.

5.7.8 Arc length

The numerical methods of integration discussed in *Sections 5.3–5.5* can be used to determine the arc length (*Eqn. 5.32*) of a function over a given interval.

$$\int_a^b ds \tag{5.32}$$

In the rectangular form, *ds* is given by *Eqn. 5.33*.

$$ds = \sqrt{1+\left(\frac{dy}{dx}\right)^2}\, dx \tag{5.33}$$

The following Scilab program determines the arc length of a curve described by a logarithmic function given in *Eqn. 5.34*.

$$y = \log x \tag{5.34}$$

The arc length of this function over an interval of 1 to 4 is given by *Eqn. 5.35*.

$$\int_1^4 \sqrt{1+\frac{1}{x^2}}\, dx \tag{5.35}$$

```
//Load the *.sci file which contains user-defined
functions of all the integration methods.
exec('integrate.sci',-1);

function alpha = f(x)                    //Define the function
```

```
alpha = 1+1/(x.*x);
endfunction


a = 1;
b = 4;
n = 31;

//Call the functions
intg(a,b,f)
integrate('f(x)','x',a,b)
trapezoidal(f,a,b,n)
simpson_1_3(f,a,b,n)
simpson_3_8(f,a,b,n)
```

The value of the arc-length determined from these methods is given in *Table 5.3*.

Table 5.3 Measurement of arc-length

Method	Arc-length
Intg	3.75
Integrate	3.75
Trapezoidal rule	3.7516373
Simpson's 1/3 rule	3.7500129
Simpson's 3/8 rule	3.7500281

5.7.9 Motion of an object

This section uses differential calculus to determine displacement, velocity, and acceleration of a particle that is executing simple harmonic motion.

- Displacement of a particle executing simple harmonic motion is given by *Eqn. 5.36*.

$$\psi(t) = A\cos(\omega t + \delta) = A\cos(2\pi vt + \delta) \tag{5.36}$$

- Velocity of this particle is given by *Eqn. 5.37*.

$$\dot{\psi}(t) = \frac{d\psi}{dt} = -\omega A\sin(2\pi vt + \delta) \tag{5.37}$$

- Velocity can also be written in the form of *Eqn. 5.38*.

$$\dot{\psi}(t) = \frac{d\psi}{dt} = \lim_{t_2 - t_1 \to 0} \frac{\psi(t_2) - \psi(t_1)}{t_2 - t_1} \tag{5.38}$$

- Acceleration of this particle is given by *Eqn. 5.39*.

$$\ddot{\psi}(t) = \frac{d\dot{\psi}}{dt} = -\omega^2 A \cos(2\pi v t + \delta) \tag{5.39}$$

- Acceleration can also be written in the form of *Eqn. 5.40*.

$$\ddot{\psi}(t) = \frac{d\dot{\psi}}{dt} = \lim_{t_2 - t_1 \to 0} \frac{\dot{\psi}(t_2) - \dot{\psi}(t_1)}{t_2 - t_1} \tag{5.40}$$

The following Scilab program determines the velocity and acceleration of the particle by using the diff function. The graph is shown in *Figure 5.10*.

```
nu = 1/(2*%pi);                        //Frequency
t = [0:0.01/nu:1/nu];                  //Time range
y = cos(2*%pi*nu*t);                   //Displacement
dy = diff(y)/(0.01/nu);                //Velocity
d2y = diff(dy)/(0.01/nu);              //Acceleration
tr1 = t(1:$-1);
tr2 = tr1(1:$-1);

//Use subplot to make three graphs in one graphic window
subplot(311)

plot2d(t,y)                            //Plot displacement

subplot(312)
plot2d(tr1,dy)                         //Plot velocity

subplot(313)
plot2d(tr2,d2y)                        //Plot acceleration
```

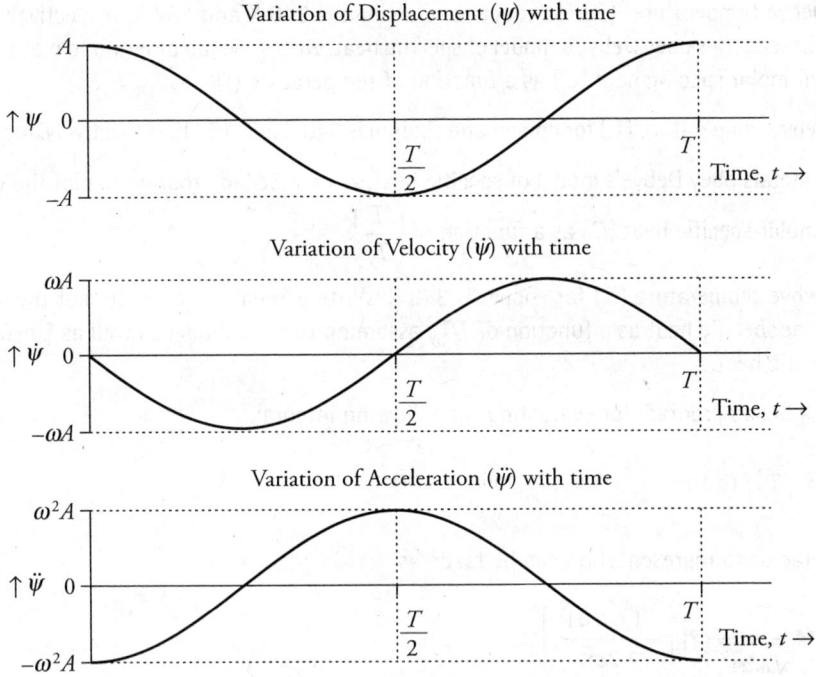

Figure 5.10 Motion of a particle executing simple harmonic motion

It is clear from the Scilab program and from *Figure 5.10*, that

- As the magnitude of velocity and acceleration depend on frequency, for clarity in the graph and for proper scaling, frequency (v) is taken as $\dfrac{1}{2\pi}$ so that angular frequency is unity. This makes the magnitude scale comparable for displacement, velocity and acceleration.

- Phase constant (δ) is taken as zero.

- Only one cycle of simple harmonic motion is shown.

5.8 Exercises

1. Use the trapezoidal method and Simpson's methods for determining the following integral and compare the results for different values of step size.

$$\int_0^{\pi/2} \sin x\,dx$$

2. With the help of a suitable function, write a Scilab program to show the significance of step size in the trapezoidal rule.

3. The Debye temperature (T_D) for copper and sodium is 340 K and 157 K, respectively. Assuming that these metals obey Debye's model of specific heat, write a Scilab program to plot the variation of their molar specific heat (C_v) as a function of temperature (T).

4. The Debye temperature (T_D) for copper and sodium is 340 K and 157 K, respectively. Assuming that these metals obey Debye's model of specific heat, write a Scilab program to plot the variation of their molar specific heat (C_v) as a function of $\left(\dfrac{T}{T_D}\right)$.

5. The Debye temperature (T_D) for copper is 340 K. Write a Scilab program to plot the variation of its molar specific heat as a function of T/T_D, assuming Debye's model as well as Einstein's model of specific heat.

6. Write a Scilab program for evaluating the following integral.

$$\int_{-\infty}^{\infty} \delta(x-a)f(x)dx$$

The Dirac delta representation can be taken as

$$f(x) = \frac{1}{\sqrt{2\pi\sigma^2}}\exp\left(-\frac{(x-3)^2}{2\sigma^2}\right)$$

Take the function to be equal to

$$f(x) = x^2$$

7. Write a Scilab program to determine Fresnel's diffraction pattern due to a slit between two straight edges kept adjacent to the wave front. Take different values for the width of the slit $(\Delta v = 1, 2, 3, 4)$.

8. Write a Scilab program to find Fresnel's diffraction pattern due to a wire kept symmetrically in front of the wave front. Take different values for the width of the wire $(\Delta v = 1, 2, 3, 4)$.

9. Determine the arc length of the curves described by the following functions in the intervals mentioned.
    a. $y = x^2$ in the interval $[0,4]$
    b. $y = e^x$ in the interval $[0,3]$
    c. $y = \sin x$ in the interval $[0,\pi]$

10. Integrate the following functions by making use of the trapezoidal method and the two Simpson's methods.
    a. $\dfrac{1}{x}$ in $[1,3]$
    b. e^{-x^2} in $[0,4]$
    c. $\sqrt{25\sin^2 x + 2\cos^2 x}$ in $[0,2\pi]$

11. Using Scilab program, show that differentiation of a triangular wave gives a square wave.

12. The displacement of an object as a function of time is given by,

$$x(t) = 1 + t^2 + 5t^3$$

Write a Scilab program to draw the displacement, velocity, and acceleration profile of the object during the time interval [0,1].

13. *Figure 5.11* shows the displacement profile of an object. Write a Scilab program to draw the velocity profile of this object.

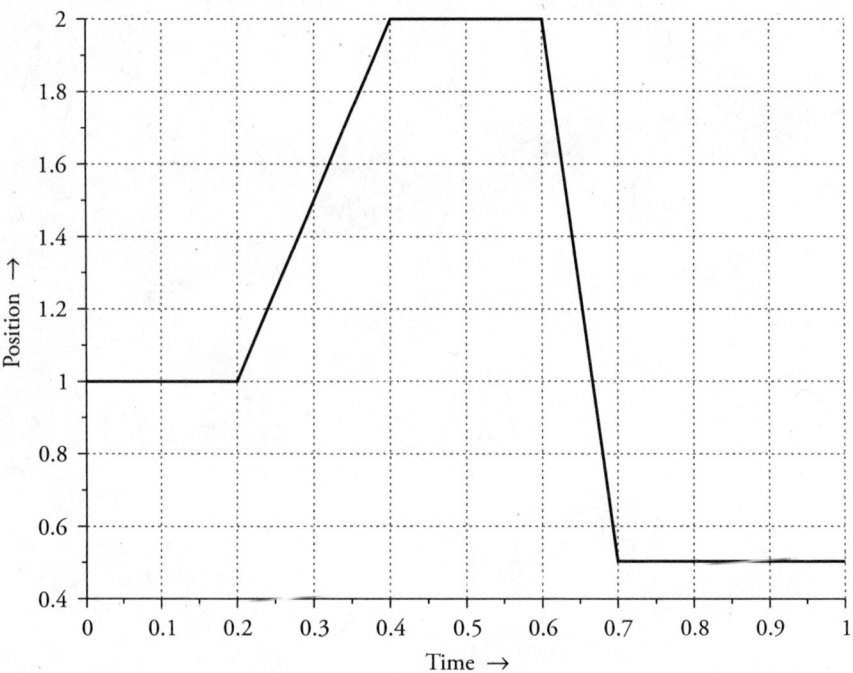

Figure 5.11   Graph for *Exercise 13*

Special Functions

After studying this chapter, the reader should be able to

◊ Write Scilab programs for generating special functions that are invariably used in Physics for solving analytical problems that are otherwise difficult to solve.

◊ Solve Bessel's differential equation and determine its solution using the finite difference method.

◊ Use the built-in Scilab functions to generate Bessel's and Legendre functions.

◊ Write user-defined functions for generating the Legendre, Laguerre, and Hermite polynomials of any order using the recursion relations.

◊ Write user-defined functions for generating the Legendre, Laguerre, and Hermite polynomials of any order by determining their power series solution.

◊ Solve improper integrals using quadrature rules

6.1 Introduction

The main objective of this chapter is familiarization with a variety of numerical methods that are essential for solving advanced problems of applied physics and engineering. With the help of suitable examples, basic skills on appropriately using these methods for various applications in physics are provided.

The chapter focuses on the following special second order differential equations, which are known to have standard functional form and/or analytical solutions.

- Bessel's equation (*Section 6.2*)
- Legendre's equation (*Section 6.3*)
- Laguerre's equation (*Section 6.4*)
- Hermite's equation (*Section 6.5*)

The solutions of these equations are referred to as 'special functions', which are significantly different from standard functions like sine/cosine, exponential and logarithmic functions.

This chapter also describes the use of quadrature methods of integration for calculating improper integrals, which are either infinite in the interval of integration, or the interval of integration has an infinite bound. The quadrature methods discussed in this chapter are as follows:

- Gauss–Legendre (*Section 6.6.1*)
- Gauss–Laguerre (*Section 6.6.2*)
- Gauss–Hermite (*Section 6.6.3*)

The chapter has been written in a manner so as to develop the necessary skills of the reader to evaluate certain integrals that are generally not discussed in introductory physics classes because they involve advanced calculations.

6.2 Bessel Function of the First Kind

Bessel functions have several applications in physics. They arise while solving Laplace's and Helmholtz equations in spherical and cylindrical coordinates. The functions are also useful while solving problems based on electromagnetic wave propagation and Schrödinger's equation.

The general features of the Bessel function are as follows.

1. Bessel functions (order n) of the first kind ($J_n(x)$) are the solutions ($y(x)$) of the differential equation given in *Eqn. 6.1*.

$$x^2 \frac{d^2y}{dx^2} + x\frac{dy}{dx} + \left(x^2 - n^2\right)y = 0 \tag{6.1}$$

A second order differential equation can be written in the form

$$\frac{d^2y}{dx^2} + f(x)\frac{dy}{dx} + g(x)y = r(x) \tag{6.2}$$

For solving *Eqn. 6.1* with the finite difference method, it is necessary to first define the functions $f(x)$, $g(x)$, and $r(x)$ in the following manner.

```
function func_r = r(x)
func_r = 0;
endfunction

function func_f = f(x)
func_f = 1/x;
endfunction

function func_g = g(x)
func_g = 1;
endfunction
```

The function for the finite difference method has already been explained in detail in *Chapter 4*. This function can be written in an executable file, 'differentiation.sci' (for example) and can be loaded using the following Scilab command.

```
exec('differentiation.sci',-1);
```

Figure 6.1 shows the zero order Bessel function of the first kind. It has been generated using the following Scilab program.

```
a = 1;                              //Initial x
b = 10;                             //Final x
ya = 0.765;                         //Initial y
yb = -0.246;                        //Final y
h = 0.1;                            //Step size

//Call the function for finite difference method and plot
the result
[x,y] = finite_diff(a,b,h,ya,yb,f,g,r);
plot2d(x,y)

//Plot the Bessel function by using the built-in SciLab
function
x = 1:0.1:10;
y = besselj(0,x);
```

Figure 6.2 shows the Bessel function of the first kind (first order). It has been generated by changing the value of '*n*' in the function for *g(x)* and the initial values.

```
function func_g = g(x)
func_g = ((x.*x)-1)/(x.*x);
endfunction
```

```
a = 1;
b = 10;
ya = 0.44;
yb = 0.0435;
h = 0.1;
```

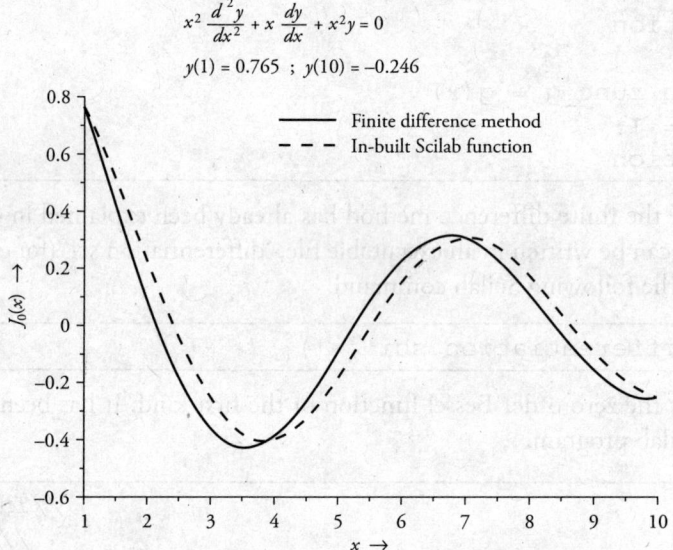

$$x^2 \frac{d^2 y}{dx^2} + x \frac{dy}{dx} + x^2 y = 0$$

$$y(1) = 0.765 \ ; \ y(10) = -0.246$$

Figure 6.1   Graph for Bessel function $J_0(x)$

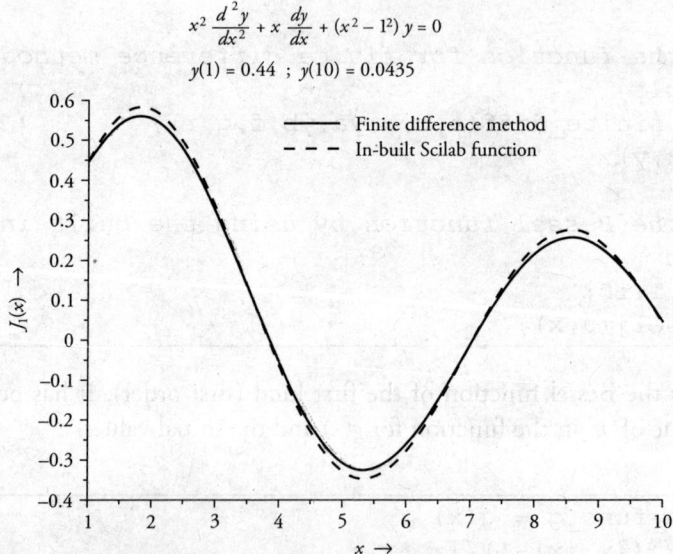

$$x^2 \frac{d^2 y}{dx^2} + x \frac{dy}{dx} + (x^2 - 1^2) y = 0$$

$$y(1) = 0.44 \ ; \ y(10) = 0.0435$$

Figure 6.2   Graph for Bessel function $J_1(x)$

Figure 6.3 shows the Bessel function $J_2(x)$. Note the changes made in the function $g(x)$ and the initial and final values of the dependent variable y.

```
function func_g = g(x)
func_g = ((x.*x)-4)/(x.*x);
endfunction
a = 1;
b = 10;
ya = 0.1149;
yb = 0.2546;
h = 0.1;
```

$$x^2 \frac{d^2 y}{dx^2} + x \frac{dy}{dx} + (x^2 - 2^2)\, y = 0$$

$$y(1) = 0.1149 \; ; \; y(10) = 0.2546$$

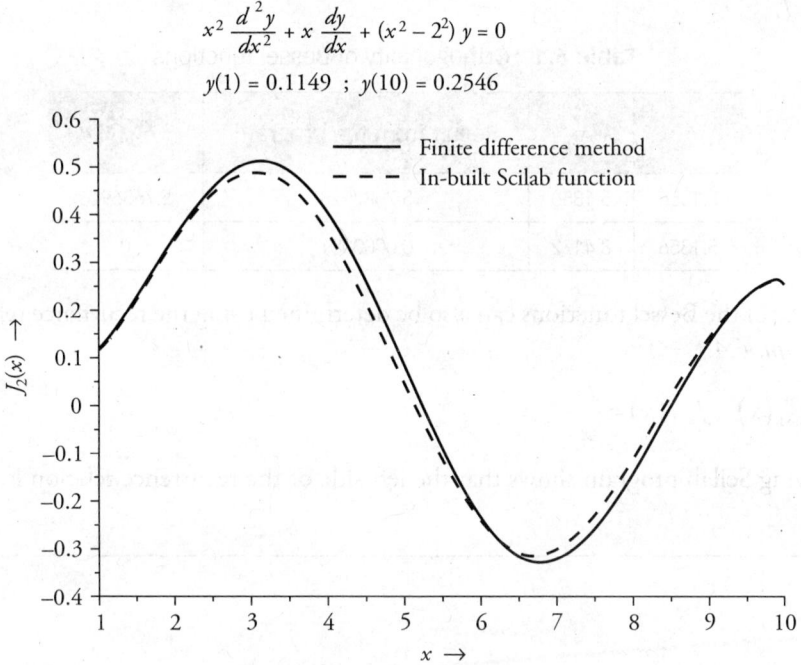

Figure 6.3 Graph for Bessel function $J_2(x)$

2. The orthogonality of Bessel functions is given by *Eqn. 6.3*.

$$\int_0^a x J_n\left(\alpha_{ni}\frac{x}{a}\right) J_n\left(\alpha_{nj}\frac{x}{a}\right) dx = \begin{cases} 0 & \text{if } i \neq j \\ \dfrac{1}{2} J^2_{n-1}(a) & \text{if } i = j \end{cases} \tag{6.3}$$

In *Eqn. 6.3*, a_{ni} and a_{nj} are the ith and j^{th} roots of $J_n(x)$ respectively. Orthogonality of Bessel functions can be proved using the following Scilab program.

```
n = 2;
x = 5;
a = 10;
root_1 = 5.1356;
root_2 = 8.4172;

function s = f(x)
s = x*besselj(n,x*root_1/a)*besselj(n,x*root_2/a);
endfunction

I = intg(0,a,f)
```

A comparison of the result from the aforementioned program and the expected result is shown in *Table 6.1*.

Table 6.1 Orthogonality of Bessel functions

α_{2i}	α_{2j}	Result from the program	$\frac{1}{2}J^2_{n-1}(a)$
5.1356	5.1356	5.7687928	5.7686926
5.1356	8.4172	0.0000081	0

3. Value of the Bessel functions can also be determined using the recurrence relation given in *Eqn. 6.4*.

$$J_{n+1}(x) + J_{n-1}(x) = \frac{2n}{x}J_n(x) \qquad (6.4)$$

The following Scilab program shows that the left side of the recurrence relation is equal to its right side.

```
x = 1:10;
n = 2;

function alpha = left_side(x)
alpha = besselj(n+1,x)+besselj(n-1,x)
endfunction

function alpha = right_side(x)
alpha = 2*n*besselj(n,x)/x
endfunction

value_of_left_side = feval(x,left_side);
value_of_right_side = feval(x,right_side);

disp([x',value_of_left_side',value_of_right_side'])
```

Result of the aforementioned program is given in *Table 6.2*.

Table 6.2 Recurring relation of Bessel functions

x	L.H.S.	R.H.S
1	0.4596139	0.4596139
2	0.7056681	0.7056681
3	0.6481217	0.6481217
4	0.3641281	0.3641281
5	0.0372521	0.0372521
6	-0.1619155	-0.1619155
7	-0.1722384	-0.1722384
8	-0.0564959	-0.0564959
9	0.0643766	0.0643766
10	0.1018521	0.1018521

6.3 Legendre Polynomial

The general features of the Legendre polynomials are as follows.

1. Legendre functions are solutions of Legendre's second order differential equation (*Eqn. 6.5*).

$$\frac{d}{dx}\left(\left(1-x^2\right)\frac{d}{dx}P_n(x)\right)+n(n+1)P_n(x)=0 \tag{6.5}$$

The solution of this ordinary differential equation for different values of n (= 0,1,2,...) along with the normalization ($P_n(1) = 1$) forms polynomial sequences, which are called the Legendre polynomials.

The following Scilab program generates the first four Legendre polynomials in the range [−1 < x < 1] by making use of the built-in function of Scilab. The corresponding graph is shown in *Figure 6.4*.

```
i = -1;
j = 0;

for n = 1:4
    x = -1.0:0.1:1.0;
    i = i+1;
    j = j+1;
    y = legendre(i,0,x);
    plot2d(x,y)
end
```

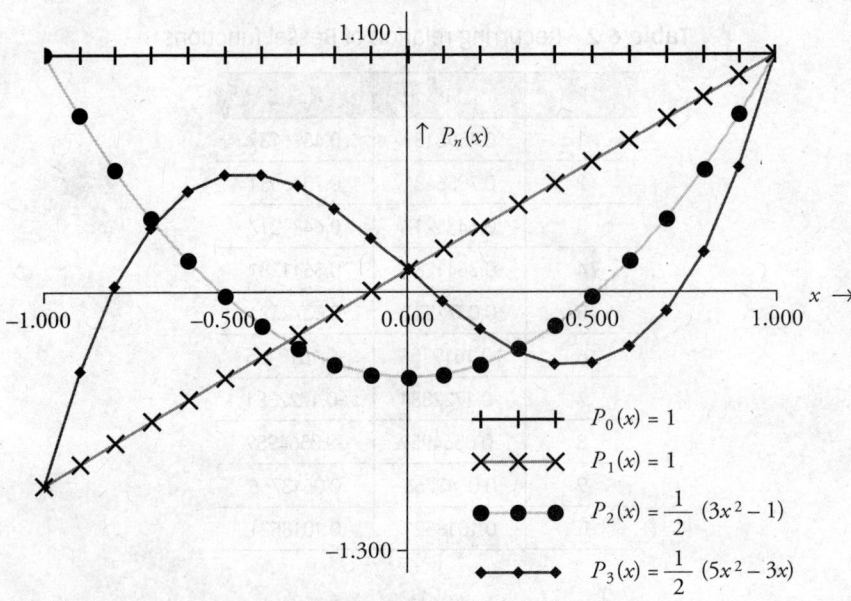

The first 4 Legendre polynomial

$P_0(x) = 1$

$P_1(x) = 1$

$P_2(x) = \dfrac{1}{2}(3x^2 - 1)$

$P_3(x) = \dfrac{1}{2}(5x^2 - 3x)$

Figure 6.4   First four Legendre polynomials

2. They are orthogonal to each other in the interval $-1 \le x \le 1$. *Eqn. 6.6* shows this orthogonality relation.

$$\int_{-1}^{1} P_m(x)P_n(x)\,dx = 0 \qquad if\ m \ne n \tag{6.6}$$

Eqn. 6.7 gives the normalization condition of Legendre polynomials.

$$\int_{-1}^{1} P_m(x)P_n(x)\,dx = \frac{2}{2n+1} \qquad if\ m = n \tag{6.7}$$

The following Scilab program shows that Legendre polynomials are orthogonal in the interval $-1 \le x \le 1$.

```
n = input("Enter value of n:");
m = input("Enter value of m:");
ans = integrate('legendre(m,0,x)*legendre(n,0,x)',
'x',-1,1,0.001);
mprintf('∫P%i(x)*P%i(x)dx = %f\n',m,n,ans);
mprintf('The value of 2/(2n+1) is %f\n',2/(2*n + 1));
```

The output of the aforementioned program is as follows.

```
Enter value of n: 1
Enter value of m: 1
```

∫ P1(x)*P1(x)dx = 0.66

The value of 2/(2n+1) is 0.66

```
Enter value of n: 1
Enter value of m: 2
```

∫ P1(x)*P1(x)dx = 0.00

The value of 2/(2n+1) is 0.66

3.   The first two Legendre polynomials are

$P_0(x) = 1$

$P_1(x) = x$

The other Legendre polynomials of order 'n' can be determined using Bonnet's recursion formula (*Eqn. 6.8*).

$$P_n(x) = \sum_{m=0}^{M} (-1)^m \frac{(2n-2m)!}{2^n \, m!(n-m)!(n-2m)!}$$ (6.8)

In *Eqn. 6.8*,

- If the order of the polynomial is even, then $M = n/2$.

- If the polynomial is of an odd order, then $M = (n-1)/2$

The following Scilab program uses this summation series to write a function for generating the Legendre polynomials of higher orders.

```
function Legendre = legendre_poly_gamma(n,var)
if n == 0 then
    cc = [1];
elseif n == 1 then
    cc = [0 1];
else
    if modulo(n,2) == 0 then
        M = n/2
```

```
    else
        M = (n-1)/2
    end;
    cc = zeros(1,M+1);
    for m = 0:M
        k = n-2*m;
        cc(k+1)=(-1)^m*gamma(2*n-2*m+1)/
 (2^n*gamma(m+1)*gamma(n-m+1)*gamma(n-2*m+1));
    end;
end;
Legendre = poly(cc,var,'coeff');
endfunction
```

As an example, the first five Legendre polynomials can be determined in the following manner.

```
p = zeros(5,1);
for order = 1:5;
    p(order) = legendre_poly_gamma(order,'x');
end;
```

Output of the aforementioned program is as follows:

```
p   =

    x

                2
  - 0.5 + 1.5x

                3
  - 1.5x + 2.5x

            2        4
  0.375 - 3.75x + 4.375x

            3        5
  1.875x - 8.75x + 7.875x
```

4. The Legendre polynomials are also related by Bonnet's recursion formula given in *Eqn. 6.9* (if $P_0(x)$ and $P_1(x)$ are known).

$$P_{n+1}(x) = \frac{(2n+1)xP_n(x) - nP_{n-1}(x)}{(n+1)} \tag{6.9}$$

The following Scilab program uses this recursion formula to write a function for generating the Legendre polynomials. The function can be called to determine and plot the Legendre polynomials of any order.

```
function y = Legendre_polynomial(n)
if (n == 0) then
    y = poly(1,"x","coeff");
elseif (n == 1) then
    y = poly([0 1],"x","coeff")
else
    polynomial_x = poly([0 1],"x","coeff")
 y_n_minus_2 = poly(1,"x","coeff")
 y_n_minus_1 = poly([0 1],"x","coeff")
 for i = 2:n;
     y = ((2*i - 1)*polynomial_x*y_n_minus_1 -
(i-1)*y_n_minus_2)/i;
     y_n_minus_2 = y_n_minus_1
     y_n_minus_1 = y
end
end
endfunction
```

As an example, the following Scilab program evaluates the first three Legendre polynomials in the interval $[-1.5 < x < 1.5]$. The graphs are shown in *Figure 6.5*.

```
x = -1.5:0.1:1.5;
y = horner(Legendre_polynomial(0),x);
plot2d(x,y)

y = horner(Legendre_polynomial(1),x);
plot2d(x,y)

y = horner(Legendre_polynomial(2),x);
plot2d(x,y)
```

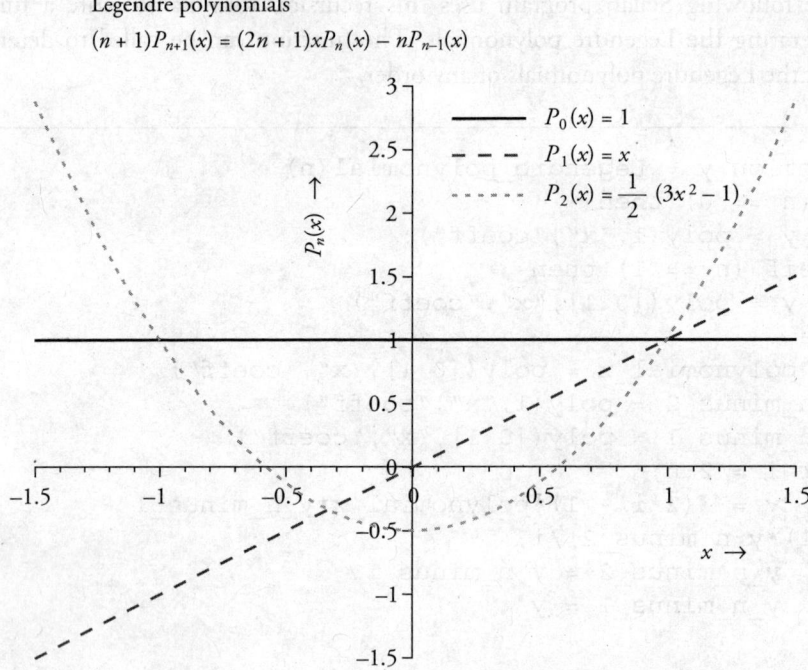

Legendre polynomials

$$(n + 1)P_{n+1}(x) = (2n + 1)xP_n(x) - nP_{n-1}(x)$$

Figure 6.5 First three Legendre polynomials

6.4 Laguerre Polynomial

The general features of the Laguerre polynomials are as follows.

1. They are the solutions of Laguerre's second order linear differential equation (*Eqn. 6.10*).

$$x\frac{d^2 y}{dx^2} + (1-x)\frac{dy}{dx} + ny = 0 \tag{6.10}$$

These polynomials ($L_0(x)$, $L_1(x)$...) are quite useful while solving radial part of the Schrödinger equation for a one electron atom and in problems involving three-dimensional harmonic oscillators.

2. The Laguerre polynomials are orthogonal to each other.

3. The Laguerre polynomials can be generated using the recursion formula given in *Eqn. 6.11* (if $L_0(x)$ and $L_1(x)$ are known).

$$L_{k+1}(x) = \frac{(2k+1-x)L_k(x) - kL_{k-1}(x)}{k+1} \tag{6.11}$$

The following Scilab program uses this recursion formula to write a function for generating the Laguerre polynomials. The function can be called to determine and plot Laguerre polynomials of any order.

```
function y = Laguerre_polynomial(n)
if (n == 0) then
    y = poly(1,"x","coeff");
elseif (n == 1) then
    y = poly([1 -1],"x","coeff")
else
    polynomial_x = poly([0 1],"x","coeff")
    y_n_minus_2 = poly(1,"x","coeff")
    y_n_minus_1 = poly([1 -1],"x","coeff")
    for i = 2:n;
        y = ((2*i - 1 - polynomial_x)*y_n_minus_1 -
(i-1)*y_n_minus_2)/i
        y_n_minus_2 = y_n_minus_1
        y_n_minus_1 = y
    end
end
endfunction
```

As an example, the following Scilab program evaluates the first three Laguerre polynomials in the interval $[-2 < x < 5]$. The graphs of these polynomials are shown in *Figure 6.6*.

```
x = -2:0.1:5;
y = horner(Laguerre_polynomial(0),x);
plot2d(x,y)

y = horner(Laguerre_polynomial(1),x);
plot2d(x,y)

y = horner(Laguerre_polynomial(2),x);
plot2d(x,y)
```

Laguerre polynomials

$$(n+1)L_{n+1}(x) = (2n+1-x)L_n(x) - nL_{n-1}(x)$$

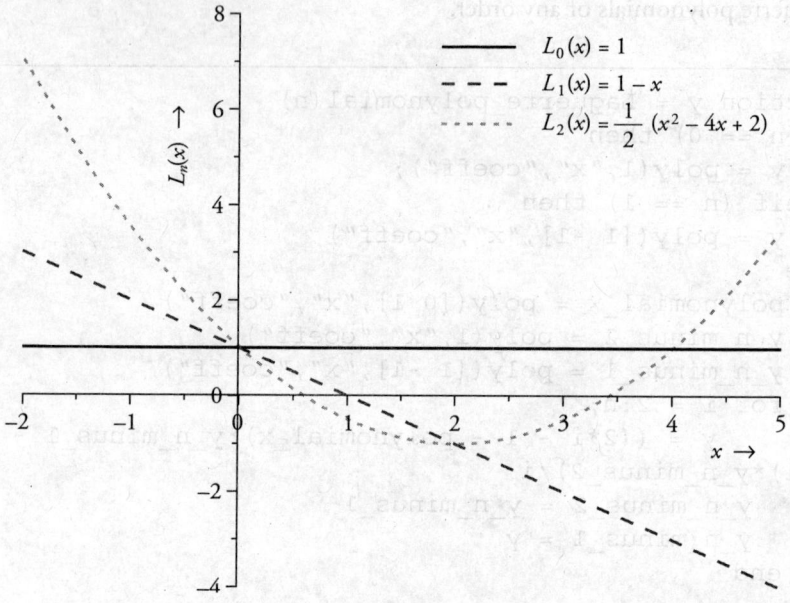

Figure 6.6 First three Laguerre polynomials

4. The Laguerre polynomials can be generated with the help of *Eqn. 6.12*.

$$L_n(x) = \sum_{m=0}^{n} \frac{(-1)^m}{m!} \frac{n!}{m!(n-m)!} x^m \qquad (6.12)$$

The following Scilab program makes use of this series to generate the Laguerre polynomials of higher orders, assuming that $L_0(x) = 0$. This function can be used to determine the Laguerre polynomials of any order.

```
function Laguerre = Laguerre_poly_gamma(n,var)
if n == 0 then
    solution = [1];
else
    solution = [];
    for m = 0:n;
     solution = [solution (-
1)^m*gamma(n+1)/((gamma(m+1))^2*gamma(n-m+1))];
    end;
end;
Laguerre = poly(solution,var,"coeff");
endfunction
```

As an example, the first few Laguerre polynomials are as follows.

```
Laguerre_poly_gamma(0,'x')
    1

Laguerre_poly_gamma(1,'x')
    1 - x

Laguerre_poly_gamma(2,'x')
                 2
    1 - 2x + 0.5x

Laguerre_poly_gamma(3,'x')
             2       3
    1 - 3x + 1.5x - 0.17x

Laguerre_poly_gamma(4,'x')
             2       3         4
    1 - 4x + 3x - 0.6667x + 0.0417x

Laguerre_poly_gamma(5,'x')
             2      3         4          5
    1 - 5x + 5x - 1.6667x + 0.2083x - 0.0083x
```

6.5 Hermite Polynomial

The general properties of the Hermite polynomials are as follows.

1. They are the solutions of the second order differential equation written in *Eqn. 6.13*.

$$\frac{d^2 y}{dx^2} - x\frac{dy}{dx} + ny = 0 \tag{6.13}$$

2. Their distribution function is given by *Eqn. 6.14*.

$$f(x) = \frac{1}{\sqrt{2\pi}} \exp\left(-\frac{x^2}{2}\right) \tag{6.14}$$

3. They can be defined by the two methods. These two definitions are scaled version of each other. The first one is referenced as 'probabilists' Hermite polynomials'. The first two Hermite (probabilists') polynomials are given by *Eqn. 6.15*.

$$H_0(x) = 1$$

$$H_1(x) = x \tag{6.15}$$

The other Hermite (probabilists') polynomials can be determined using the recursion formula given in *Eqn. 6.16* for $n \geq 1$.

$$H_{n+1}(x) = xH_n(x) - nH_{n-1}(x) \tag{6.16}$$

The following Scilab program uses this recursion relation for generating the Hermite (probabilists') polynomials of higher orders. This function can be called to determine and plot these polynomials of any order over any interval.

```
function y = Hermite_polynomial_prob(n)
if (n == 0) then
    y = poly(1,"x","coeff");
elseif (n == 1) then
    y = poly([0 1],"x","coeff")
else
    polynomial_x = poly([0 1],"x","coeff")
    y_n_minus_2 = poly(1,"x","coeff")
    y_n_minus_1 = poly([0 1],"x","coeff")
    for i = 2:n;
        y = polynomial_x*y_n_minus_1 - (i-1)*y_n_minus_2;
        y_n_minus_2 = y_n_minus_1
        y_n_minus_1 = y
    end
end
endfunction
```

As an example, the following Scilab program evaluates the first three Hermite polynomials ($H_0(x)$, $H_1(x)$, and $H_2(x)$) in the interval $[-4 < x < 4]$. The graphs of these polynomials are shown in *Figure 6.7*.

```
x = -4:0.1:4;
y = horner(Hermite_polynomial_prob(0),x);
plot2d(x,y)


y = horner(Hermite_polynomial_prob(1),x);
plot2d(x,y)


y = horner(Hermite_polynomial_prob(2),x);
plot2d(x,y)
```

Hermite polynomials

$$H_{n+1}(x) = xH_n(x) - nH_{n-1}(x)$$

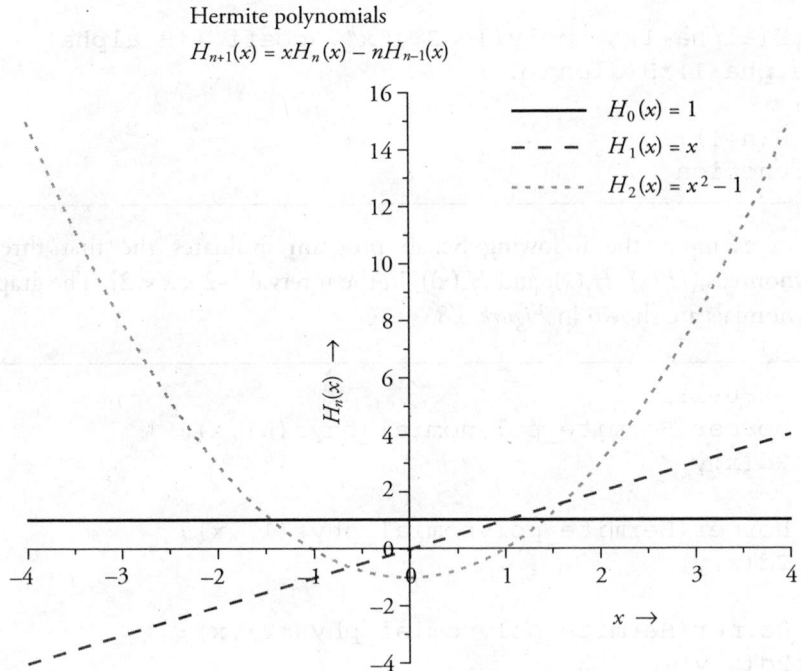

Figure 6.7   First few Hermite (probabilists') polynomials

4. The second type of polynomials is the 'physicists' Hermite polynomials'. The first two Hermite (physicists') polynomials are given by *Eqn. 6.17*.

$$H_0(x) = 1$$

$$H_1(x) = 2x \tag{6.17}$$

The other Hermite (physicists') polynomials can be determined using the recursion formula given in *Eqn. 6.18* for $n \geq 1$.

$$H_{n+1}(x) = 2xH_n(x) - 2nH_{n-1}(x) \tag{6.18}$$

The following Scilab program uses this recursion relation for generating the Hermite (physicists') polynomials of higher orders. This function can be called to determine and plot these polynomials of any order over any interval.

```
function y = Hermite_polynomial_phys(n)
H = zeros(1,n+1);
H(1) = poly([1],"x",'coeff');
H(2) = poly([0 2],"x",'coeff');
for alpha = 2:n
```

```
    H(alpha+1) = poly([0 2],"x",'coeff')*H(alpha) -
2*(alpha-1)*H(alpha-1);
end;
y = H(n+1);
endfunction
```

As an example, the following Scilab program evaluates the first three Hermite polynomials $(H_0(x), H_1(x),$ and $H_2(x))$ in the interval $[-2 < x < 2]$. The graphs of these polynomials are shown in *Figure 6.8*.

```
x = -2:0.1:2;
y = horner(Hermite_polynomial_phys(0),x);
plot2d(x,y)

y = horner(Hermite_polynomial_phys(1),x);
plot2d(x,y)

y = horner(Hermite_polynomial_phys(2),x);
plot2d(x,y)
```

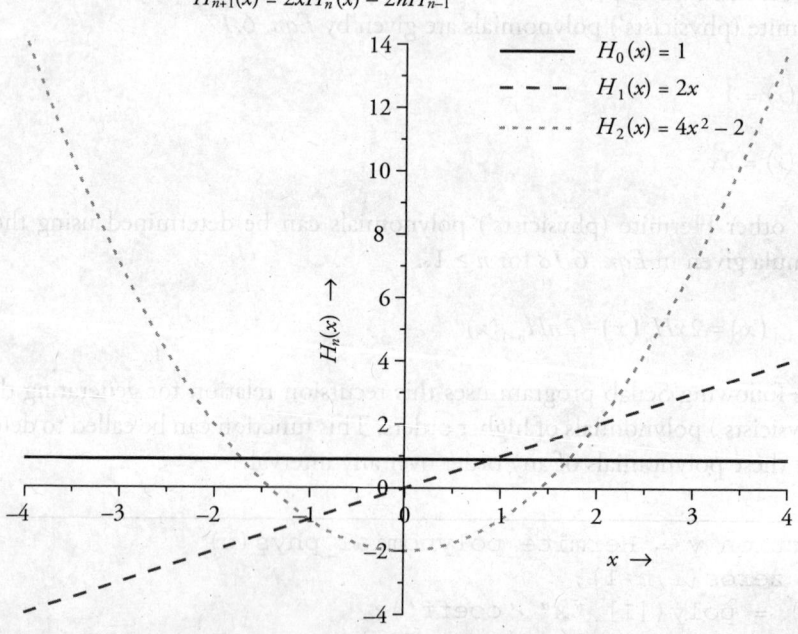

Figure 6.8   First few Hermite (physicists') polynomials

5. The Hermite polynomials are orthogonal such that for, $n, m \geq 0$,

$$\left(H_n, H_m\right) = \sqrt{2\pi}\, n!\, \delta_{nm} \tag{6.19}$$

6.6 Improper Integrals – Quadrature Methods

Suppose $f(x)$ is a real valued function of x, which is defined over the interval $[a < x < b]$. The integral of this function is called improper if

- The function $f(x)$ is undefined in the interval $[a < x < b]$. Or,

- Either or both the bounds, a and b are infinite.

It has been discussed in *Chapter 5* that the methods of integration, such as the trapezoidal method and the Simpson's methods, determine the value of the integrand at the endpoints of the interval of integration. Therefore, if the function is not defined in this interval and/or the bounds are infinite, then these methods cannot be used.

The quadrature methods are an important numerical technique for determining an approximation of the improper integrals. They are better suited for evaluating improper integrals because they use only interior points of the domain. According to the quadrature rule (*Eqn. 6.20*),

$$\int_{-1}^{1} f(x)\, dx = \sum_{i=1}^{n} w_i\, f(x_i) \tag{6.20}$$

Eqn. 6.20 is the Gaussian quadrature rule of order n. It yields exact results for polynomials of degree $(2n - 1)$ or less by a suitable choice of x_i and w_i. The common weight functions are as follows.

- Gauss–Legendre: $w(x) = 1$
- Gauss–Laguerre: $w(x) = e^{-x}$
- Gauss–Hermite: $w(x) = e^{-x^2}$

The following sub-sections will describe the use of the Gauss quadrature methods to solve some definite integrals with the help of Scilab programs.

6.6.1 Gauss–Legendre quadrature

The key features of this method are as follows.

1. The Gauss–Legendre quadrature is a numerical integration method over the interval $[-1, 1]$. Therefore, any other arbitrary limits of integration are mapped onto $[-1, 1]$.

2. The abscissas (x_i) of the quadrature for order n are roots of the Legendre polynomial $P_n(x)$.

3. The weight coefficients (w_i) are given by *Eqn. 6.21*.

$$w_i = \frac{2}{\left(1 - x_i^2\right)\left(P_n'(x)\right)^2} \tag{6.21}$$

For any other arbitrary limits of integration [a, b], the integral is calculated in the following manner (*Eqn. 6.22*).

$$\int_a^b f(x)\,dx \approx \frac{1}{2}(b-a)\sum_{i=1}^n w_i f\left(\frac{b-a}{2}x_i + \frac{b+a}{2}\right) \tag{6.22}$$

The user-defined Scilab function to calculate the Gauss–Legendre quadrature is as follows. It makes use of the user-defined recursion function of Legendre, which was explained in *Section 6.3*.

```
function y = gauss_legendre(f,a,b,n)
p  = legendre_poly_gamma(n,'x');
xroots = roots(p);
w  = [];
for j = 1:n
    poly_deriv = derivat(p);
    w = [w 2/((1- xroots(j)^2)*(horner(poly_deriv,
xroots(j)))^2)];
end;
arg = ((b-a)/2.* xroots)+((b+a)/2);
y = (b-a)/2*w*f(arg);
endfunction
```

It is now extremely trivial to evaluate definite integrals using this function. The following Scilab program shows one such example for evaluating the integral given in *Eqn. 6.23*.

$$\int_0^1 \frac{1}{\left(1 + x^3\right)}\,dx \tag{6.23}$$

```
function alpha = f(x)
alpha = (1+x.^3).^(-1);
endfunction

gauss_legendre(f,0,1,20)
```

The answer is equal to 0.8356488. This matches very well with the analytical solution, which is equal to 0.8357.

6.6.2 Gauss–Laguerre quadrature

The key features of this method are described as follows.

1. Gauss–Laguerre quadrature is an integration method for calculating improper integrals over the interval $[0, \infty]$.

2. The abscissas (x_i) of the quadrature for order n are roots of the Laguerre polynomial $L_n(x)$.

3. The weight coefficients (w_i) are given by *Eqn. 6.24*.

$$w_i = \frac{x_i}{(n+1)^2 \left(L_{n+1}(x_i)\right)^2} \tag{6.24}$$

4. The integral is calculated in the following manner (*Eqn. 6.25*).

$$\int_0^\infty f(x)\,dx = \sum_{i=1}^n w_i\, e^{x_i}\, f(x_i) \tag{6.25}$$

The user-defined Scilab function to calculate the Gauss–Laguerre quadrature is as follows. It makes use of the user-defined recursion function of Laguerre, which was explained in *Section 6.4*.

```
function y = gauss_laguerre(f,n)
p = Laguerre_poly_gamma(n,'x');
p_n_plus_1 = Laguerre_poly_gamma(n+1,'x');
xroots = roots(p);
w = [];
for i = 1:n
    w = [w xroots(i)/((n+1)^2*(horner(p_n_plus_1,
xroots(i)))^2)];
end;
y = w*(exp(xroots).*f(xroots));
endfunction
```

It is now extremely trivial to evaluate improper integrals using this function. The following Scilab program shows one such example for evaluating the integral given in *Eqn. 6.26*.

$$\int_0^\infty \frac{1}{\left(1+x^2\right)}\,dx \tag{6.26}$$

```
function alpha = f(x)
alpha = (1+x.^2).^(-1);
endfunction
gauss_laguerre(f,30)                    //Order is 30
```

The answer will be equal to 1.567

6.6.3 Gauss–Hermite quadrature

The key features of this method are described as follows.

1. The Gauss–Hermite quadrature is an integration method for calculating improper integrals over the interval $[-\infty,\infty]$.

2. The abscissas (x_i) for the quadrature of order n are roots of the Hermite (physicists') polynomial $H_n(x)$.

3. The weight coefficients (w_i) are given by *Eqn. 6.27*.

$$w_i = \frac{2^{n-1} n! \sqrt{\pi}}{n^2 \left(H_{n-1}(x_i) \right)^2} \tag{6.27}$$

4. The integral is calculated in the following manner (*Eqn. 6.28*).

$$\int_{-\infty}^{\infty} f(x)\,dx = \sum_{i=1}^{n} w_i\, e^{x_i^2} f(x_i) \tag{6.28}$$

The user-defined Scilab function to calculate the Gauss–Hermite quadrature is given as follows. It makes use of the user-defined recursion function of Hermite polynomials, which was explained in *Section 6.5*.

```
function y = gauss_hermite(f,n)
p = Hermite_polynomial_phys(n);
p_n_minus_1 = Hermite_polynomial_phys(n-1);
xroots = roots(p);
w = [];
for j = 1:n
    w = [w 2^(n-1)*gamma(n+1)*sqrt(%pi)/(n^2*horner(p_n_
minus_1, xroots(j))^2)];
end;
y = w*(exp(xroots.^2).*f(xroots));
endfunction
```

It is now extremely trivial to evaluate improper integrals using this function. The following Scilab program shows one such example for evaluating the integral given in *Eqn. 6.29*.

$$\int_{-\infty}^{\infty} e^{-|x|}\,dx \tag{6.29}$$

```
function alpha = f(x)
alpha = exp(-abs(x));
endfunction
gauss_hermite(f,20)          //Order is 20
```

The answer turns out to be equal to 1.9747

6.7 Applications

6.7.1 Simple pendulum

Pendulums are one of the most well studied systems in the mechanics practical course of the physics curriculum. However, in the physics laboratories, the amplitude is often restricted to small angles and only linear harmonic solutions are analyzed. This restriction of small angle approximation takes away the beauty and understanding of the real world behavior of pendulums.

Gauss–Legendre quadrature method is a handy tool to investigate the variation of the time period of a simple pendulum with its angular amplitude (both small and large).

The time period (T) of a simple pendulum is given by *Eqn. 6.30*.

$$T = 4\sqrt{\frac{L}{2g}} \int_0^{\theta_{max}} \frac{d\theta}{\sqrt{\cos\theta - \cos\theta_{max}}} \tag{6.30}$$

In *Eqn. 6.30*,

- L is length of the pendulum
- g is acceleration due to gravity
- θ is the angular amplitude
- θ_{max} is the maximum angular amplitude

It should be noted that,

- This is an improper integral because the integrand is undefined at the upper limit of integration (at $\theta = \theta_{max}$).
- At small values of θ_{max} (i.e., $\theta_{max} \to 0$), the time period approaches the small angle approximation.

The following Scilab program calculates the time period of a simple pendulum for a wide range ($10° - 90°$) of maximum angular amplitude of the swinging pendulum. The time period is compared with the time period from the small angle approximation $\left(T_0 = 2\pi\sqrt{\dfrac{L}{g}} \right)$ and the result is shown in *Figure 6.9*. It is clear from the figure that the time period increases with an increase in the initial angular displacement.

```
//Load the *.sci file which contains Gauss-Legendre
functions
exec(' special_func.sci ',-1);

function y = f(theta)                  //Define the integral
y = (cos(theta)-cos(theta_m)).^(-0.5);
endfunction
```

```
T_0 = 2*%pi*sqrt(1/9.81);

i = 1;
for theta = 10:2:90;
    theta_max(i) = theta*%pi/180;
    theta_m = theta*%pi/180;
    timeperiod(i) = gauss_legendre(f,0,theta_m,39);
    timeperiod(i) = (4*sqrt(1/(2*9.81)))*timeperiod(i);
    i=i+1;
end
plot2d(theta_max*180/%pi,(timeperiod-T_0)/T_0)
```

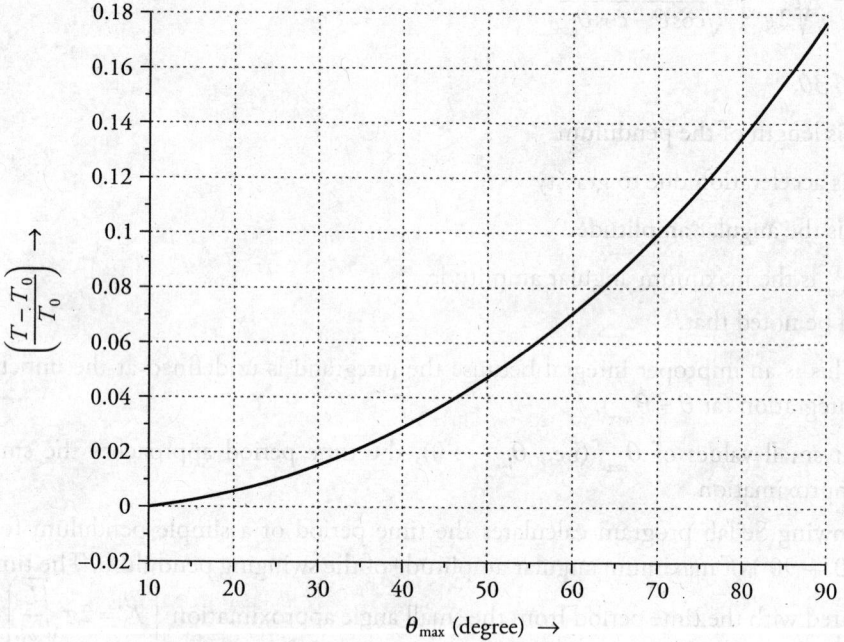

Figure 6.9 Application of the Gauss–Legendre quadrature

6.8 Exercises

1. Write a Scilab program to plot the first three Bessel function of the first kind.

2. Write a Scilab program to determine the value of the Bessel function of the first kind ($J_\alpha(x)$) at x = 1 and 2. Use the following definition.

$$J_\alpha(x) = \sum_{n=0}^{\infty} \frac{(-1)^n}{n!(n+\alpha)!}\left(\frac{x}{2}\right)^{2n+\alpha}$$

3. Write a Scilab program to interpolate the values of the Bessel function and determine the value at $x = 9.95$ and 25.2.

4. Write a Scilab program to determine the roots of the first five Legendre polynomials.

5. Write a Scilab program to determine the roots of the first five Laguerre polynomials.

6. Write a Scilab program to determine the expression for the first five Hermite polynomials.

7. Write a Scilab program to determine the roots of the first five Hermite polynomials.

8. Write a Scilab program to evaluate the following integrals using the quadrature methods of integration.

a) $\int_0^2 (1+x^2)\,dx$

b) $\int_0^4 xe^{2x}\,dx$

c) $\int_1^{1.5} x^2\ln x\,dx$

d) $\int_0^{10} \frac{e^x}{(1+x^2)^2}\,dx$

e) $\int_0^1 x^2 e^{-x}\,dx$

f) $\int_0^1 \frac{2}{x^2-4}\,dx$

g) $\int_0^1 \sin\pi x\,dx$

h) $\int_0^\pi \frac{\sin x}{x}\,dx$

i) $\int_0^{10} \frac{\sin x}{1+x^2}\,dx$

j) $\int_0^1 \frac{1}{\sqrt{x}}\,dx$

k) $\int_0^3 \frac{1}{\sqrt{3-x}}\,dx$

l) $\int_0^\infty \frac{\sin x}{1+x^2}\,dx$

m) $\displaystyle\int_0^\infty e^{-x}dx$

n) $\displaystyle\int_0^\infty x^2 e^{-x}dx$

o) $\displaystyle\int_0^\infty \frac{e^{-|x|}}{1+x^2}dx$

p) $\displaystyle\int_{-\infty}^\infty \frac{\sin x}{1+x^2}dx$

q) $\displaystyle\int_{-\infty}^\infty \frac{e^{-|x|}}{1+x^2}dx$

7

Fourier Analysis

After studying this chapter, the reader should be able to

◊ Write Scilab programs for generating periodic functions

◊ Appreciate the significance of representing a continuous periodic function in the form of a Fourier series

◊ Understand the concept of representing functions as a sum of harmonics

◊ Generate Fourier series for different periodic functions.

◊ Perform integral Fourier transform of common functions such as square, sine–cosine, and Gaussian functions.

◊ Visualize how the Fourier transform determines the presence of periodicity in a particular signal.

7.1 Introduction

The theory of Fourier series and Fourier integrals is of great importance for a wide range of scientific applications, such as in acoustics, optics and signal processing. The Fourier series is a mathematical representation of a continuous periodic function as an infinite sum of sinusoidal waves. The Fourier analysis is an excellent method to decompose an arbitrary function into sinusoidal components and solve it to get analytical solutions that are otherwise difficult to obtain.

Fourier transform is the frequency domain representation of continuous time signals. It results when the period of a time signal is stretched and allowed to approach infinity.

This chapter introduces the reader to write Scilab programs for computation of Fourier series of various periodic functions. An outline of the chapter is as follows. *Section 7.2* discusses the generation of periodic functions. A quick recapitulation of the Fourier series and the significance of harmonics are done in *Sections 7.3* and *7.4*, respectively. The method of writing Scilab programs for determining Fourier series of various periodic functions has been explained in *Section 7.5*. In *Section 7.6*, Fourier transform of some commonly used functions has been discussed. *Section 7.7* winds up the chapter with a brief summary of the Fourier analysis. Some practice questions have been given in *Section 7.8*.

7.2 Periodic Functions

A periodic function is a function whose value repeats itself after a regular interval which is called as 'period' of that function. As shown in *Eqn. 7.1*, if function $f(x)$ is periodic and periodicity is δ (a non-zero constant number), then for all the values of 'x',

$$f(x) = f(x + \delta) \tag{7.1}$$

Trigonometric functions are common examples of periodic functions. For example, consider the function in *Eqns. 7.2*. It has a base period equal to $\pi/2$.

$$f(x) = \cos(4x) \tag{7.2a}$$

$$f(x + \pi/2) = \cos\{4(x + \pi/2)\} = f(x) \tag{7.2b}$$

The following Scilab function generates periodic functions over a given interval. If periodicity of the function is $2T$, then, there can be two cases for generating a periodic function.

Case (I) The function is defined within the range $[-T, T]$.

```
function alpha = periodic1(f,T,x)
if (x <= T) & (x >= -T) then
    alpha = f(x);
elseif x < -T then
    new_x = x + 2.*T;
    alpha = periodic1(f,T,new_x);
elseif x > T then
    new_x = x - 2.*T;
    alpha = periodic1(f,T,new_x);
end
endfunction
```

Case (II) The function is defined within the range $[0, 2T]$.

```
function alpha = periodic2(f,T,x)
if (x >= 0) & (x <= 2.*T) then
    alpha = f(x);
elseif x < 0 then
    new_x = x + 2.*T;
    alpha = periodic2(f,T,new_x);
elseif x > T then
    new_x = x -2.*T;
    alpha = periodic2(f,T,new_x);
end
endfunction
```

The usefulness of these functions is explained with the help of the following examples.

Example 1: Suppose $f(x)$ is a periodic function in the interval $[-2, 2]$ such that,

$$f(x) = \begin{cases} -x & \text{for } -2 < x < 0 \\ x & \text{for } 0 < x < 2 \end{cases} \tag{7.3}$$

The function in *Eqn. 7.3* has a periodicity of 4. The following Scilab code generates this periodic function in the interval $[-8, 8]$. The graph is shown in *Figure 7.1*.

```
//Load the *.sci file which contains the function for
periodicity
exec('fourier.sci',-1);
period = 4;
//Periodicity

function func = f(x)
//Define function
if x > 0 then
   func = x;
else
   func = -x
end
endfunction

//Range for plotting the periodic function
x = [-2*period:0.01:2*period];
```

```
//Determine value of the function for all values of x.The
function is periodic in the interval (-T to T).Therefore,
'periodic1' function has been called with half the
periodicity
for i =1:length(x)
y(i)=periodic1(f,0.5*period,x(i));
end

plot2d(x,y')                              //Plot the function
```

Example 2: Suppose $f(x)$ is a periodic function in the interval such that,

$$f(x) = \begin{cases} x & \text{for } 0 < x < 2 \\ -x & \text{for } 2 < x < 4 \end{cases} \tag{7.4}$$

The function in *Eqn. 7.4* has a periodicity of 4. The following Scilab code generates this periodic function in the interval [–8, 8]. The graph is shown in *Figure 7.2*.

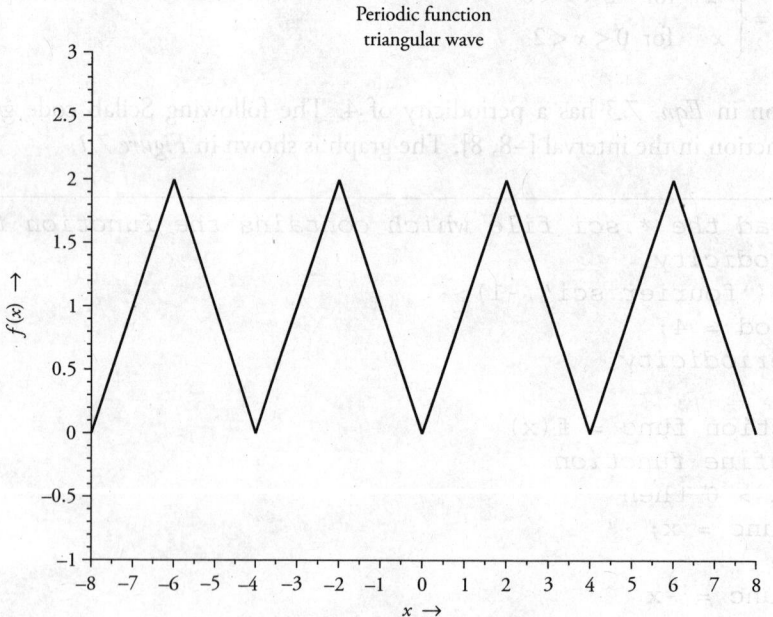

Figure 7.1 Generation of a periodic function

```
//Load the *.sci file which contains function for
periodicity
exec('fourier.sci',-1);
period = 4;                              //Periodicity
```

```
function func = f(x)
if x < period*0.5 then
    func = x;
else
    func = period-x
end
endfunction

//Range for plotting the periodic function
x = [-2*period : 0.01 : 2*period];

//Determine value of the function for all values of x.
The function is periodic in the interval (0 to 2T).
Therefore, 'periodic2' function has been called with
periodicity given above.
for i =1:length(x)
y(i)=periodic2(f,0.5*period,x(i));
end

plot2d(x,y')                        //Plot the function
```

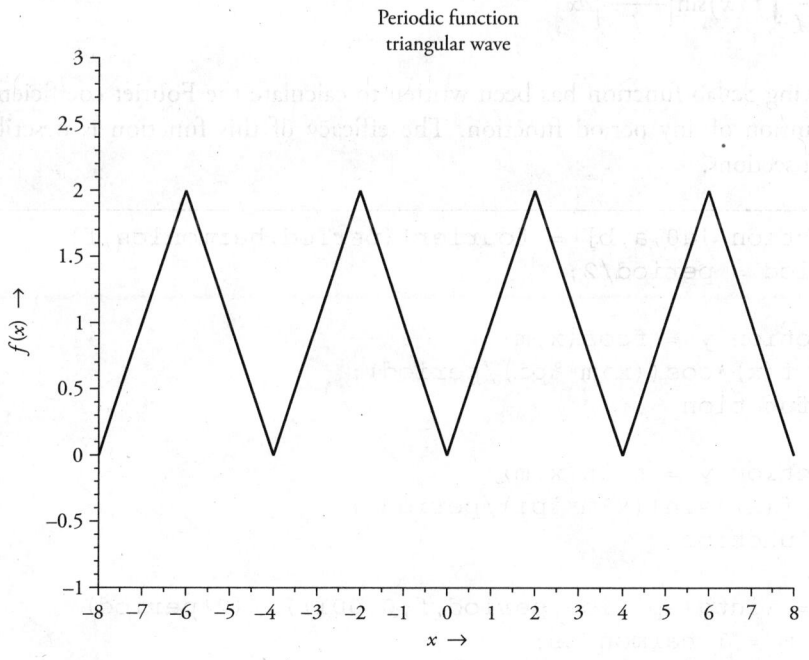

Figure 7.2 Generation of a periodic function

7.3 Fourier Series

As discussed in the previous section, there can be two ways of defining a function that has periodicity equal to '$2L$'.

Case (I)

Suppose $f(x)$ is periodic in the interval $[-L, L]$. The Fourier series expansion of $f(x)$ is given in *Eqn. 7.5*.

$$f(x) = a_0 + \sum_{n=1}^{\infty} a_n \cos\left(\frac{n\pi x}{L}\right) + \sum_{n=1}^{\infty} b_n \sin\left(\frac{n\pi x}{L}\right) \tag{7.5}$$

In *Eqn. 7.5*, a_0, a_n, and b_n are called coefficients of the Fourier series expansion, and their values are (*Eqns. 7.6*),

$$a_0 = \frac{1}{2L} \int_{-L}^{L} f(x)\, dx \tag{7.6a}$$

$$a_n = \frac{1}{L} \int_{-L}^{L} f(x)\cos\left(\frac{n\pi x}{L}\right) dx \tag{7.6b}$$

$$b_n = \frac{1}{L} \int_{-L}^{L} f(x)\sin\left(\frac{n\pi x}{L}\right) dx \tag{7.6c}$$

The following Scilab function has been written to calculate the Fourier coefficients and the series expansion of any period function. The efficacy of this function is described in the subsequent sections.

```
function [a0,a,b] = fourier1(period,harmonics,f)
period = period/2;

function y = fcos(x,m)
y = f(x)*cos((x*m*%pi)/period);
endfunction

function y = fsin(x,m)
y = f(x)*sin((x*m*%pi)/period);
endfunction

a0 = (intg(-period,period,f,0.001))./(2*period)
for m = 1:harmonics;
    a(m) = (1/period)*(intg(-period,period,fcos,0.001));
    b(m) = (1/period)*(intg(-period,period,fsin,0.001));
```

```
end

sum = a0;
for i = 1:harmonics;
sum = sum + (a(i).*cos(i.*x.*%pi/period)) +
    (b(i).*sin(i.*x.*%pi/period));
end

plot2d(x,sum,5)
endfunction
```

Case (II)

Suppose $f(x)$ is periodic in the interval $[0, 2L]$. Fourier series expansion of $f(x)$ is given in *Eqn. 7.7*.

$$f(x) = a_0 + \sum_{n=1}^{\infty} a_n \cos\left(\frac{n\pi x}{L}\right) + \sum_{n=1}^{\infty} b_n \sin\left(\frac{n\pi x}{L}\right)$$

(7.7)

In *Eqn. 7.7*, a_0, a_n and b_n are called coefficients of Fourier series expansion, and their values are (*Eqns. 7.8*),

$$a_0 = \frac{1}{2L}\int_0^{2L} f(x)\,dx$$

(7.8a)

$$a_n = \frac{1}{L}\int_0^{2L} f(x)\cos\left(\frac{n\pi x}{L}\right)dx$$

(7.8b)

$$b_n = \frac{1}{L}\int_0^{2L} f(x)\sin\left(\frac{n\pi x}{L}\right)dx$$

(7.8c)

The following Scilab function has been written to calculate the Fourier coefficients and series expansion of any periodic function. The efficacy of this function is described in the subsequent sections.

```
function [a0,a,b] = fourier2(period,harmonics,f)
function y = fcos(x,m)
y = f(x)*cos((x*m*(%pi))/(0.5*period));
endfunction

function y = fsin(x,m)
y = f(x)*sin((x*m*(%pi))/(0.5*period));
endfunction
```

```
a0 = (intg(0,period,f,0.001)).*(1/(period))
for m = 1:harmonics;
a(m) = (1/(0.5*period))*(intg(0,period,fcos,0.001));
b(m) = (1/(0.5*period))*(intg(0,period,fsin,0.001));
end

sum = a0;
for i = 1:harmonics;
sum = sum + (a(i).*cos(i.*x.*%pi/(0.5*period))) +
(b(i).*sin(i.*x.*%pi/(0.5*period)));
end

plot2d(x,sum,5)
endfunction
```

Application of these functions for determining the Fourier series of continuous periodic functions will be explained in subsequent sections.

7.4 Harmonics

In the Fourier series expansion of a periodic function whose period is equal to 'T', the fundamental frequency is given by sine and cosine terms corresponding to the period 'T'. Harmonics refers to all the other sine and cosine terms of the series having higher frequencies (and smaller period). Superposition of all these sine and cosine terms (assumed to be infinite in number) gives an approximation of the original periodic function.

For example, consider a periodic function having periodicity 2π (*Eqn. 7.9*).

$$f(x) = \sin x + \sin 2x + \sin 3x \text{ for } -\pi < x < \pi \tag{7.9}$$

The following Scilab program plots this function in the interval $[-3\pi, 3\pi]$. The graph is shown in *Figure 7.3*.

```
//Load the *.sci file which contains function for
periodicity
exec('fourier.sci',-1);

period = 2*%pi;

function func = f(x)
func = sin(x)+sin(2*x)+sin(3*x);
endfunction
```

```
//Give range for x-variable
x = [-1.5*period:0.01:1.5*period];

//Determine value of the function
for i =1:length(x)
y(i)=periodic1(f,0.5*period,x(i));
end

plot2d(x,y')                          //Plot the function
```

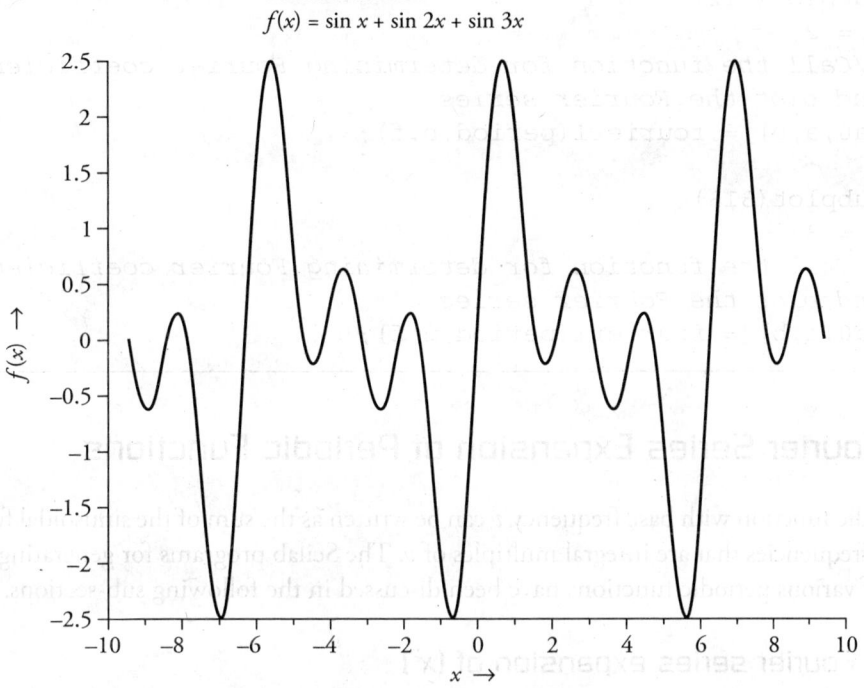

Periodic function
$f(x) = \sin x + \sin 2x + \sin 3x$

Figure 7.3 Graph of a periodic function

The following Scilab program generates the Fourier series of this function for different harmonics. The result is shown in *Figure 7.4*. It should be noticed that the plots in the first and second panel do not reproduce the original signal. However, the superposition of three harmonics gives a reasonably good representation of the original signal. This is to be expected because the signal itself consists of a superposition of three harmonics.

```
//Load *.sci file which contains the function for Fourier
series
exec('fourier.sci',-1);

subplot(311)
n = 1;
//Call the function for determining Fourier coefficients
and to plot the Fourier series
[a0,a,b] = fourier1(period,n,f);

subplot(312)
n = 2;
//Call the function for determining Fourier coefficients
and plot the Fourier series
[a0,a,b] = fourier1(period,n,f);

subplot(313)
n = 3;
//Call the function for determining Fourier coefficients
and plot the Fourier series
[a0,a,b] = fourier1(period,n,f);
```

7.5 Fourier Series Expansion of Periodic Functions

A periodic function with base frequency v can be written as the sum of the sinusoidal functions having frequencies that are integral multiples of v. The Scilab programs for generating Fourier series of various periodic functions have been discussed in the following sub-sections.

7.5.1 Fourier series expansion of (x^2)

Suppose a function has a periodicity equal to 2π and is defined in the interval $[-\pi, \pi]$ such that

$$f(x) = x^2 \tag{7.10}$$

The following Scilab program plots the function given in *Eqn. 7.10* in the range $[-3\pi, 3\pi]$ determines its Fourier series expansion for the fundamental harmonic. This is shown in *Figure 7.5* where the original function is overlaid with the Fourier series representation.

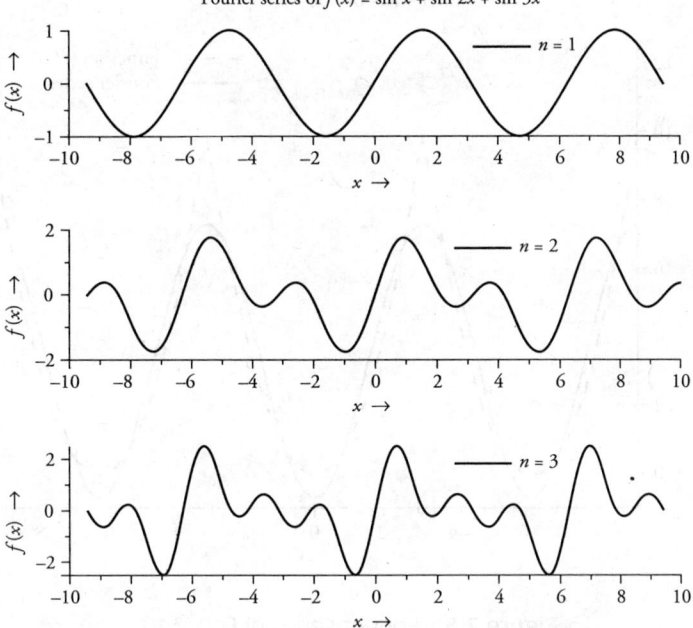

Figure 7.4   Fourier series of the given periodic function

```
//Load the *.sci file which contains functions for
periodicity and the Fourier series
exec('fourier.sci',-1);
period = 2*%pi;

function func = f(x)
func = x.^2;
endfunction

//Range for x-variable
x = [-1.5*period : 0.01 : 1.5*period];

//Determine value of the function
for i = 1:length(x)
y(i) = periodic1(f,0.5*period,x(i));
end

plot2d(x,y')    //Plot the function
n = 1;     //Number of harmonics

//Call the function for determining Fourier coefficients
and to plot the Fourier series
[a0,a,b] = fourier1(period,n,f);
```

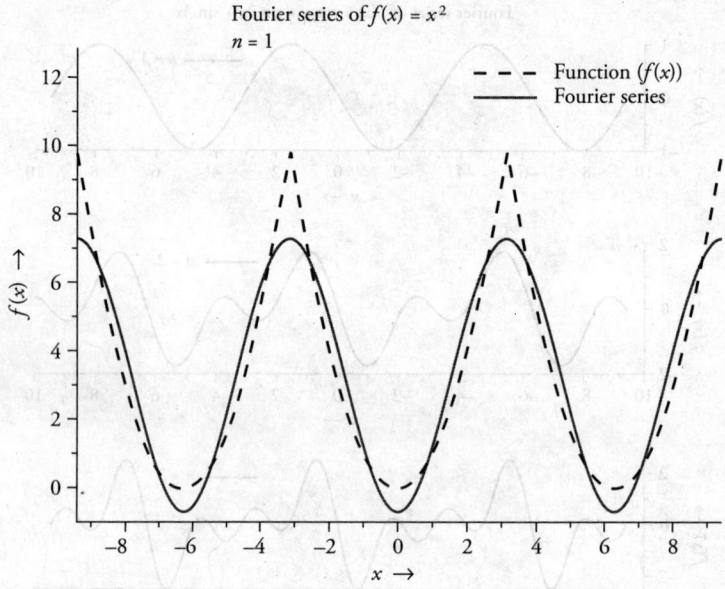

Figure 7.5   Fourier series of *Eqn. 7.10*

Figure 7.6 shows the Fourier series of *Eqn. 7.10*. In this case, ten harmonics have been taken to determine the Fourier series. It is clear from the graph that superposition of a large number of harmonics gives a better representation of the function.

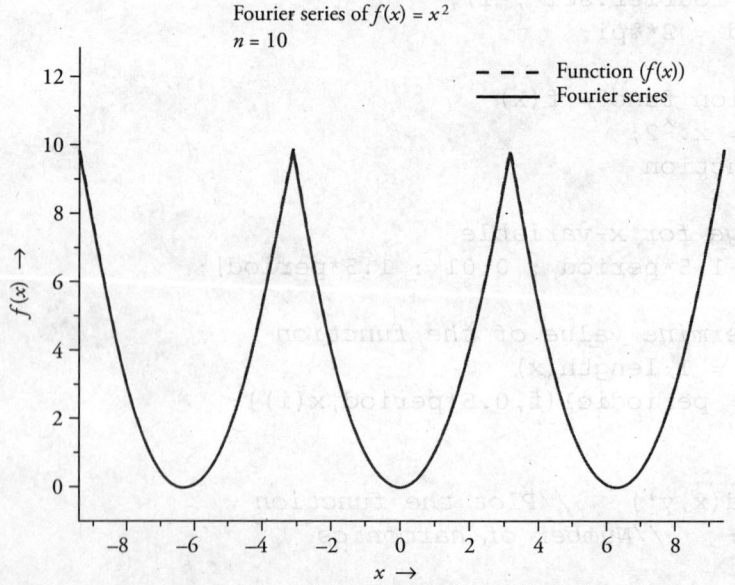

Figure 7.6   Fourier series of *Eqn. 7.10*

7.5.2 Fourier series expansion of saw-tooth wave

Consider a saw-tooth wave having a periodicity equal to 'p' and amplitude varying between -1 and 1. The general expression of this wave is given by *Eqn. 7.11*.

$$y(x) = 2\left(\frac{x}{p} - \left\lfloor \frac{1}{2} + \frac{x}{p} \right\rfloor\right)$$

(7.11)

The Scilab program for generating a saw-tooth wave and its Fourier series is given as follows.

```
//Load the *.sci file which contains the function for
generating the Fourier series
exec('fourier.sci',-1);

period = 2;                      //Periodicity of the wave
n=1;                            //Number of harmonics
x = 0:0.01:2*period;            //Range of x-variable

//Function for generating the saw-tooth wave
function alpha = f(x);
alpha = 2*((x/period)-floor(0.5+(x/period)));
endfunction

plot2d(x,f(x))                  //Plot the function

//Call the function for determining the Fourier
coefficients and to plot the Fourier series
[a0,a,b] = fourier2(period,n,f);
```

Figure 7.7 shows the plot of *Eqn. 7.11* and its Fourier series generated by using the aforementioned code. The saw-tooth shaped signal can be reconstructed reasonably well if the number of harmonics is increased. This is shown in *Figure 7.8* for three harmonics. The original saw-tooth shaped signal can be well retrieved if the number of harmonics is increased further.

The saw-tooth wave is an odd function. Therefore, Fourier coefficients a_0 and a_n are equal to zero. The values of the first few b_n determined from the aforementioned Scilab program have been tabulated in *Table 7.1*.

For a saw-tooth wave of amplitude one, these values are clearly in accordance with theoretically expected values as given in *Eqn. 7.12*.

$$b_n = -\frac{2A}{\pi n}(-1)^n$$

(7.12)

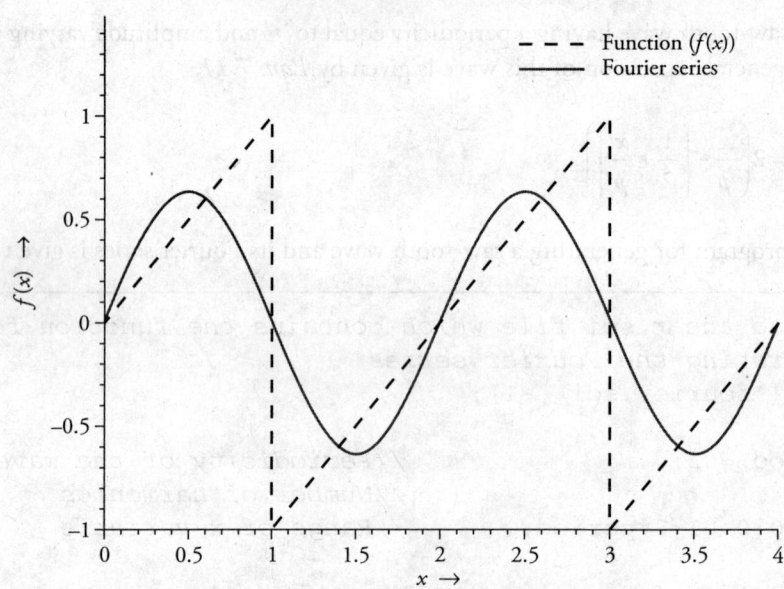

Fourier series of Saw-tooth wave
$n = 1$

Figure 7.7 Fourier series of *Eqn. 7.11*

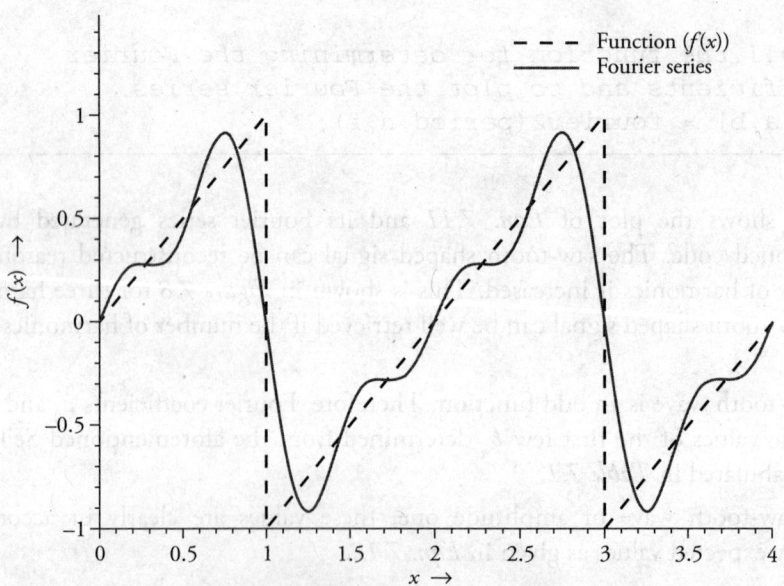

Fourier series of Saw-tooth wave
$n = 3$

Figure 7.8 Fourier series of *Eqn. 7.11*

Table 7.1   Fourier coefficients of saw-tooth wave

n	b_n
1	0.6366
2	-0.3183
3	0.2122
4	-0.1591
5	0.1273
6	-0.1061
7	0.0909

7.5.3 Fourier series expansion of a square wave

Consider a square wave having a periodicity equal to 'p' (or frequency $f = \dfrac{1}{p}$) and an amplitude varying between 0 and 1. The general expression for a square wave is (*Eqn. 7.13*).

$$y(x) = \text{sgn}(2\pi fx) \tag{7.13}$$

The Scilab program for generating a square wave and its Fourier series is as follows.

```
//Load the *.sci file which contains the function for
generating Fourier series
exec('fourier.sci',-1);
period = 2*%pi;              //Periodicity of the wave
n = 1;                       //Number of harmonics
x = 0:0.01:2*period;         //Range of x-variable

//Function for generating the square wave
function func = f(x)
func = sign(sin(2*%pi*(1/(2*%pi))*x));
endfunction

plot2d(x,f(x))              //Plot the square wave

//Call the function for determining Fourier coefficients
and to plot the Fourier series
[a0,a,b] = fourier2(period,n,f);
```

Figure 7.9 shows the square wave and its Fourier series generated using the aforementioned program. A better approximation of the signal will be obtained if the number of harmonics is increased. This is shown in *Figure 7.10* for three harmonics. The reader can increase the number of harmonics to appreciate its importance.

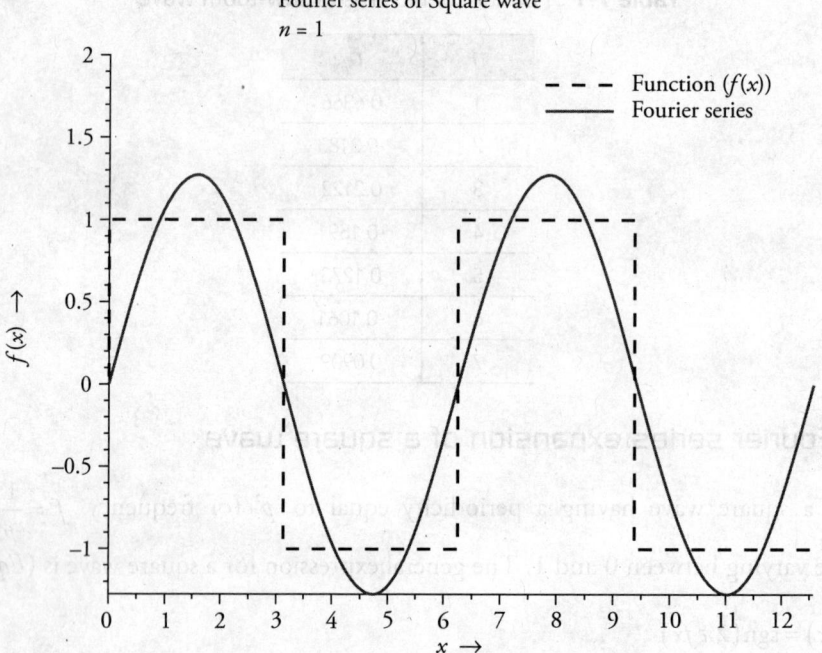

Figure 7.9 Fourier series of *Eqn. 7.13*

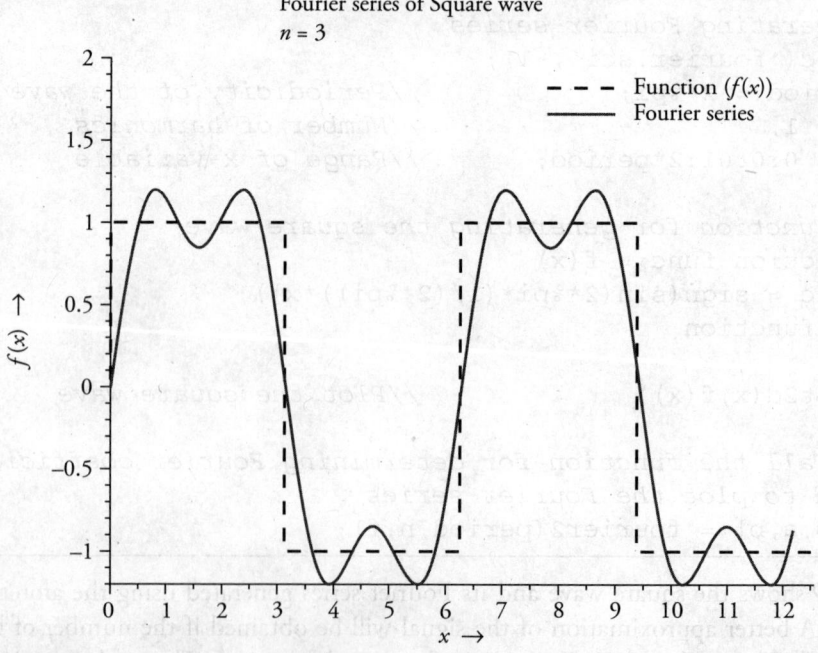

Figure 7.10 Fourier series of *Eqn. 7.13*

7.5.4 Fourier series expansion of a triangular wave

Consider a triangular wave having a periodicity equal to 'p' and an amplitude varying between 0 and 1. The expression for the triangular wave is the absolute value of the expression for the saw-tooth wave. It is given by *Eqn. 7.14*.

$$y(x) = \left| 2\left(\frac{x}{p} - \left\lfloor \frac{1}{2} + \frac{x}{p} \right\rfloor \right) \right| \tag{7.14}$$

The Scilab program for generating a triangular wave and its Fourier series is as follows.

```
//Load the *.sci file which contains the function for
generating the Fourier series
exec('fourier.sci',-1);

period = 2*%pi;              //Periodicity of the wave
n = 1;                       //Number of harmonics
x = 0:0.01:2*period;         //Range of x-variable

//Function for generating a triangular wave
function func = f(x)
func = abs(2*((x/period)-floor(0.5+(x/period))));
endfunction
plot2d(x,f(x))               //Plot the triangular wave

//Call the function for determining Fourier coefficients
and to plot the Fourier series
[a0,a,b] = fourier2(period,n,f);
```

Figure 7.11 shows the triangular wave and its Fourier series generated using the aforementioned program. As compared to the saw-tooth and square-shaped signals, which have finite discontinuities, the convergence of Fourier series is faster in case of the triangular wave. In the former, the nth coefficient falls as $1/n$. The nth coefficient in case of the triangular function decreases as $1/n^2$. Therefore, the triangular wave can be reasonably approximated even at the fundamental frequency. Inclusion of higher harmonics rolls off faster in the case of a triangular wave. This is shown in *Figure 7.12* for three harmonics. The triangular wave is an even function of x. The Fourier coefficients b_n are equal to zero. The value of a_0 is 0.5. *Table 7.2* gives the values of other a_n determined from the aforementioned Scilab program. For a triangular wave of amplitude (A) equal to 0.5, these values are clearly in accordance with theoretically expected values, which are given by *Eqn. 7.15*.

$$a_n = \begin{cases} 4A\dfrac{1-(-1)^n}{\pi^2 n^2} & \text{for odd } n \\ 0 & \text{for even } n \end{cases} \tag{7.15}$$

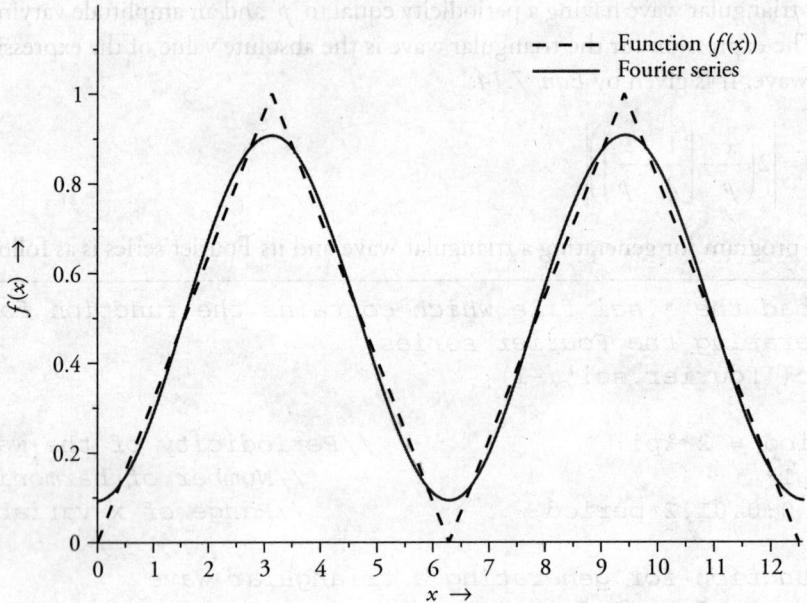

Figure 7.11 Fourier series of *Eqn. 7.14*

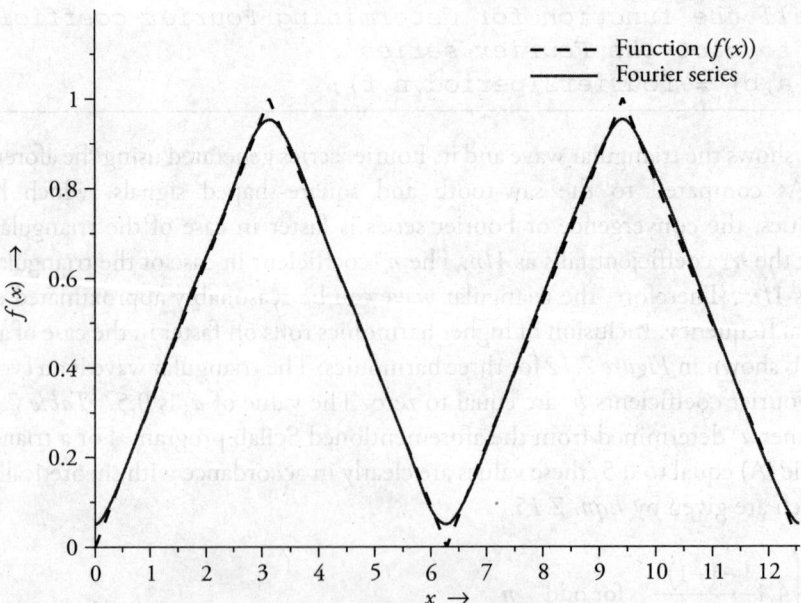

Figure 7.12 Fourier series of *Eqn. 7.14*

Table 7.2 Fourier coefficients of *Eqn. 7.14*

n	a_n
1	0.4053
2	0
3	0.0450
4	0
5	0.0162
6	0
7	0.0083

7.5.5 Fourier series expansion of output of half wave rectifier

If the periodicity of the output of a half wave rectifier is equal to '2π', then its functional form is given by *Eqn. 7.16*.

$$f(t) = \begin{cases} 0 & \forall & \pi \le t \le 2\pi \\ \sin \omega_0 t & \forall & 0 \le t \le \pi \end{cases} \tag{7.16}$$

The Scilab program for generating the half wave rectifier output and its Fourier series is as follows. The graph is shown in *Figures 7.13* and *7.14* for different number of harmonics.

```
//Load the *.sci file which contains functions for
generating the periodic function and Fourier series
exec('fourier.sci',-1);

period = 2*%pi;              //Periodicity of the wave
omega = 2*%pi/period;              //Angular frequency
n = 10;                           //Number of harmonics

//Function for generating half wave rectifier output
function func = f(x)
if (x < 0.5*period) & (x > 0) then
func = sin(omega*x);
else
func = 0
end
endfunction

x = [-2*period:0.01:2*period];     //Range of x-variable
for i=1:length(x)
y(i) = periodic2(f,0.5*period,x(i));
```

```
end

plot2d(x,y')                              //Plot the wave

//Call the function for determining the Fourier
coefficients and to plot the Fourier series
[a0,a,b] = fourier2(period,n,f);
```

Fourier series of Half wave rectifier (*n* = 3)

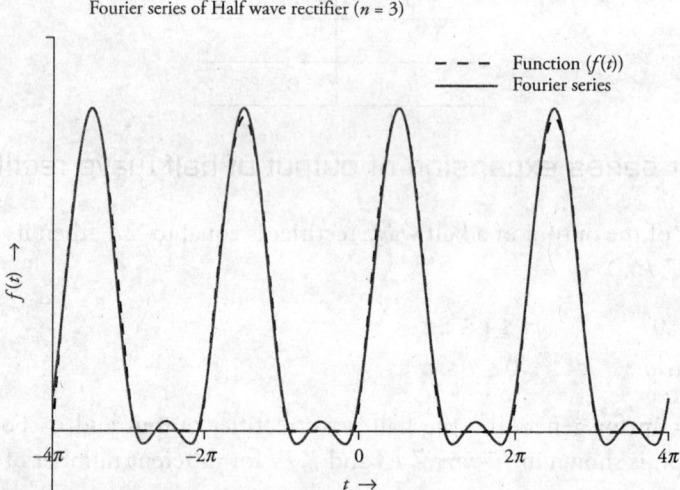

Figure 7.13 Fourier series of *Eqn. 7.16*

Fourier series of Half wave rectifier (*n* = 10)

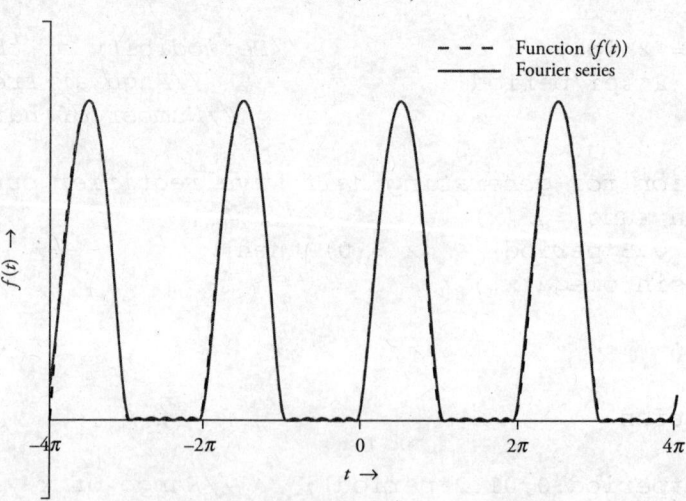

Figure 7.14 Fourier series of *Eqn. 7.16*

7.6 Fast Fourier Transform

A Fourier transform is used for decomposing a time signal into its frequency constituents. It is also called the 'frequency domain representation' of the time function. If a function is localized in time domain, then its Fourier transform will be a spread out function in the frequency domain. The utility of the Fourier transform lies in the fact that sometimes it is easier to perform an operation in one of the domains. For example, sometimes it might be difficult to solve differential equations in the time domain. Therefore, in such cases, it has been repeatedly observed that it is easier to perform multiplicative operation in the frequency domain. *Eqn. 7.17* gives the expression for the Fourier transform ($\mathcal{F}$) of a function $f(t)$.

$$\mathcal{F}\{f(t)\} = \int_{-\infty}^{\infty} f(t)e^{-2\pi i v t}\,dt \tag{7.17}$$

7.6.1 FFT of a sine wave

The following Scilab program determines the Fourier transform of a sine wave having frequency 50 Hz. The fast Fourier transform can be performed by using the built-in function (fft) of Scilab. The graph is shown in *Figure 7.15*. The peak (at 50 Hz) in its bottom panel corresponds to the frequency of the signal.

```
sample_rate = 1000;
t = 0:1/sample_rate:0.1;
func = sin(2*%pi*50*t) ;
subplot(211)
plot2d(t,func)
X = fft(func);  //FFT of the given function
N = length(t);
f = sample_rate*(0:(N/2))/N;
subplot(212)
plot2d(f,abs(X(1:length(f))))    //Plot FFT of the signal
```

7.6.2 FFT of the sum of two cosine wave signals

The following Scilab program determines the Fourier transform of the sum of two cosine wave signals having frequencies 100 Hz and 350 Hz. The graph is shown in *Figure 7.16*. The peaks (at 100 Hz and 350 Hz) in its bottom panel correspond to the frequency of the two waves.

```
sample_rate = 1000;
```

```
t = 0:1/sample_rate:0.1;
func = cos(2*%pi*100*t) + cos(2*%pi*350*t);
subplot(211)
plot2d(t,func)

X = fft(func);                           //FFT of the signal
N = length(t);
f = sample_rate*(0:(N/2))/N;
subplot(212)
plot2d(f,abs(X(1:length(f)))) //Plot FFT of the signal
```

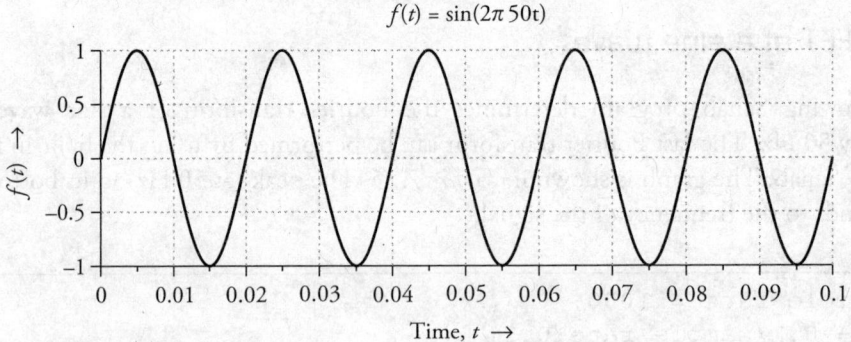

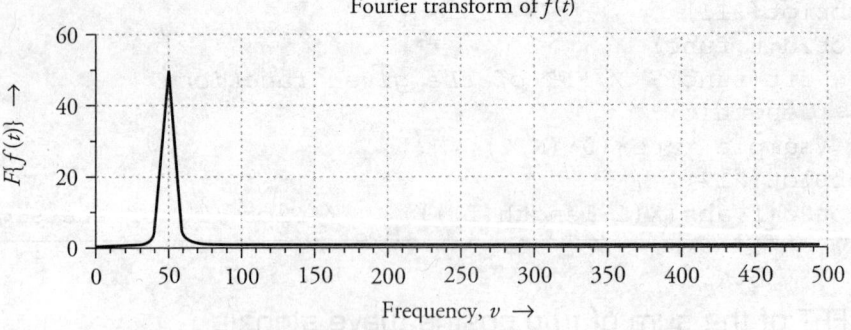

Figure 7.15   FFT of a sinusoidal wave

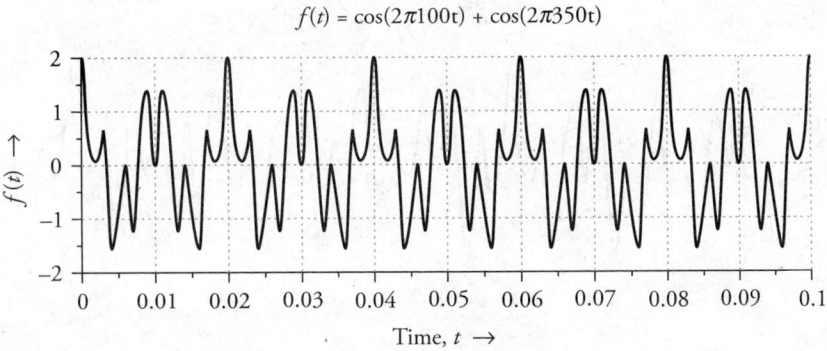

$$f(t) = \cos(2\pi 100t) + \cos(2\pi 350t)$$

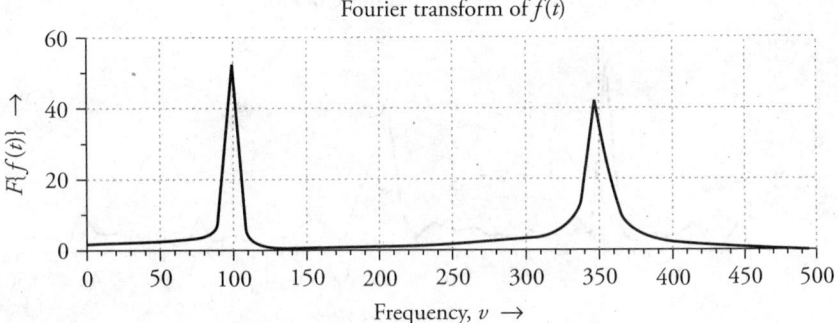

Fourier transform of $f(t)$

Figure 7.16   FFT of the sum of two cosine wave signals

7.6.3 FFT of a noisy signal

The following Scilab program determines the Fourier transform of a noisy signal. The graph is shown in *Figure 7.17*.

```
sample_rate = 100;
t = 0:1/sample_rate:1;
func=sin(2*%pi*10*t)+cos(2*%pi*25*t)+0.5*grand(1,length
(t),'nor',0,1);
subplot(211)
plot2d(t,func)
X = fft(func);
N = length(t);
f = sample_rate*(0:(N/2))/N;
subplot(212)
plot2d(f,abs(X(1:length(f))))
```

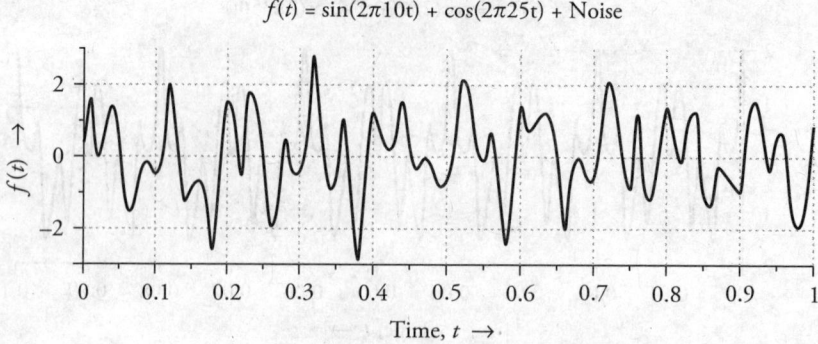

$$f(t) = \sin(2\pi10t) + \cos(2\pi25t) + \text{Noise}$$

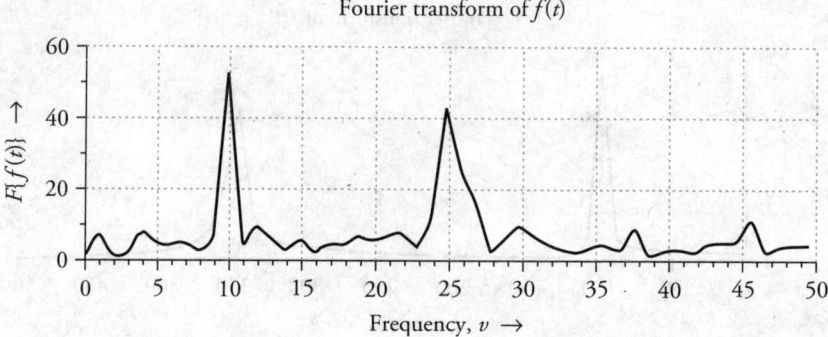

Fourier transform of $f(t)$

Figure 7.17 FFT of a noisy signal

7.6.4 FFT of a square wave

The following Scilab program determines the Fourier transform of the square wave signal. The frequency domain representation is a 'sinc' function, which is shown in *Figure 7.18*.

```
// Load the *.sci file which contains the function for
generating periodic signals
exec('fourier.sci',-1)
period = 8;      //Periodicity
function func = f(x)
if (x > -0.25*period) & (x < 0.25*period) then
func = 1;
else
func = 0
end
endfunction
sample_rate = 100;
x = -0.75*period:1/sample_rate:0.75*period;
for i =1:length(x)
```

```
y(i)=periodic1(f,0.5*period,x(i));
end
subplot(211)
plot2d(x,y)
X = (fft(y))/length(y);
f = 0:0.005:1/period;
subplot(212)
plot2d(f,X(1:length(f)))
plot2d(-f,X(1:length(f)))
```

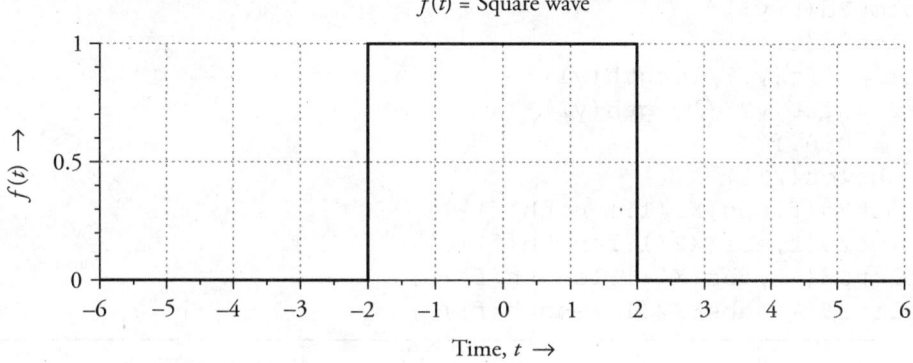

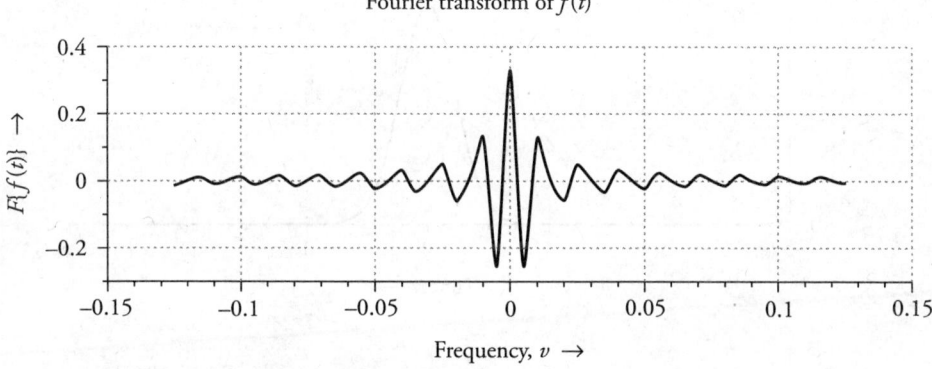

Figure 7.18   FFT of a square wave

The Fourier transform of a square wave (a box function) has an important application in physical optics. A diffracting slit is described by a box function and the amplitude of the corresponding diffraction pattern is described by its Fourier transform.

7.6.5 FFT of a Gaussian curve

The following Scilab program determines the Fourier transform of a Gaussian curve. The graph is shown in *Figure 7.19*.

```
sample_rate = 100;
t = -3:1/sample_rate:3;
for i =1:length(t)
y1(i) = exp(-10.*t(i).*t(i));
end
t = -3:1/sample_rate:3;
for i =1:length(t)
y2(i) = exp(-1.*t(i).*t(i));
end
subplot(211)
plot2d(t,y1)
plot2d(t,y2)
X1 = fft(y1)/length(y1);
X2 = fft(y2)/length(y2);
f = 0:0.1:3;
subplot(212)
plot2d(f,abs(X1(1:length(f))))
plot2d(f,abs(X2(1:length(f))))
plot2d(-f,abs(X1(1:length(f))))
plot2d(-f,abs(X2(1:length(f))))
```

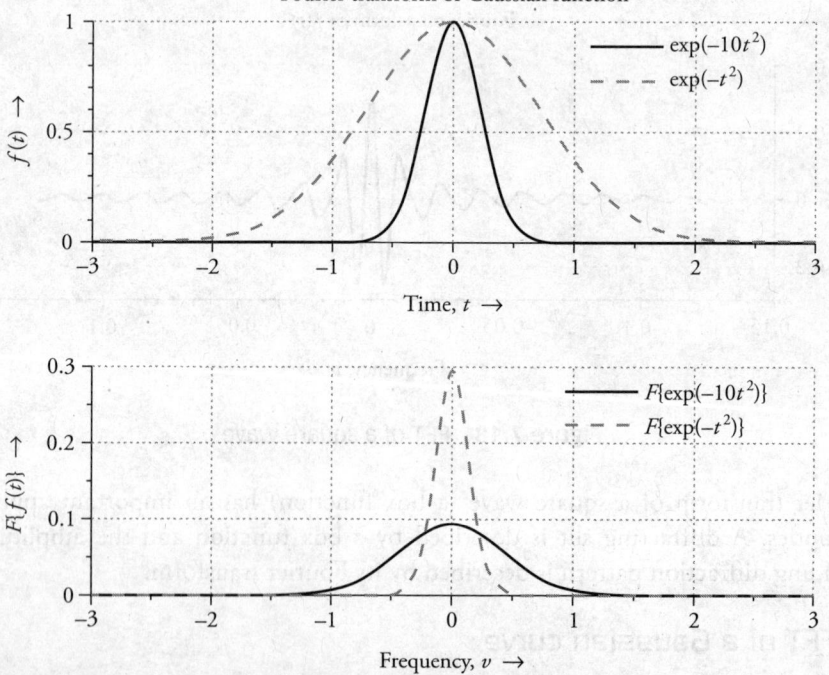

Figure 7.19   FFT of a Gaussian function

From *Figure 7.19*, it is clear that

- Fourier transform of a Gaussian curve is a Gaussian.

- If width of a Gaussian curve is more in the time domain, then the Fourier transform will have a narrower width.

7.7 Summary

The Fourier analysis is an interesting technique that involves several mathematical methods such as trigonometry, integration, and summation. Some of the important points that should be taken care of while performing the Fourier analysis in Scilab are as follows:

- If a function has periodicity $2L$, then always use its correct interval of periodicity. It can either be $[0\ 2L]$ or $[-L\ +L]$. Therefore, make use of appropriate formulae.

- The Fourier series is used for analysing periodic functions. It is an infinite series expansion of a function and involves sinusoidal functions.

- The frequencies of these sinusoidal functions are called harmonics.

- The Fourier series converges to the function if more and more terms involving higher harmonics are included in the series expansion.

- The convergence is slow if the original function has discontinuities and summation of several terms is required to approximate it.

- The Fourier transform decomposes a time-domain signal into its constituent frequencies and it deals with non-periodic functions also.

7.8 Exercises

1. Suppose $f(x)$ is a periodic function in the interval (-2 to 2) such that

   $$f(x) = \begin{cases} x & \text{for} \quad -2 < x < 0 \\ -x & \text{for} \quad 0 < x < 2 \end{cases}$$

   Write a Scilab program to plot this function in the interval (-8 to 8).

2. Write a Scilab program to generate the graph shown in *Figure 7.20*.

3. Suppose $f(x)$ is a periodic function in the interval (-2, 2) such that

   $$f(x) = x \text{ for} -2 < x < 2$$

   Write a Scilab program to plot this function in the interval (-8, 8).

4. What are the Fourier series coefficients of the function discussed in *Section 7.5.5?*

5. Write a Scilab program for finding the Fourier series coefficients of

$$f(x) = \cos(\pi x) + \cos(2\pi x)$$

Plot the Fourier series expansion. Take the number of harmonics to be 5.

6. Write a Scilab program for finding the Fourier series coefficients of

$$f(x) = \cos(5\pi x) + \sin(10\pi x)$$

Plot the Fourier series expansion of this function. Take the number of harmonics to be 5

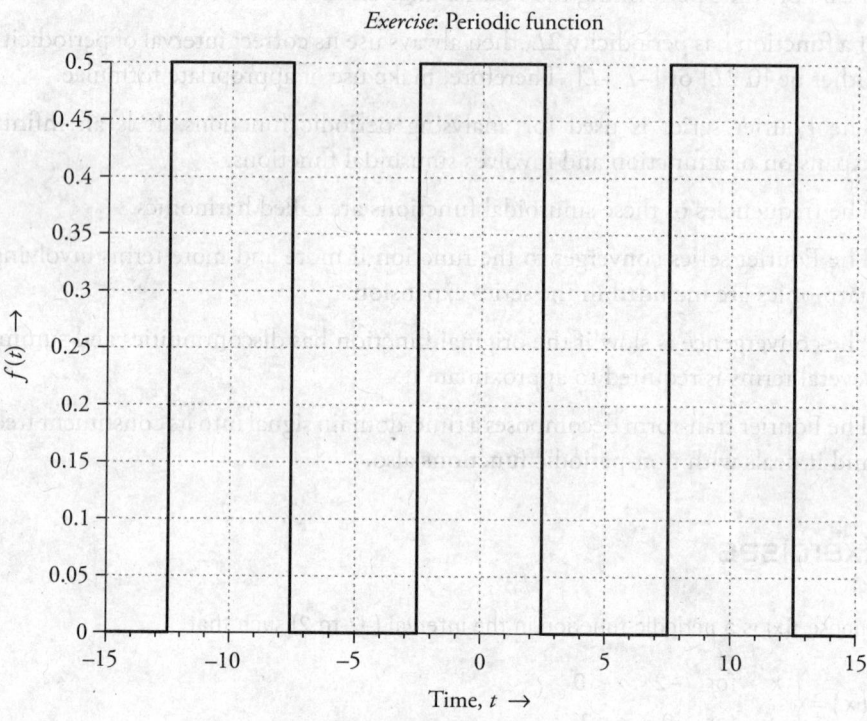

Exercise: Periodic function

Figure 7.20 Graph for *Exercise 2*

7. Write a Scilab program to show the effect of the first and third harmonic in Fourier series expansion of square wave.

8. Write a Scilab program for finding the Fourier series expansion of a full wave rectifier output in the interval (0, π).

$$f(t) = |\sin \omega t|$$

Show that coefficients of Fourier series expansion can be written in the following form

$$|\sin \omega t| = \frac{2}{\pi} - \frac{4}{\pi} \sum_{n-2,4,\ldots}^{\infty} \frac{\cos(n\omega t)}{n^2 - 1}$$

9. Write a Scilab program to determine the Fourier series expansion of a trapezoidal function defined in the interval [0, 2]. Take the number of harmonics to be 7.

$$f(x) = \begin{cases} 5x & \text{for} & x < 0.4 \\ 2 & \text{for} & 0.4 < x < 1.6 \\ 2 - 5\left(x - \dfrac{8}{5}\right) & \text{for} & x > 1.6 \end{cases}$$

10. Write a Scilab program for finding the Fourier series expansion of

$$f(x) = \sqrt{1 + x^2}$$

It is defined in the interval [−1, 1].

11. Write a Scilab program for finding the Fourier series expansion of

$$f(x) = \ln(2 + x)$$

It is defined in the interval [−1, 1].

12. Write a Scilab program to determine the periodicity in the following time function by performing its Fourier transform.

$$f(t) = \sin(20\pi t)e^{-2\pi t^2}$$

13. Write a Scilab program to determine the Fourier transform of $e^{-2|t|}$.

Algebraic and Transcendental Equations

After studying this chapter, the reader should be able to

◊ Write Scilab programs for solving algebraic and transcendental equations using the built-in Scilab functions.

◊ Write Scilab programs to solve systems of linear equations by making use of the iterative Gauss–Seidel method.

◊ Identify the benefits of the Gauss–Seidel method and the conditions under which it converges to give a solution.

◊ Write Scilab programs to solve systems of linear equations by making use of the Gaussian elimination method.

◊ Use the Gaussian elimination method for writing Scilab programs to find the determinant and inverse of a matrix.

◊ Write Scilab programs to solve non-linear and transcendental equations by making use of the bisection, secant, false position and Newton–Raphson methods.

◊ Apply these numerical methods to solve interesting problems like trajectory of a particle, bearing angle of a boat, binding energy of deuteron and single slit diffraction pattern.

8.1 Introduction

In mathematics, an algebraic expression is a combination of integers, variables, and algebraic operations such as summation, subtraction, product, division, and exponentiation. Equating two such algebraic expressions gives rise to an algebraic equation. For example, *Eqns. 8.1–8.2* are algebraic expressions, whereas *Eqns. 8.3–8.4* are algebraic equations.

$$2x^3 + 5xy - 6 \tag{8.1}$$

$$\sqrt[3]{\frac{x+y}{x-y}} \tag{8.2}$$

$$\sqrt{x} + 5x = 32 \tag{8.3}$$

$$3^x = 5 \tag{8.4}$$

There are different kinds of algebraic equations, such as, linear equations, non-linear equations (quadratic, cubic, and higher order polynomial equations) and exponential equations. The aim of this chapter is to find solutions for these equations by making use of various numerical techniques and to determine a set of values of the unknown variables contained in these equations. In this chapter, the Gauss–Seidel and the Gaussian elimination methods have been used for solving systems of linear equations, which involve '*n*' equations and an equal number of unknown variables.

Transcendental equations are encountered ubiquitously in theoretical physics. They consist of transcendental functions that cannot be expressed in the form of finite sequences of algebraic operations. Elementary transcendental functions are logarithmic, trigonometric, exponential, and hyperbolic functions. These equations commonly occur while solving heat transfer problems and studying diffraction patterns. This chapter focuses on Scilab programs for determining the approximate numerical solution of non-linear and transcendental equations using the following iterative numerical techniques,

- Bracketing methods
  - Bisection method
  - Method of false position
- Open methods
  - Secant method
  - Newton–Raphson method

This chapter is arranged such that *Section 8.2* recapitulates the built-in functions of Scilab that are often used for solving linear, quadratic, and polynomial equations. The reader is advised to revise the chapter on matrices before attempting to go through this section. In *Section 8.3*, the Gauss–Seidel method has been discussed along with its pitfalls. In *Section 8.4*, a brief overview of the Gaussian elimination method for solving system of linear equations has been provided. It is followed by an explicit explanation of the Scilab program based on this numerical technique. *Section 8.5* discusses a variant of the Gaussian elimination method. This version re-orients the pivot elements of the Gaussian elimination method. The technique of bisection method has been explained in *Section 8.6*. The other techniques for solving non-linear and transcendental equations, such as the Regula Falsi method, the Secant method, and the Newton–Raphson method have been explained in *Sections 8.7–8.9* respectively. Some interesting applications of these methods have been given in *Section 8.10*, for example determination of the trajectory of a particle, determination of the bearing angle of a boat, determinant and inverse of matrices by making use of the Gaussian elimination method; determination of binding energy of deuteron and single slit diffraction pattern.

8.2 Equation Solver in Scilab

A linear equation consists of a polynomial of degree one. An example of a linear equation is given in *Eqn. 8.5*. It involves only one unknown variable x. Similarly, *Eqn. 8.6* is also a linear equation. It involves an additional unknown variable y.

$$2x + 3 = 5 \tag{8.5}$$

$$9x + y = 13 \tag{8.6}$$

A non-linear equation consists of a polynomial of degree greater than one. Trigonometric functions and exponential functions also constitute a non-linear equation. An example of a non-linear equation is given in *Eqn. 8.7*.

$$2x^2 - x = 7 \tag{8.7}$$

There are several built-in functions and commands in Scilab for solving linear and non-linear equations. Some of them have been discussed in the following sub-sections.

8.2.1 Division operator

Consider the system of linear equations given in *Eqns. 8.8*.

$$3x_1 + x_2 + x_3 = 8 \tag{8.8a}$$

$$2x_1 - x_2 - x_3 = -3 \tag{8.8b}$$

$$x_1 + 2x_2 + 3x_3 = 0 \tag{8.8c}$$

The Scilab program based on the division operator method is given next. The main steps of this program are as follows.

- Generate a coefficient matrix corresponding to all the unknown variables, x_1, x_2 and x_3.

- Let this matrix be A such that,

$$A = \begin{bmatrix} 3 & 1 & 1 \\ 2 & -1 & -1 \\ 1 & 2 & 3 \end{bmatrix}$$

- Generate a column matrix corresponding to the numbers given on the right side of these equations.

- Let this matrix be B such that

$$B = \begin{bmatrix} 8 \\ -3 \\ 0 \end{bmatrix}$$

- The aforementioned two points imply that $AX = B$. The solution matrix X is

$$X = \begin{bmatrix} x_1 \\ x_2 \\ x_3 \end{bmatrix}$$

- Therefore, the solution matrix will be equal to

$$X = A \backslash B$$

```
A = [3 1 1 ; 2 -1 -1 ; 1 2 3];
B = [8 ; -3 ; 0]
A\B
```

Values of the variables will be equal to, 1, 16 and -11.

8.2.2 Built-in Scilab function – 'linsolve'

Consider the same example as that taken in *Section 8.2.1 (Eqn. 8.8)*.

$$3x_1 + x_2 + x_3 - 8 = 0$$

$$2x_1 - x_2 - x_3 + 3 = 0$$

$$x_1 + 2x_2 + 3x_3 = 0$$

The Scilab program based on the function 'linsolve' is given next. The main steps of this program are as follows.

- Generate a coefficient matrix corresponding to all the unknown variables, x_1, x_2 and x_3.
- Let this matrix be A such that,

$$A = \begin{bmatrix} 3 & 1 & 1 \\ 2 & -1 & -1 \\ 1 & 2 & 3 \end{bmatrix}$$

- Generate a column matrix corresponding to the numbers in these equations.
- Let this matrix be B such that

$$B = \begin{bmatrix} -8 \\ 3 \\ 0 \end{bmatrix}$$

- The aforementioned two points imply that $AX + B = 0$. The solution matrix X is

$$X = \begin{bmatrix} x_1 \\ x_2 \\ x_3 \end{bmatrix}$$

- The function, '`linsolve`' takes the matrices A and B as its input and gives the values of all the variables of this equation.

```
A = [3 1 1 ; 2 -1 -1 ; 1 2 3];
B = [-8 ; 3 ; 0]
linsolve(A,B)
```

Value of the variables will be equal to, 1, 16, and -11.

8.2.3 Built-in Scilab function – 'fsolve'

The Scilab function, '`fsolve`' is used for solving systems of 'n' linear or non-linear equations that involve 'n' unknown variables. For example, suppose a non-linear equation involving only one variable is given by *Eqn. 8.9*.

$$x^2 - 5 = 0 \tag{8.9}$$

The Scilab program for solving *Eqn. 8.9* is given next. The main steps of this program are as follows.

- Define a function for the polynomial of given equation.

- Give an initial trial value for the unknown variable.

- Call the 'fsolve' function. Its arguments are the unknown variable and the function which is already defined.

```
function[alpha] = F(x)
alpha = x.^2 - 5;
endfunction
x = 1;
fsolve(x,F)
```

The answer will be equal to 2.2360679

Consider another example consisting of two unknown variables. The equations are given by *Eqns. 8.10.*

$$x^2 + y^2 = 13 \qquad\qquad (8.10a)$$

$$x^3 + y^3 = 35 \qquad\qquad (8.10b)$$

The Scilab program is given next. The main steps of this program are as follows.

- Define the polynomials of both the equations in one function, such that the variables are elements of a column matrix. In the program given next,

$$z = \begin{bmatrix} x \\ y \end{bmatrix}$$

- Give an initial trial value of both the variables in the form of a vector.

- Invoke the 'fsolve' command.

```
function z = f(x_y)
z(1) = x_y(1).^2 + x_y(2).^2 - 13;
z(2) = x_y(1).^3 + x_y(2).^3 - 35
endfunction

x_y = [1 2]
fsolve(x_y,f)
```

The answer will be equal to 2 and 3.

8.3 Gauss–Seidel Method

The Gauss–Seidel method is an iterative technique; it may or may not converge to give a solution, but a judicious initial estimate in the iterative method can lead to a faster convergence. As a thumb rule, this method converges for those systems of linear equations where the coefficient matrix of unknown variables is diagonally dominant, i.e., if A is $n \times n$ matrix, then for a diagonally dominant matrix,

- For all k,

$$|A_{kk}| \geq \left\{ \sum_{j=1}^{n} |A_{kj}| \right\}_{k \neq j}$$

- For at least one k,

$$|A_{kk}| > \left\{ \sum_{j=1}^{n} |A_{kj}| \right\}_{k \neq j}$$

- It should be noted that the diagonal dominance of the coefficient matrix is not a necessary condition. It is just a sufficient condition.

- It should also be noted that if a system of linear equations is not diagonally dominant, then a small re-arrangement of the equations can form a diagonally dominant matrix.

The user-defined Scilab function for using the Gauss–Seidel iterative method to solve a system of n equations having n unknown variables is given next.

The steps of this function are as follows.

- This method is used for solving a square system of 'n' linear equations involving an equal number of unknown variables, such that,

  $AX = B$

Here,

- A is the coefficient matrix
- X is the solution vector, such that,

  $X = \left[x_1, x_2, \ldots x_n \right]$

- B is the column matrix corresponding to the constants in the equation

- The input for this function is the augmented matrix formed from the coefficient matrix of the unknown variables and the column matrix corresponding to the constants in the equations.

- If a system of linear equations is given by *Eqns. 8.11*, then its augmented matrix will be given by *Eqn. 8.12*.

$$a_{11}x_1 + a_{12}x_2 = b_1 \tag{8.11a}$$

$$a_{21}x_1 + a_{22}x_2 = b_2 \tag{8.11b}$$

$$\begin{bmatrix} a_{11} & a_{12} & b_1 \\ a_{21} & a_{22} & b_2 \end{bmatrix} \tag{8.12}$$

- The program determines the number of rows in this matrix. It will obviously be equal to the number of unknown variables.

- The trial initial guess value for all the unknown variables is taken as unity. Readers can also change these numbers. The change in numbers will also change the number of iterations required to generate the solution vector.

- The initial limit for tolerance is taken as any number greater than 10^{-10}. In this case, it is taken to be unity. Readers can change this value also and examine the effect.

- This program algebraically solves each linear equation for every variable. The first equation is used for solving the first variable; the second equation is used for the second variable and similarly for others. It thus calculates a new estimate for every variable.

- For example, if the equations are given by

$$a_{11}x_1 + a_{12}x_2 + a_{13}x_3 + \ldots a_{1n}x_n = b_1$$

$$a_{21}x_1 + a_{22}x_2 + a_{23}x_3 + \ldots a_{2n}x_n = b_2$$

$$\vdots \qquad \vdots \qquad \vdots$$

$$a_{n1}x_1 + a_{n2}x_2 + a_{n3}x_3 + \ldots a_{nn}x_n = b_n$$

Then,

$$x_1 = \frac{b_1 - a_{12}x_2 - a_{13}x_3 - \ldots - a_{1n}x_n}{a_{11}} = \frac{b_1 - \sum_{\substack{j=1 \\ j \neq 1}}^{n} a_{1j}x_j}{a_{11}}$$

$$x_2 = \frac{b_2 - a_{21}x_1 - a_{23}x_3 - \ldots - a_{2n}x_n}{a_{22}} = \frac{b_2 - \sum_{\substack{j=1 \\ j \neq 2}}^{n} a_{2j}x_j}{a_{22}}$$

$$x_n = \frac{b_n - a_{n1}x_1 - a_{n2}x_2 - a_{n3}x_3 - \ldots - a_{nn-1}x_{n-1}}{a_{nn}} = \frac{b_n - \sum_{\substack{j=1 \\ j \neq n}}^{n} a_{nj}x_j}{a_{nn}}$$

- These equations can be summarized as in *Eqn. 8.13*.

$$x_i = \frac{b_i - \sum_{\substack{j=1 \\ j \neq i}}^{n} a_{ij} x_j}{a_{ii}}$$

(8.13)

- This program determines the absolute relative error after every iteration. It repeats this process until for every x_i, this error is not more than a pre-defined tolerance for all unknown variables.

```
function [solution] = gauss_seidel(A)
[rows_A,columns_A] = size(A);
n = rows_A;
for i = 1:n;
    x(i,1) = 1;
end
tolerance = 1;
while(tolerance >= 1d-10)
    y = x;
    for i = [1:n];
        sum = 0;
        for j = [1:n];
            if i ~= j then
                sum = sum + A(i,j)*x(j);
            else
                end
        end
        x(i) = (A(i,n+1)-sum)/A(i,i);
    end
    tolerance = abs((x-y)./x);
end
solution = x;
endfunction
```

The usefulness of this function is shown with the help of the following example. Suppose *Eqn. 8.14* gives a system of linear equations,

$$x - y = 5$$

(8.14a)

$$x + 2y = 11$$

(8.14b)

```
//Load the *.sci file which contains function for Guass-
Seidel method.
exec('numerical_techniques.sci',-1)
A = [1 -1 5 ; 1 2 11];
gauss_seidel(A)
```

The solution of *Eqn. 8.14* is 7 and 2.

8.4 Gaussian Elimination Method

In this method, a square matrix of coefficients of 'n' unknown variables, belonging to 'n' linear equations is reduced to upper triangular matrix (row echelon form) through a series of row operations.

The fundamental idea behind the Gaussian elimination method comprises of two steps,

- Forward elimination: Multiples of one equation are added to other equations for eliminating a variable. The process is continued until just one variable is left.

- Back substitution: Value of the final variable is substituted in other equations for evaluating the remaining unknown variables.

The method of Gaussian elimination is explained next with the help of an example. Suppose *Eqns. 8.15* corresponds to a system of linear equations.

$$x - y = 4 \qquad\qquad (8.15a)$$

$$3x + 2y = 22 \qquad\qquad (8.15b)$$

A step wise explanation of this method is as follows.

- *Step 1*: Generate a coefficient matrix for *Eqn. 8.15*. This is done by writing the coefficients of unknown variables in the form of a matrix.

$$A = \begin{bmatrix} 1 & -1 \\ 3 & 2 \end{bmatrix}$$

- *Step 2*: The right-hand side of *Eqn. 8.15* is represented in the form of a column matrix.

$$B = \begin{bmatrix} 4 \\ 22 \end{bmatrix}$$

- *Step 3*: Generate the augmented matrix for *Eqn. 8.15*. This is done by writing the constants on the right side of *Eqn. 8.15* in the form of an extra column in the coefficient matrix. Thus, each row of this matrix will correspond to an equation of the system of equations.

$$C = \begin{bmatrix} 1 & -1 & 4 \\ 3 & 2 & 22 \end{bmatrix}$$

- The aforementioned steps imply the following:
  - The linear system of equations is represented by

    $$AX = B$$

  - The augmented matrix is represented by

    $$C = \begin{bmatrix} A | B \end{bmatrix}$$

- *Step 4*: This method reduces the augmented matrix into an upper triangular matrix $(C_{ij} = 0$ for $i > j)$, or the row echelon form in which, the left most nonzero entry of a row is called the *pivot* (leading coefficient) of that row. Moreover, the leading coefficient of subsequent rows (example second row, second column) lies to the right of the leading coefficient of the previous row (example first row, first column).

- *Step 5*: The variables can be eliminated by performing row operations, as follows:
  - Swapping of any two rows.
  - Multiplication of elements of a row by a non-zero number.
  - Addition of multiples of a row to another row.

- For the system of equations given in *Eqn. 8.15*, the following row operations will result in an upper triangular matrix.

$$\begin{bmatrix} 1 & -1 & 4 \\ 3 & 2 & 22 \end{bmatrix} \xrightarrow{R_2 \to R_2 - 3R_1} \begin{bmatrix} 1 & -1 & 4 \\ 0 & 5 & 10 \end{bmatrix}$$

This implies that

$$5y = 10 \Rightarrow y = 2$$

- *Step 6*: Substituting this value into the first row gives $x = 6$.

The following Scilab function has been written to accomplish the aforementioned method of Gaussian elimination for solving *Eqn. 8.15*. The steps in this program are as follows.

- The inputs to the function are the coefficient matrix and the column matrix corresponding to the constants in the equations.
- Determine the dimension of the coefficient matrix.

- Generate the augmented matrix.

$$\begin{bmatrix} 1 & -1 & 4 \\ 3 & 2 & 22 \end{bmatrix}$$

- Perform the forward elimination method and create an upper triangular matrix by taking $A(1,1)$ as the pivot of the first row.

- If coefficient matrix is 3×3 matrix, then the pivot elements will be $A(1,1)$ followed by $A(2,2)$.

- After the aforementioned step, the function performs backward substitution to obtain the solution set.

- Value of the last variable is determined first. It is followed by the values of its antecedent variables.

```
function [solution] = gauss_elimination(A,B)
[rows_A,columns_A] = size(A);
C = [A B];

//Forward elimination
n = rows_A;
for i = 1:n-1;
    for j = i+1:n;
        for k = i+1:n+1;
            C(j,k) = C(j,k) - C(i,k)*C(j,i)/C(i,i);
        end;
    end;
end;

//Backward substitution
solution(n) = C(n,n+1)/C(n,n);

for i = n-1:-1:1;
    value = 0;
    for j = i+1:n;
        value = value + C(i,j)*solution(j);
    end
    solution(i) = (C(i,n+1)-value)/C(i,i);
end

endfunction
```

Now again refer to the equations mentioned in *Eqns. 8.15*.

$$x - y = 4$$

$$3x + 2y = 22$$

The Scilab program for determination of solution of these equations is written as follows.

```
//Load the *.sci file which contains function for gauss
elimination method
exec('numerical_techniques.sci',-1)
A = [1 -1 ; 3 2];
B = [4 ; 22];

gauss_elimination(A,B)
```

The solution is equal to (6,2).

8.5 'pivoting' Gaussian Elimination Method

In *Section 8.4*, it has been observed that row operations in the Gaussian elimination method are based on dividing the respective elements by their pivot element, which lies on the main diagonal of the augmented matrix. However, if the pivot element is zero, then this method will fail and give an error because of its inability to perform division by zero. In such a condition, it is better to perform *'pivoting'*, wherein before the forward elimination, the rows and columns in the augmented matrix are exchanged such that pivot element is the largest absolute number in a given row and column. Thus, the *'pivoting'* Gaussian elimination method is executed in three main steps,

- *Step 1*: Compare the pivot of the augmented matrix with all the elements of its column. When the largest absolute number in the column is found, the corresponding row and its pivot are exchanged. This will generate a new augmented matrix that differs from the original matrix because of the re-orientation of the linear equations.

- *Step 2*: The new augmented matrix should be reduced to an upper triangular matrix as done earlier.

- *Step 3*: Perform back substitution for obtaining the values of all unknown variables.

The following Scilab user-defined function is based on the *pivoting* Gaussian elimination method explained earlier.

```
function [solution] = gauss_elimination_pivot(A,B)
[rows_A,columns_A] = size(A);
C = [A B];
n = rows_A;
```

```
for i = 1:n-1;
    pivot_new = i;
    max_C = abs(C(i,i));
    for j = i+1:n;
        if abs(C(j,i)) > max_C then
            pivot_new = j;
            max_C = C(i,j);
        end;
    end;
    new_row = C(pivot_new,:);
    C(pivot_new,:) = C(i,:);
    C(i,:) = new_row;
    for j = i+1:n;
        for k = i+1:n+1;
            C(j,k) = C(j,k) - C(i,k)*C(j,i)/C(i,i);
        end;
    end;
end;

solution(n) = C(n,n+1)/C(n,n);
    for i = n-1:-1:1;
        value = 0;
        for j = i+1:n;
            value = value + C(i,j)*solution(j);
        end;
        solution(i) = (C(i,n+1)-value)/C(i,i);
    end;
endfunction
```

The following Scilab program shows the execution of *pivoting* Gaussian elimination method for solving the equations given in *Eqns. 8.16*.

$$3x - 2y + 4z = 39 \tag{8.16a}$$

$$5x + y - 3z = 6 \tag{8.16b}$$

$$x - 4y = -3 \tag{8.16c}$$

```
//Load the *.sci file which contains function for pivot-
ing gauss elimination method
exec('numerical_techniques.sci',-1)
A = [3 -2 4 ; 5 1 -3 ; 1 -4 0];
B = [39 ; 6 ; -3];
gauss_elimination_pivot(A,B)
```

The answer is as follows:

$$x = 5, y = 2, z = 7$$

8.6 Bracketing Method: Bisection Method

The numerical techniques for solving transcendental equations are broadly categorized into *bracketing methods* and *open methods*. The first method is based on the intermediate value theorem, according to which, a continuous function changes its sign on either side of its root. In this method, an initial guess of the solution is made so as to bracket the root. In successive iterations, the width of the bracket (interval) is reduced until the desired accuracy in the estimation of root is reached. This section focuses on the *bisection method*. It is the simplest root-finding *bracketing* algorithm and is applicable to equations of the form

$$f(x) = 0 \tag{8.17}$$

In *Eqn. 8.17*,

- The function $f(x)$ is a continuous function of real variable x.

- The function $f(x)$ is assumed to be defined in interval $[a, b]$. $f(a)$ and $f(b)$ have opposite signs. Therefore, the function has at least one root in the interval $[a,b]$.

The following Scilab function evaluates the approximate solution of an equation by using the bisection method. The algorithm of this function is as follows.

- The function `Bisection_Method(a,b,N,tolerance)` is called inside the main program for a given range of the interval $[a,b]$, a reasonable number of cycles of iterations (N) and tolerance level (difference) between the successive approximations of the solution.

- The function executes the methodology of the bisection method as long as the number of iterations is less than N, or the required tolerance is achieved (whichever is earlier).

- At every step, this method divides (bisects) the interval into two equal sub-intervals by determining its mid-point $\left(\dfrac{a+b}{2}\right)$.

- If $f\left(\dfrac{a+b}{2}\right) = 0$, then the function prints its roots.

- If $f\left(\dfrac{a+b}{2}\right) \neq 0$ & $f(a) * f\left(\dfrac{a+b}{2}\right) > 0$, then the root lies between $f\left(\dfrac{a+b}{2}\right)$ and $f(b)$.
  Therefore the program replaces a by $\left(\dfrac{a+b}{2}\right)$.

- If $f\left(\dfrac{a+b}{2}\right) \neq 0$ & $f(a) * f\left(\dfrac{a+b}{2}\right) < 0$, then the root lies between $f(a)$ and $f\left(\dfrac{a+b}{2}\right)$.
  Therefore, the program replaces b by $\left(\dfrac{a+b}{2}\right)$.

- The process continues until $f\left(\dfrac{a+b}{2}\right) = 0$ or the tolerance level is achieved.

This method converges slowly to the solution because it repeatedly bisects the search interval in every cycle of iteration.

```
function root = Bisection_Method(a,b,N,tolerance)
    i = 1;
while(i <= N)
    approx_root = (a+b)/2
    if((b-a)/2 < tolerance | f(approx_root) == 0)
then
    root = approx_root;
    mprintf("\n Root = %f",root);
    mprintf("\n Number of Iterations = %i",i);
    break
    end
    if(f(a)* f(approx_root) > 0) then
a = approx_root
        else
            b = approx_root
        end
        i = i + 1
    end
endfunction
```

As an example for application of this method, the following Scilab program determines the two roots of the transcendental equation given in *Eqn. 8.18*, that lie in the intervals [0,1] and [1,4].

$$f(x) = \sin x - e^{-x} \tag{8.18}$$

```
//Load the *.sci file which contains function for bisec-
tion method
exec('numerical_techniques.sci',-1)

function func = f(x)              //Define function
func = sin(x) - exp(-x)
endfunction

Result = [];
initial_x = 0; //Initial value of first interval
final_x = 1;            //Final value of first interval
```

```
Root = Bisection_Method(initial_x,final_x,100,1d-4);
Result = [Result, Root];

initial_x = 1;  //Initial value of second interval
final_x = 4;    //Final value of second interval

Root = Bisection_Method(initial_x,final_x,100,1d-4);
Result = [Result, Root];
x = 0:0.05:4;

plot2d(x,f(x))                   //Plot the function
plot2d(Result,f(Result))         //Mark the roots
```

Roots of the given equation are equal to 0.588562 and 3.096283. They have been marked in *Figure 8.1*.

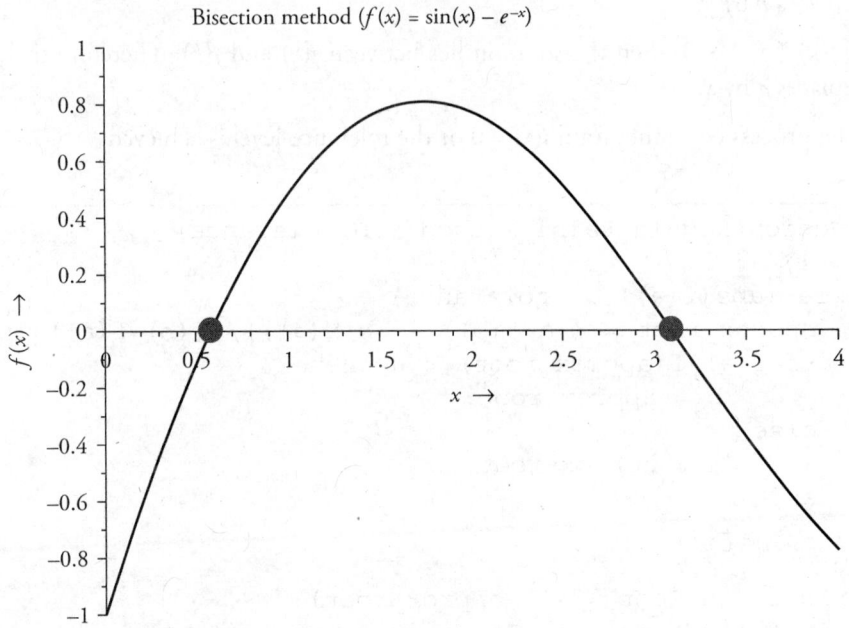

Figure 8.1 Example of Bisection Method

8.7 Bracketing Method: Regula Falsi Method

This method also falls under the category of bracketing methods. It requires two initial approximations a and b, such that, the value of the function ($f(a)$ and $f(b)$) has opposite signs at these two initial guess points. The root of the equation is determined by determining a secant line.

The method of false position is based on the recursive relation given in *Eqn. 8.22*.

$$x = b - f(b) \left\{ \frac{b-a}{f(b)-f(a)} \right\} = \frac{a f(b) - b f(a)}{f(b) - f(a)} \tag{8.22}$$

The following user-defined Scilab function determines the roots of an equation by recursively determining a new approximation until the desired tolerance is achieved.

The algorithm of this method is as follows.

- The function `Regula_Falsi_Method(a,b,tolerance)` is called inside the main program for a given range of the interval [*a*,*b*], a reasonable number of cycles of iteration (N) and tolerance level (difference) between successive approximations of the solution.

- The first approximation (*x*) is determined using *Eqn. 8.22*.

- If *f*(*a*) * *f*(*x*) < 0, then the solution lies between *f*(*a*) and *f*(*x*). Therefore the program replaces *b* by *x*.

- If *f*(*a*) * *f*(*x*) > 0, then the solution lies between *f*(*x*) and *f*(*b*). Therefore the program replaces *a* by *x*.

- The process continues until *f*(*x*) = 0 or the tolerance level is achieved.

```
function Regula_Falsi_Method(a,b,tolerance)
i = 1;
while (abs(f(a)) >= tolerance)
    approx_root = ((a*f(b)) - (b*f(a)))/(f(b)-f(a))
    if f(a)*f(approx_root) < 0 then
        b = approx_root;
    else
        a = approx_root;
    end
    i = i+1;
end
mprintf("\n Root = %f",approx_root);
mprintf("\n Number of Iterations = %i",i);
endfunction
```

The following Scilab program shows the use of the Regula Falsi method for determining the solution of *Eqn. 8.23* in the range [−1, 0].

$$x + e^x = 0 \tag{8.23}$$

```
//Load the *.sci file which contains function for regula
falsi method.
exec('numerical_techniques.sci',-1)

function func = f(x)        //Define the function
func = x + exp(x)
endfunction


Regula_Falsi_Method(-1,0,1e-4)
```

The root of the given equation is -0.567206

8.8 Open Method: Secant Method

Open methods do not rely on changing the sign of a continuous function in the vicinity of a root. They are based on algorithms that begin iteration with an initial guess of the root and then improvise it in successive cycles. This is based on approximating the roots of a given equation by determining secant lines. Similar to the bisection method, this method also requires two trial approximations x_0 and x_1, which are preferably near the expected root (x) of the equation. However in this case, the value of the function at these two initial guess points need not be of opposite sign. If $f(x_0)$ and $f(x_1)$ are values of the given function at these two initial points, then the equation of secant line is given by *Eqn. 8.19*.

$$f(x) - f(x_1) = \frac{f(x_1) - f(x_0)}{x_1 - x_0}(x - x_1)$$ (8.19)

It is obvious that if x is solution of the given equation, then $f(x) = 0$. This gives the recursion relation in *Eqn. 8.20*.

$$x = x_1 - f(x_1)\frac{x_1 - x_0}{f(x_1) - f(x_0)}$$ (8.20)

The user-defined Scilab function for the secant method is as follows. The approximate value of the root is determined recursively until a pre-defined tolerance is achieved.

```
function Secant_Method(x0,x1,tolerance)
i = 1;
while (abs(f(x1)-f(x0)) >= tolerance)
    approx_root = x1 - (f(x1)*(x1 - x0))/(f(x1)-f(x0))
    x0 = x1;
    x1 = approx_root;
```

```
        i = i+1;
    end
    mprintf("\n Root = %f",approx_root);
    mprintf("\n Number of Iterations = %i",i);
    endfunction
```

The following Scilab program shows the use of the secant method for determining the solution of transcendental equation given in *Eqn. 8.21* in the range [–3, –4].

$$\sin(x) + xe^x = 0 \tag{8.21}$$

```
//Load the *.sci file which contains function for secant
method.
exec('numerical_techniques.sci',-1)

function func = f(x)        //Define the function
func = sin(x) + x.*exp(x)
endfunction

Secant_Method(-3,-4,1e-4)
```

Root of the given equation is equal to -3.266500

8.9 Open Method: Newton–Raphson Method

This method is based on determining a line tangent to the curve,

$$y = f(x)$$

Suppose x_0 is an initial guess for the solution of a given function. A more accurate approximation (x_1) is determined using *Eqn. 8.24*.

$$x_1 = x_0 - \frac{f(x_0)}{f'(x_0)} \tag{8.24}$$

The process of determining an approximate solution is repeated until a pre-defined accuracy is achieved. If h is the step size for every iteration, then the derivative of the function is determined by *Eqn. 8.25*.

$$f'(x_0) = \frac{f(x_0 + h) - f(x_0)}{h} \tag{8.25}$$

```
function Newton_Raphson(x0,h,epsilon)
j = 1;
while (abs(f(x0)) > epsilon)
    f_prime = (f(x0+h)-f(x0))/h;
    approx_root = x0-f_prime\f(x0);
    x0 = approx_root;
    j = j + 1;
end
mprintf("\n Root = %f",approx_root);
mprintf("\n Number of Iterations = %i",j);
endfunction
```

The following Scilab program shows the use of the Newton–Raphson method for determining the solution of the transcendental equation given in *Eqn. 8.26*.

$$\cos(x) - 5x^4 = 0 \tag{8.26}$$

```
//Load the *.sci file which contains function for Newton
Raphson method.
exec('numerical_techniques.sci',-1)

function func = f(x)          //Define the function
func = cos(x) - 5*x.^4
endfunction

Newton_Raphson(0.5,1e-4,1e-4)
```

Root of the given equation is 0.633617.

8.10 Applications

8.10.1 Trajectory of a particle

Height (h) of a body thrown in air is a function of time (t) and is given by *Eqn. 8.27* which is quadratic in variable t.

$$h = a + bt + ct^2 \tag{8.27}$$

The value of constants (a, b, c) can be determined using the Gaussian elimination method if height of this object at different times is known. Suppose the data given in *Table 8.1* is known.

Table 8.1 Data for trajectory of a particle

Time (t)	Height (h)
1	2
2	3
3	6

Therefore,

- At $t = 1$,

  $a + b + c = 2$

- At $t = 2$,

  $a + 2b + 4c = 3$

- At $t = 3$,

  $a + 3b + 9c = 6$

Value of the unknown variables can be determined using the following Scilab program.

```
//Load the *.sci file which contains function for gauss
elimination method
exec('numerical_techniques.sci',-1)

A = [1 1 1 ; 1 2 4 ; 1 3 9];
B = [2 ; 3 ; 6];

gauss_elimination(A,B)
```

The answer will be

$a = 3$

$b = -2$

$c = 1$

8.10.2 Matrix inverse

Determination of inverse of matrices is an interesting application of the method of Gaussian elimination. It is based on the fact that if A is a coefficient matrix, X is a solution matrix and B is identity matrix, then, $AX = B$ implies $X = A^{-1}B = A^{-1}$

Therefore in this application of the Gaussian elimination method, the augmented matrix is created by adding an identity matrix (of the same dimension as the coefficient matrix) on the right side of matrix A.

The basic concept behind the following Scilab function is similar to that of the *pivoting* Gaussian elimination method, except the fact that an identity matrix is used to generate the augmented matrix.

```
function [inverse] = gauss_inverse(A)
[rows_A,columns_A] = size(A);
C = [A eye(rows_A,rows_A)];
n = rows_A;
m = columns_A;

for i = 1:n-1;
    pivot_new = i;
    max_C = abs(C(i,i));
    for j = i+1:n;
        if abs(C(j,i)) > max_C then
            pivot_new = j;
            max_C = C(i,j);
        end;
    end;
    new_row = C(pivot_new,:);
    C(pivot_new,:) = C(i,:);
    C(i,:) = new_row;
    for j = i+1:n;
        for k = i+1:n+m;
            C(j,k) = C(j,k) - C(i,k)*C(j,i)/C(i,i);
        end;
    end;
end;

for i = 1:m;
    inverse(n,i) = C(n,n+i)/C(n,n);
    for j = n-1:-1:1;
        value = 0;
        for k = j+1:n;
            value = value + C(j,k)*inverse(k,i);
        end;
        inverse(j,i) = (C(j,n+i) - value)/C(j,j);
    end;
end;

endfunction
```

The execution of the aforementioned user-defined function is shown here with the help of the following matrix.

$$A = \begin{bmatrix} 1 & 2 \\ 3 & 4 \end{bmatrix}$$

```
//Load the *.sci file which contains function for gauss
elimination method
exec('numerical_techniques.sci',-1)

A = [1 2 ; 3 4];
gauss_inverse(A)
```

The answer will be

$$A^{-1} = \begin{bmatrix} -2 & 1 \\ 1.5 & -0.5 \end{bmatrix}$$

8.10.3 Determinant of a matrix

The method of Gaussian elimination for computing the determinant of a square matrix, is based on the following rules of elementary row operations:

- Value of the determinant changes sign when elements of two rows are swapped.

- Value of the determinant is scaled by a non-zero scalar number when elements of a row are multiplied by that number.

- Value of the determinant is unchanged when scalar multiple of one row is added to another.

In the method of Gaussian elimination, a square matrix A is reduced to a row echelon matrix B by forward elimination. The product of diagonal elements of this matrix gives the determinant of square matrix A. In the case of *pivoting* Gaussian elimination method, the value of the determinant is based on row operations because its sign changes after every row operation. The following Scilab function can be used to determine the determinant of a square matrix.

```
function [determinant] = gauss_determinant(A)
[rows_A,columns_A] = size(A);
n = rows_A;

counter = 0;
for i = 1:n-1;
    pivot_new = i;
```

```
        max_A = abs(A(i,i));
        for j = i+1:n;
            if abs(A(j,i)) > max_A then
                pivot_new = j;
                max_A = A(i,j);
                counter = counter+1;
            end;
        end;
        new_row = A(pivot_new,:);
        A(pivot_new,:) = A(i,:);
        A(i,:) = new_row;
        for j = i+1:n;
            for k = i+1:n;
                A(j,k) = A(j,k) - A(i,k)*A(j,i)/A(i,i);
            end;
        end;
    end;

    determinant = (-1)^counter*prod(diag(A))
endfunction
```

The execution of the aforementioned user-defined function is shown herewith the help of the following square matrix.

$$A = \begin{bmatrix} 2 & -3 & 1 \\ 2 & 0 & -1 \\ 1 & 4 & 5 \end{bmatrix}$$

```
//Load the *.sci file which contains function for
determinant of a matrix
exec(' numerical_techniques.sci',-1)
A = [2 -3 1 ; 2 0 -1 ; 1 4 5];
gauss_determinant(A)
```

The answer will be 49.

8.10.4 Fraunhofer diffraction pattern

In Fraunhofer diffraction through a single slit, the secondary maxima occur at positions that are given by roots of the transcendental equation given in *Eqn. 8.28*.

$$x = \tan x \tag{8.28}$$

The solution of *Eqn. 8.28* can be easily determined using any of the numerical methods discussed in *Sections 8.6–8.9*.

The following Scilab program plots these functions (*Figure 8.2*) and calculates approximate solution of the equation lying in the x-range $[-2\pi, 2\pi]$ by using the bisection method. The tolerance level is taken to be equal to 10^{-4}.

```
//Load the *.sci file containing user defined functions
for the bisection method
exec('numerical_techniques.sci',-1)

function func = f(x)                    //Define function
func = x - tan(x)
endfunction

x = -2*%pi:0.1: 2*%pi;                  //Plot the two functions
plot2d(x,x)
plot2d(x,tan(x));

answer = [];
initial_x = -2*%pi;
final_x = -%pi;
for i = 1:3;
Root = Bisection_Method(initial_x,final_x,100,1d-4);
answer(i) = Root
initial_x = initial_x+%pi;
final_x = final_x+%pi;
end

initial_x = %pi/2;
final_x = 3*%pi/2;
for i = 4:5
Root = Bisection_Method(initial_x,final_x,100,1d-4);
answer(i) = Root
initial_x = initial_x+%pi;
final_x = final_x+%pi;
end

plot2d(answer,answer)                   //Mark the roots
```

The roots occur at - 4.7124849, - 1.5708922, 0, 1.5708922 and 4.7124849. The root at $x = 0$, is ignored because it coincides with the primary maxima.

8.10.5 Bound state of proton and neutron

Deuteron is the bound state of a proton and neutron. The one-dimensional quantum mechanical model can be approximated by choosing a finite square potential well,

$$V(x) = \begin{cases} 0 & \text{for } |x| \le a \\ V_0 & \text{for } |x| > a \end{cases} \tag{8.29}$$

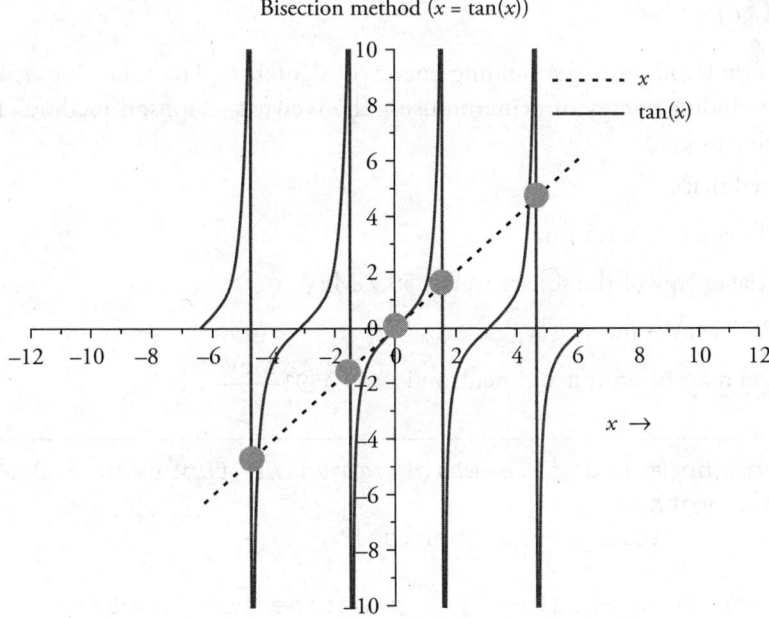

Figure 8.2 Application of numerical methods

In *Eqn. 8.29*,

- a is the size of the deuteron
- V_0 is the depth of the well and its strength depends on the neutron–proton potential

The one-dimensional Schrödinger's wave equation is given by *Eqn. 8.30*.

$$-\frac{\hbar^2}{2m}\frac{d^2 u}{dx^2} + V(x)u(x) = Eu(x) \tag{8.30}$$

Here, m is the reduced mass of proton and neutron. Analytical solution for the bound states is given by *Eqn. 8.31*.

$$u(x) = \begin{cases} A\sin(k_1 x) & \text{for } |x| \le a \\ Be^{-k_2 x} & \text{for } |x| > a \end{cases} \tag{8.31}$$

In *Eqn. 8.31*,

- $k_1 = \dfrac{\sqrt{2m(V_0 - E)}}{\hbar}$

- $k_2 = \dfrac{\sqrt{2mE}}{\hbar}$

The continuity condition at $x = a$ leads to the transcendental equation,

$$k_1 a \cot(k_1 a) = -k_2 a \tag{8.32}$$

The root of *Eqn. 8.32* gives the binding energy of deuteron. The following Scilab program calculates the binding energy of deuteron using the Newton–Raphson method. The graph is shown in *Figure 8.3*.

It is assumed that

- Size of deuteron is 1.5 fm.

- Potential energy of the square well is 59.7 MeV

- $\hbar c = 197.3 \text{MeV} - \text{fm}$

- Reduced mass of proton and neutron is $469.4591 \dfrac{\text{MeV}}{c^2}$

```
//Load the *.sci file which contains function for Newton
Raphson method
exec('numerical_techniques.sci',-1)

function func = f(E)        //Define the function
k = sqrt(2*m.*(V-E))/hbar;
kappa = sqrt(2*m.*E)/hbar;
func = k.*cotg(k*a) + kappa;
endfunction

E = 0.1:0.1:5;
V = 59.7;                   //Potential energy (in MeV)
hbar = 197.3;
a = 1.5;                                //Size of deuteron
                                        //        MeV
m = 469.4591;                           //Reduced (in ---- )
                                        //         c^2
plot2d(E,f(E))                          //Plot the function

Newton_Raphson(1.5,1e-4,1e-4)

plot2d(2.242206,0)
```

The binding energy of deuteron will be 2.242206 MeV.

8.10.6 Central angle of an elliptical orbit

At any time t, Kepler's equation for an elliptical orbit having a period P, eccentricity e and central angle θ is given by *Eqn. 8.33*.

$$2\pi t = P\theta - Pe \sin\theta \tag{8.33}$$

The numerical techniques discussed in *Sections 8.6–8.9* can be used to find the central angle from the periapsis position at different elapsed times.

For example, suppose the eccentricity of the elliptical orbit is 0.4 and the period of revolution is 50 days, then for time $t = 20$ days, Kepler's equation reduces to *Eqn. 8.34*.

$$40\pi = 50\theta - 20\sin\theta \tag{8.34}$$

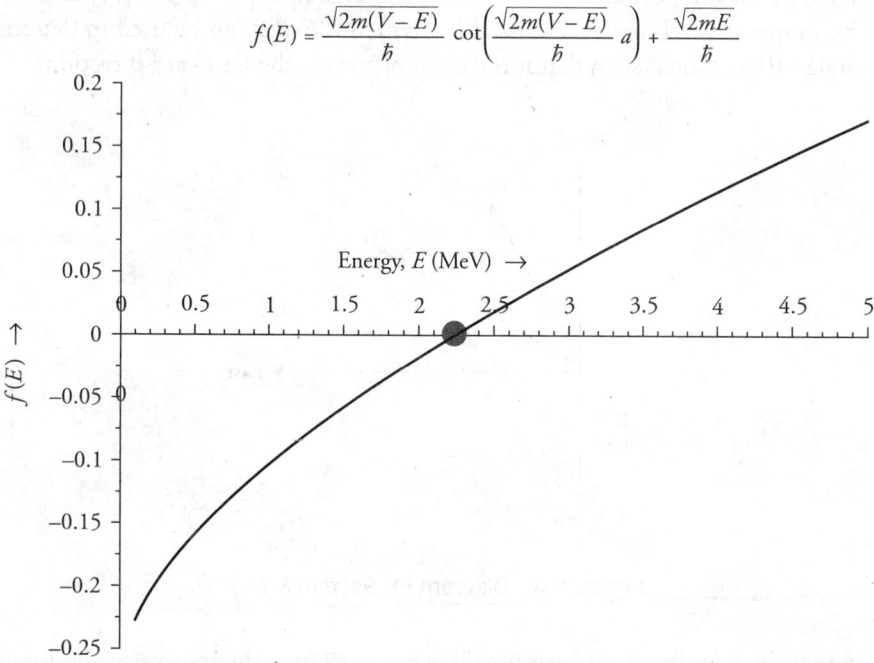

$$f(E) = \frac{\sqrt{2m(V-E)}}{\hbar} \cot\left(\frac{\sqrt{2m(V-E)}}{\hbar}a\right) + \frac{\sqrt{2mE}}{\hbar}$$

Figure 8.3   Estimation of binding energy of deuteron

The following Scilab program determines the value of the central angle for the given elliptical orbit after 20 days from the periapsis.

```
//Load the *.sci file which contains functions for all
the numerical techniques
exec('numerical_techniques.sci',-1)
```

```
function alpha = f(x)
alpha = 40.*%pi - 50.*x + 20.*sin(x)
endfunction

Bisection_Method(2,3,100,1d-5);
Newton_Raphson(1,1e-4,1e-5)
Secant_Method(1,2,1e-4)
Regula_Falsi_Method(0,1,1e-6)
```

The answer will be $2.688408^{rad}\left(=154.11256°\right)$.

8.10.7 Bearing angle of a boat

Suppose a river is flowing towards the South direction (as shown in *Figure 8.4*) at a velocity of 3 m/s. The numerical techniques discussed in *Sections 8.6–8.9* can be used to determine the bearing angle (θ) of a boat such that it travels at 30° w.r.t. the Eastward direction.

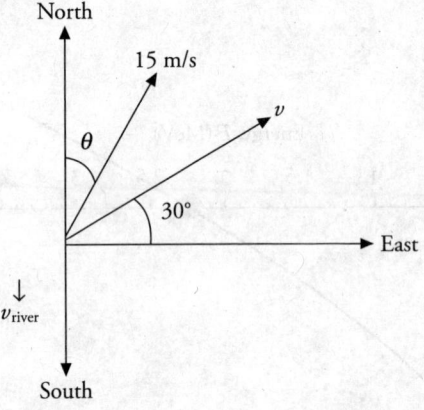

Figure 8.4 Diagram for *Section 8.10.7*

Assuming that the velocity of the boat in still water is 15 m/s, the transverse and longitudinal components of velocities are given by *Eqns. 8.35*.

$$v\cos 30 = 15\sin\theta \tag{8.35a}$$

$$v\sin 30 = -3 + 15\cos\theta \tag{8.35b}$$

After elimination of v, these equations will give

$$15\sin\theta = -3\sqrt{3} + 15\sqrt{3}\cos\theta \tag{8.36}$$

The following Scilab program determines the value of the bearing angle by solving *Eqn. 8.36.*

```
//Load the *.sci file which contains functions for all
the numerical techniques
exec('numerical_techniques.sci',-1)

function alpha = f(x)
alpha = 15.*sin(x) + 3.*sqrt(3) - 15.*sqrt(3).*cos(x)
endfunction

Bisection_Method(0,1,100,1d-5);
Newton_Raphson(1,1e-4,1e-5)
Secant_Method(1,2,1e-4)
Regula_Falsi_Method(0,1,1e-6)
```

The bearing angle will be equal to $0.873115^{rad} \left(= 50.02586^{\circ} \right)$.

8.11 Exercises

1. Write a Scilab program to find solutions of the following equations by making use of the Gauss–Seidel method. Discuss the problems in using this method if any.
    a) $x + 2y = 11$
        $x - y = 5$
    b) $3x + y + z = 8$
        $x + 5y - 3z = 2$
        $2x - y + 4z = 12$
    c) $2x + 4y + 6z = 14$
        $3x - 2y + z = -3$
        $4x + 2y - z = -4$

2. Write a Scilab program to find solutions of the following equations by making use of the method of Gaussian elimination.
    a) $4x - 2y = 6$
        $2x + y = 45$
    b) $6x - 3y = -21$
        $x + 5y = 46$
    c) $x + y - z = 6$
        $4x - y + 5z = 8$
        $3x + 2y - 2z = 14$
    d) $2y + 3z = 13$
        $x - 2z = -5$

$$4x + 3y = 10$$

e) $x + 3y = 9$

$$x + 3z = -3$$
$$2y + z = 2$$

f) $z + x = 4$

$$z + y = 5$$
$$x + y = 3$$

3. Find the inverse of matrix A by making use of the Gaussian elimination technique.

$$A = \begin{bmatrix} 3 & 4 & 0 \\ 6 & 8 & 2 \\ 1 & 1 & 3 \end{bmatrix}$$

4. Write a Scilab program to solve the following equations by making use of the Newton–Raphson method.

   a) $5x + \log x = 100$

   b) $e^x = x^3$

   c) $2^x = x^2$

   d) $x \cos x + \sin x = 0$

5. Write a Scilab program to plot the following function and determine its roots by making use of all the four numerical methods discussed in this chapter.

   $x \tan x - 1 = 0$

6. Write a Scilab program to plot the following functions. Also solve and plot the transcendental equation, $f(x) = g(x)$ and mark the root on the graph.

   • $f(x) = x$

   • $g(x) = 2 \sin x$

7. Write a Scilab program to find the value of following expressions.

   a) $\sqrt{3}$

   b) $\sqrt{\dfrac{3}{4}}$

   c) $\sqrt[3]{2}$

   d) $\sqrt[4]{0.8}$

   e) $\sqrt[5]{5}$

8. Write a Scilab program to solve the following equations by making use of the bisection and the Newton–Raphson method.

   a) $2x^3 - x^2 - 5x + 1 = 0$

   b) $5x^4 - 13x^3 = 1$

9. Write a Scilab program to find the positive roots of the following Legendre polynomials using the bisection method.

   a) $\dfrac{3x^2 - 1}{2}$

   b) $\dfrac{5x^3 - 3x}{2}$

   c) $\dfrac{35x^4 - 30x^2 + 3}{8}$

   d) $\dfrac{63x^5 - 70x^3 + 15x}{8}$

10. The equation of motion of an object is given by $s(t) = t^3 - 81t$. Write a Scilab program to determine the moment when the velocity of the object becomes zero.

Appendix

1 Matrices and Vector Spaces (vectors.sci)

1.1 Distance between two points

```
function distance = distance_between_points(A,B)
AB = A - B;
distance = (AB(1)^2 + AB(2)^2 + AB(3)^2)^0.5;
endfunction
```

1.2 Coordinate conversion

```
function [coordinates] = cartesian_to_cylindrical(A)
x = A(1);
y = A(2);
z = A(3);
r = (x.^2 + y.^2)^0.5;
// theta = atan(y/x);   //angle in radian
theta = atand(y/x);   //angle in degree
if theta < 0 then
theta = theta + 180
end
coordinates = [r theta z]
endfunction
```

```
function [coordinates] = cartesian_to_spherical(A)
x = A(1);
y = A(2);
z = A(3);
r = (x.^2 + y.^2 + z.^2)^0.5;
theta = atand(((x.^2 + y.^2)^0.5)/(z));
phi = atand(y/x);
coordinates = [r theta phi]
endfunction
```

2 Plotting and Graphics Design (plot.sci)

2.1 Formatting of coordinate axes

```
function set_my_axes(fontsize, font_color, fontstyle)
    a = get("current_axes")
    a.font_size = fontsize;
    a.labels_font_color = font_color;
    a.font_style = fontstyle;
    a.y_location = "origin";
    a.x_location = "origin";
endfunction
```

2.2 Formatting of the line styles

```
function set_my_line_styles(style, thickness)
    e = gce();
    e.children.thickness = thickness;
    e.children.line_style = style;
endfunction
```

2.3 Formatting of the markers

```
function set_my_line_styles_mark(thickness,markstyle,mar
ksize,color)
    e = gce();
    e.children.thickness = thickness;
    e.children.mark_style = markstyle;
    e.children.mark_size = marksize;
    e.children.mark_foreground = color;
    e.children.line_mode ='off';
endfunction
```

2.4 Formatting of the legend

```
function set_my_legend(size, style, color, mode)
    a=get("current_axes");
    legend=a.children(1);
    legend.font_size = size;
    legend.font_style = style;
    legend.font_color = color;
    legend.line_mode = mode
endfunction
```

3 Least Square Curve Fitting

3.1 Exponential Fitting

```
function [f,m,c] = exponential_fit(x,y)
n = length(x);
for i = 1:n
    z(i) = log(y(i));
end

x1 = sum(x);
x2 = sum(x.*x);
x1y1 = sum(x'.*z);
y1 = sum(z);

A = [x2 x1; x1 n];
B = [x1y1; y1];
C = A\B;
m = C(1);
c = C(2);
c = exp(c);

for i = 1:n
    f(i) = c*exp(m*x(i));
end
endfunction
```

3.2 Polynomial Fitting

```
function alpha = polynomial_fit(x,y,k)
format(6);
n = length(x);
X(1) = n;

for i = 1:2*k
X(i+1) = sum(x.^i);
end


for i = 1:k+1
    for j = 1:k+1
        A(i,j) = X(i+j-1);
    end
end


for i = 0:k
    B(i+1) = sum(x.^i.*y);
end
B = -B;


alpha = linsolve(A,B);
endfunction
```

4 Ordinary Differential Equation (differentiation.sci)

4.1 Euler's Method (for first order differential equation)

```
function [x,y] = euler(initial_x,initial_y,h,final);
i = 1;
x(i) = initial_x;
y(i) = initial_y;
while(x(i) < final);
x(i+1) = x(i) + h;
y(i+1) = y(i) + (f(x(i),y(i)).*h);
disp('Euler Method : Value of y at x = ' +string(x(i))+ '
is : ' +string(y(i)));
i = i+1;
end
endfunction
```

4.2 Euler's Method (for second order differential equation)

```
function [t,x,y] = euler2(initial_t,initial_x,initial_y,
h,final);
i = 1;
t(i) = initial_t;
x(i) = initial_x;
y(i) = initial_y;
while (t(i) < final);
t(i+1) = t(i) + h;
x(i+1) = x(i) + (f1(t(i), x(i), y(i)).*h);
y(i+1) = y(i) + (f2(t(i), x(i), y(i)).*h);
i = i+1;
end
endfunction
```

4.3 Modified Euler's Method

```
function[x,y] = modeuler(initial_x,initial_y,h,final)
i = 1;
x(i) = initial_x;
y(i) = initial_y;
while(x(i) < final);
x(i+1) = x(i) + h;
y(i+1) = y(i) + (f(x(i),y(i)).*h);
z = y(i) + (h/2).*(f(x(i),y(i)) + f(x(i+1),y(i+1)));
// disp('Modified Euler Method : Value of y at x = '
+string(x(i+1))+ ' is : ' +string(z));
i = i+1;
end
endfunction
```

4.4 Second order Runge–Kutta Method (for first order differential equation)

```
function [x,y] = rk2(initial_x,initial_y,h,final);
i = 1;
x(i) = initial_x;
y(i) = initial_y;
while (x(i) < final);
x(i+1) = x(i) + h;
k1 = h.*(f(x(i),y(i)));
```

```
k2 = h.*(f(x(i)+h,y(i)+k1));
y(i+1) = y(i)+((k1+k2)./2);
// disp('Second Order Runge-Kutta : Value of y at x = '
+string(x(i))+ ' is : ' +string(y(i)));
i = i+1;
end
endfunction
```

4.5 Fourth Order Runge–Kutta Method (for first order differential equation)

```
function [x,y] = rk4(initial_x,initial_y,h,final);
i = 1;
x(i) = initial_x;
y(i) = initial_y;
while (x(i) < final);
x(i+1) = x(i) + h;
k1 = h.*(f(x(i),y(i)));
k2 = h.*(f(x(i) + (h/2),y(i) + (k1/2)));
k3 = h.*(f(x(i) + (h/2),y(i) + (k2/2)));
k4 = h.*(f(x(i) + h, y(i) + k3));
y(i+1) = y(i) + ( ((k1 + (2.*(k2+k3)) + k4))./6 );
// disp('Fourth Order Runge-Kutta : Value of y at x = '
+string(x(i))+ ' is : ' +string(y(i)));
i = i+1;
end
endfunction
```

4.6 Fourth Order Runge–Kutta Method (for second order differential equation)

```
function [t,x,y] = rk42(initial_t,initial_x,initial_y,h,
final);
i = 1;
t(i) = initial_t;
x(i) = initial_x;
y(i) = initial_y;
while (t(i) < final);
t(i+1) = t(i) + h;
k1 = h.*(f1(t(i),x(i),y(i)));
l1 = h.*(f2(t(i),x(i),y(i)));

k2 = h.*(f1(t(i) + (h/2),x(i) + (k1/2),y(i) + (l1/2)));
l2 = h.*(f2(t(i) + (h/2),x(i) + (k1/2),y(i) + (l1/2)));
```

```
k3 = h.*(f1(t(i) + (h/2),x(i) + (k2/2),y(i) + (12/2)));
13 = h.*(f2(t(i) + (h/2),x(i) + (k2/2),y(i) + (12/2)));

k4 = h.*(f1(t(i) + h, x(i) + k3, y(i) + 13));
14 = h.*(f2(t(i) + h, x(i) + k3, y(i) + 13));

x(i+1) = x(i) + ( ((k1 + (2.*(k2+k3)) + k4))./6 );
y(i+1) = y(i) + ( ((11 + (2.*(12+13)) + 14))./6 );

// disp('Fourth Order Runge-Kutta : Value of y at x = '
+string(x(i))+ ' is : ' +string(y(i)));
i = i+1;
end
endfunction
```

4.7 Finite Difference Method

```
function [x, Y] = finite_diff(a,b,h,ya,yb,f,g,r)
n = ((b-a)/h)+1;
A = zeros(n,n)
A(1,1) = 1;
A(n,n) = 1;

x0 = a;
x = a;
for i = 2:n-1;
    x = x + h;
    x0 = [x0 x];
    A(i,i) = g(x) - (2/(h*h));
    A(i,i+1) = (1 + (0.5*h*f(x)))/(h*h);
    A(i,i-1) = (1 - (0.5*h*f(x)))/(h*h);
    B(i,1) = r(x);
end

B(1,1) = ya;
B(n,1) = yb;

Y = A\B;
x = [x0 b];
endfunction
```

5 Integration and Differentiation (integrate.sci)

5.1 Trapezoidal Rule

```
function Y = trapezoidal(f,a,b,N)
h = (b - a)/(N-1);
x = linspace(a,b,N);
y = feval(x,f);
Y = (y(1) + 2.*sum(y(2:N-1)) + y(N)).*(h/2);
endfunction
```

5.2 Simpson's 1/3 – Rule

```
function Y = simpson_1_3(f,a,b,N)
h = (b - a)/(N-1);
x = linspace(a,b,N);
y = feval(x,f)
Y = (y(1) + 2.*sum(y(3:2:N-2)) + 4.*sum(y(2:2:N-1)) +
y(N)).*(h/3.0);
endfunction
```

5.3 Simpson's 3/8 – Rule

```
function Y = simpson_3_8(f,a,b,N)
h = (b-a)/(N-1);
x = linspace(a,b,N);
y = feval(x,f)
Y = (y(1) + 3.*sum(y(2:3:N-2)) + 3.*sum(y(3:3:N-1)) +
2.*sum(y(4:3:N-3)) + y(N)).*(3.*h/8.0);
endfunction
```

5.4 Line Integral in Cylindrical Coordinates

```
function line_int = line_integral(radius_1,radius_2,thet
a_1,theta_2,height_1,height_2)
if (radius_1 == radius_2) & (height_1 == height_2) then
line_int = radius_1.*integrate('1','phi',theta_1,the
ta_2)

elseif (radius_1 == radius_2) & (theta_1 == theta_2)
then
```

```
line_int = integrate('1','z',height_1,height_2)

elseif (theta_1 == theta_2) & (height_1 == height_2)
then
line_int = integrate('1','r',radius_1,radius_2);
end

endfunction
```

5.5 Surface Integral in Cylindrical Coordinates

```
function surface_int = surface_integral(radius_1,radius_
2,theta_1,theta_2,height_1,height_2)
if (radius_1 == radius_2) then
alpha_z = integrate('1','z',height_1,height_2)
surface_int = radius_1.*alpha_z.*integrate('1','phi',the
ta_1,theta_2)
elseif (theta_1 == theta_2) then
alpha_z = integrate('1','z',height_1,height_2)
surface_int = alpha_z.*integrate('1','r',radius_1,radi
us_2)
elseif (height_1 == height_2) then
alpha_phi = integrate('1','phi',theta_1,theta_2);
surface_int = alpha_phi.*integrate('r','r',radius_1,radi
us_2);
end
endfunction
```

5.6 Volume Integral in Cylindrical Coordinates

```
function volume_int = volume_integral(radius_1,radius_2,
theta_1,theta_2,height_1,height_2)
radial = integrate('r','r',radius_1,radius_2)
angular = integrate('1','phi',theta_1,theta_2)
z_direction = integrate('1','z',height_1,height_2);
volume_int = radial.*angular.*z_direction
endfunction
```

6 Special Functions (special_func.sci)

6.1 Legendre Polynomials

6.1.1 Function for the recursion relation

```
function y = Legendre_polynomial(n)
    if (n == 0) then
        y = poly(1,"x","coeff");
    elseif (n == 1) then
        y = poly([0 1],"x","coeff")
    else
        polynomial_x = poly([0 1],"x","coeff")
y_n_minus_2 = poly(1,"x","coeff")
y_n_minus_1 = poly([0 1],"x","coeff")
for i = 2:n;
y = ((2*i - 1)*polynomial_x*y_n_minus_1 - (i-1)*y_n_
minus_2)/i;
y_n_minus_2 = y_n_minus_1
y_n_minus_1 = y
end
end
endfunction
```

6.1.2 Function for the summation series

```
function Legendre = legendre_poly_gamma(n,var)
if n == 0 then
        cc = [1];
elseif n == 1 then
        cc = [0 1];
else
        if modulo(n,2) == 0 then
                M = n/2
        else
                M = (n-1)/2
        end;
        cc = zeros(1,M+1);
        for m = 0:M
            k = n-2*m;
            cc(k+1)=(-1)^m*gamma(2*n-2*m+1)/
(2^n*gamma(m+1)*gamma(n-m+1)*gamma(n-2*m+1));
        end;
end;
Legendre = poly(cc,var,'coeff');
endfunction
```

6.2 Laguerre Polynomials

6.2.1 Function for the recursion relation

```
function y = Laguerre_polynomial(n)
    if (n == 0) then
        y = poly(1,"x","coeff");
    elseif (n == 1) then
        y = poly([1 -1],"x","coeff")
    else
        polynomial_x = poly([0 1],"x","coeff")
y_n_minus_2 = poly(1,"x","coeff")
y_n_minus_1 = poly([1 -1],"x","coeff")
for i = 2:n;
y = ((2*i - 1 - polynomial_x)*y_n_minus_1 - (i-1)*y_n_
minus_2)/i
y_n_minus_2 = y_n_minus_1
y_n_minus_1 = y
end
end
endfunction
```

6.2.2 Function for the summation series

```
function Laguerre = Laguerre_poly_gamma(n,var)
if n == 0 then
    solution = [1];
else
    solution = [];
    for m = 0:n;
        solution = [solution (-1)^m*gamma(n+1)/
((gamma(m+1))^2*gamma(n-m+1))];
    end;
end;
Laguerre = poly(solution,var,"coeff");
endfunction
```

6.3 Hermite Polynomials

6.3.1 Function for the recursion (probabilists')

```
function y = Hermite_polynomial_prob(n)
    if (n == 0) then
        y = poly(1,"x","coeff");
```

```
      elseif (n == 1) then
          y = poly([0 1],"x","coeff")
      else
          polynomial_x = poly([0 1],"x","coeff")
y_n_minus_2 = poly(1,"x","coeff")
y_n_minus_1 = poly([0 1],"x","coeff")
for i = 2:n;
y = polynomial_x*y_n_minus_1 - (i-1)*y_n_minus_2;
y_n_minus_2 = y_n_minus_1
y_n_minus_1 = y
end
end
endfunction
```

6.3.2 Function for the recursion (physicists')

```
function y = Hermite_polynomial_phys(n)
H = zeros(1,n+1);
H(1) = poly([1],"x",'coeff');
H(2) = poly([0 2],"x",'coeff');
for alpha = 2:n
H(alpha+1) = poly([0 2],"x",'coeff')*H(alpha) - 2*(al-
pha-1)*H(alpha-1);
end;
y = H(n+1);
endfunction
```

6.4 Gauss–Legendre Quadrature

```
function y = gauss_legendre(f,a,b,n)
p  = legendre_poly_gamma(n,'x');
xroots = roots(p);
w  = [];
for j = 1:n
    poly_deriv = derivat(p);
    w = [w 2/((1- xroots(j)^2)*(horner(poly_deriv,
xroots(j)))^2)];
end;
arg = ((b-a)/2.* xroots)+((b+a)/2);
y = (b-a)/2*w*f(arg);
endfunction
```

6.5 Gauss–Laguerre Quadrature

```
function y = gauss_laguerre(f,n)
p = Laguerre_poly_gamma(n,'x');
p_n_plus_1 = Laguerre_poly_gamma(n+1,'x');
xroots = roots(p);
w = [];
for i = 1:n
w = [w xroots(i)/((n+1)^2*(horner(p_n_plus_1,
xroots(i)))^2)];
end;
y = w*(exp(xroots).*f(xroots));
endfunction
```

6.6 Gauss–Hermite Quadrature

```
function y = gauss_hermite(f,n)
p = Hermite_polynomial_phys(n);
p_n_minus_1 = Hermite_polynomial_phys(n-1);
xroots = roots(p);
w = [];
for j = 1:n
w = [w 2^(n-1)*gamma(n+1)*sqrt(%pi)/(n^2*horner(p_n_mi-
nus_1, xroots(j))^2)];
end;
y = w*(exp(xroots.^2).*f(xroots));
endfunction
```

7 Fourier analysis (fourier.sci)

7.1 Periodic Functions

Case (I) A function is defined in the interval $[-T, T]$.

```
function alpha = periodic1(f,T,x)
    if (x >= -T) & (x <= T) then
        alpha = f(x);
    elseif x < -T then
        new_x = x + 2.*T;
        alpha = periodic1(f,T,new_x);
    elseif x > T then
        new_x = x - 2*T;
```

```
        alpha = periodic1(f,T,new_x);
    end
endfunction
```

Case (II) The function is defined in the interval $[0, 2T]$.

```
function alpha = periodic2(f,T,x)
    if (x >= 0) & (x <= 2.*T) then
        alpha = f(x);
    elseif x < 0 then
        new_x = x + 2.*T;
        alpha = periodic2(f,T,new_x);
    elseif x > T then
        new_x = x - 2.*T;
        alpha = periodic2(f,T,new_x);
    end
endfunction
```

7.2 Fourier Series

Case (I) A function is defined in the interval $[-L, L]$.

```
function [a0,a,b] = fourier1(period,harmonics,f)
    period = period/2;

    function y = fcos(x,m)
        y = f(x)*cos((x*m*%pi)/period);
    endfunction

    function y = fsin(x,m)
        y = f(x)*sin((x*m*%pi)/period);
    endfunction

    a0 = (intg(-period,period,f,0.001))./(2*period)

    for m = 1:harmonics;
        a(m) = (1/period)*(intg(-
period,period,fcos,0.001));
        b(m) = (1/period)*(intg(-
period,period,fsin,0.001));
    end

    sum = a0;
```

```
      for i = 1:harmonics;
        sum = sum + (a(i).*cos(i.*x.*%pi/period)) +
          (b(i).*sin(i.*x.*%pi/period));
      end
      plot2d(x,sum,5)
endfunction
```

Case (II) The function is defined in the interval [0, 2L].

```
function [a0,a,b] = fourier2(period,harmonics,f)
        function y = fcos(x,m)
        y = f(x)*cos((x*m*(%pi))/(0.5*period));
        endfunction
    function y = fsin(x,m)
        y = f(x)*sin((x*m*(%pi))/(0.5*period));
    endfunction

    a0 = (intg(0,period,f,0.001)).*(1/(period))
    for m = 1:harmonics;
        a(m) = (1/(0.5*period))*(intg(0,period,fc
os,0.001));
        b(m) = (1/(0.5*period))*(intg(0,period,fs
in,0.001));
end

    sum = a0;
    for i = 1:harmonics;
        sum = sum + (a(i).*cos(i.*x.*%pi/(0.5*period)))
+                            (b(i).*sin(i.*x.*%pi/
(0.5*period)));
    end

    plot2d(x,sum,5)
endfunction
```

8 Algebraic and Transcendental Equations (numerical_ techniques.sci)

8.1 Gauss–Seidel Method

```
function [solution] = gauss_seidel(A)
[rows_A,columns_A] = size(A);
```

```
n = rows_A;
for i = 1:n;
    x(i,1) = 1;
end
tolerance = 1;
while(tolerance >= 1d-10)
    y = x;
    for i = [1:n];
        sum = 0;
        for j = [1:n];
            if i ~= j then
                sum = sum + A(i,j)*x(j);
            else
                end
            end
        x(i) = (A(i,n+1)-sum)/A(i,i);
    end
    tolerance = abs((x-y)./x);
end
solution = x;
endfunction
```

8.2 Gaussian Elimination Method

```
function [solution] = gauss_elimination(A,B)
[rows_A,columns_A] = size(A);
C = [A B];

//Forward elimination
n = rows_A;
for i = 1:n-1;
    for j = i+1:n;
        for k = i+1:n+1;
            C(j,k) = C(j,k) - C(i,k)*C(j,i)/C(i,i);
        end;
    end;
end;

//Backward substitution
solution(n) = C(n,n+1)/C(n,n);

for i = n-1:-1:1;
    value = 0;
```

```
    for j = i+1:n;
         value = value + C(i,j)*solution(j);
    end
    solution(i) = (C(i,n+1)-value)/C(i,i);
end
endfunction
```

8.3 'pivoting' Gaussian Elimination Method

```
function [solution] = gauss_elimination_pivot(A,B)
[rows_A,columns_A] = size(A);
C = [A B];
n = rows_A;

for i = 1:n-1;
    pivot_new = i;
    max_C = abs(C(i,i));
    for j = i+1:n;
        if abs(C(j,i)) > max_C then
             pivot_new = j;
             max_C = C(i,j);
        end;
    end;
    new_row = C(pivot_new,:);
    C(pivot_new,:) = C(i,:);
    C(i,:) = new_row;
    for j = i+1:n;
        for k = i+1:n+1;
             C(j,k) = C(j,k) - C(i,k)*C(j,i)/C(i,i);
        end;
    end;
end;

solution(n) = C(n,n+1)/C(n,n);
    for i = n-1:-1:1;
        value = 0;
        for j = i+1:n;
             value = value + C(i,j)*solution(j);
        end;
        solution(i) = (C(i,n+1)-value)/C(i,i);
    end;
endfunction
```

8.4 Inverse of a Matrix

```
function [inverse] = gauss_inverse(A)
[rows_A,columns_A] = size(A);
C = [A eye(rows_A,rows_A)];
n = rows_A;
m = columns_A;

for i = 1:n-1;
    pivot_new = i;
    max_C = abs(C(i,i));
    for j = i+1:n;
        if abs(C(j,i)) > max_C then
            pivot_new = j;
            max_C = C(i,j);
        end;
    end;
    new_row = C(pivot_new,:);
    C(pivot_new,:) = C(i,:);
    C(i,:) = new_row;
    for j = i+1:n;
        for k = i+1:n+m;
            C(j,k) = C(j,k) - C(i,k)*C(j,i)/C(i,i);
        end;
    end;
end;

for i = 1:m;
    inverse(n,i) = C(n,n+i)/C(n,n);
    for j = n-1:-1:1;
        value = 0;
        for k = j+1:n;
            value = value + C(j,k)*inverse(k,i);
        end;
        inverse(j,i) = (C(j,n+i) - value)/C(j,j);
    end;
end;

endfunction
```

8.5 Determinant of a Matrix

```
function [determinant] = gauss_determinant(A)
[rows_A,columns_A] = size(A);
n = rows_A;

counter = 0;
for i = 1:n-1;
    pivot_new = i;
    max_A = abs(A(i,i));
    for j = i+1:n;
        if abs(A(j,i)) > max_A then
            pivot_new = j;
            max_A = A(i,j);
            counter = counter+1;
        end;
    end;
    new_row = A(pivot_new,:);
    A(pivot_new,:) = A(i,:);
    A(i,:) = new_row;
    for j = i+1:n;
        for k = i+1:n;
            A(j,k) = A(j,k) - A(i,k)*A(j,i)/A(i,i);
        end;
    end;
end;

determinant = (-1)^counter*prod(diag(A))
endfunction
```

8.6 Bisection Method

```
function root = Bisection_Method(a,b,N,tolerance)
    i = 1;
    while(i <= N)
        approx_root = (a+b)/2
        if((b-a)/2 < tolerance | f(approx_root) == 0)
then
            root = approx_root;
            mprintf("\n Root = %f",root);
            mprintf("\n Number of Iterations = %i",i);
            break
```

```
        end

        if(f(a)* f(approx_root) > 0) then
            a = approx_root
        else
            b = approx_root
        end
        i = i + 1
    end
endfunction
```

8.7 Regula Falsi Method

```
function Regula_Falsi_Method(a,b,tolerance)
i = 1;
while (abs(f(a)) >= tolerance)
    approx_root = ((a*f(b)) - (b*f(a)))/(f(b)-f(a))
if f(a)*f(approx_root) < 0 then
    b = approx_root;
else
    a = approx_root;
end
i = i+1;
end
mprintf("\n Root = %f",approx_root);
mprintf("\n Number of Iterations = %i",i);
endfunction
```

8.8 Secant Method

```
function Secant_Method(x0,x1,tolerance)
i = 1;
while (abs(f(x1)-f(x0)) >= tolerance)
approx_root = x1 - (f(x1)*(x1 - x0))/(f(x1)-f(x0))
x0 = x1;
x1 = approx_root;
i = i+1;
end
mprintf("\n Root = %f",approx_root);
mprintf("\n Number of Iterations = %i",i);
endfunction
```

8.9 Newton–Raphson Method

```
function Newton_Raphson(x0,h,epsilon)
j = 1;
while (abs(f(x0)) > epsilon)
f_prime = (f(x0+h)-f(x0))/h;
approx_root = x0-f_prime\f(x0);
x0 = approx_root;
j = j + 1;
end
mprintf("\n Root = %f",approx_root);
mprintf("\n Number of Iterations = %i",j);
endfunction
```

References

Bajaj, N. K. (1988). *The Physics of Waves and Oscillations*. Noida: New Delhi: Tata McGraw Hill Education.

Baudin, M. and Jean-Marc Martinez, J. (2013). *Introduction to Polynomials Chaos with NISP*, Version 0.4.

Beiser, A. (2002). *Concepts of Modern Physics*. Noida: New Delhi: Tata McGraw Hill Education.

Bleany, B. I. and Bleany, B. (1978). *Electricity and Magnetism*. Oxford: Oxford University Press.

Born, M. and Wolf, E. *Principles of Optics: Electromagnetic Theory of Propagation, Interference and Diffraction of Light,* 6th ed. New York: Pregamon Press.

Coddington, E. A. (2009). *An Introduction to Ordinary Differential Equations*. Delhi: PHI Learning Pvt. Ltd.

Fausett, L. V. (2012). *Applied Numerical Analysis – Using MATLAB*. (2nd Edition). Pearson Education.

Folland, G. B. (1992). *Fourier Analysis and Its Applications*. Pacific Grove, California: Wadsworth & Brooks/Cole Advanced Books & Software.

Garg, S. C., Bansal, R. M. and Ghosh, C. K. *Thermal Physics: Kinetic Theory, Thermodynamics and Statistical Mechanics*, 2nd ed. Tata McGraw-Hill Education.

Griffiths, D. J. (2005). *Introduction to Quantum Mechanics*, 2nd ed. London: Pearson Education.

Hjorth-Jensen, M. (2011). *Computational Physics: Lecture Notes Fall 2011*.

Jain, M. C. (2014). *Vector Spaces and Matrices in Physics*, 2nd ed. Delhi: Narosa Publishing House.

Jain, M. K., Iyengar, S. R. K. and Jain, R. K. (2012). *Numerical Methods for Scientific and Engineering Computation*, 6th ed. New Delhi: New Age International Pub.

Jenkins, F. A. and White, H. E. (2011). *Fundamentals of Optics*, 4th ed. New York: McGraw-Hill International Editions.

Kittel, C. (2004). *Introduction to Solid State Physics*, 8th ed. Delhi: Wiley India Pvt. Ltd.

Lokanathan, S. and Gambhir, R. S. (1991). *Statistical and Thermal Physics: An Introduction*, Eastern Economy Edition. Delhi: PHI Learning Pvt. Ltd.

Sadiku, M. N. O. (2006). *Elements of Electromagnetics*, 4th ed. Oxford: Oxford University Press.

Sastry, S. S. (2000). *Introductory Methods of Numerical Analysis*, 3rd ed. Delhi: Prentice Hall of India Pvt. Ltd.

Satya Prakash. (2018). *Electromagnetic Theory and Electrodynamics*. Meerut: Kedar Nath Ram Nath and Co.

Sharma, M. (2016). *Scilab Codes and Programs for Physics as well as Mathematical Problems*. Retrieved from https://www.bragitoff.com/.

Urroz, G. E. (2001). *Introduction to SCILAB*. Retrieved from https://www.scilab.org.

Urroz, G. E. (2001). *ODEs with SCILAB*. Retrieved from https://www.scribd.com

Urroz, G. E. (2001). *Ordinary differential equations with SCILAB*. Retrieved from https://www.math.utah.edu.

Urroz, G. E. (2001). *Orthogonal Functions, Gaussian Quadrature, and Fourier analysis with SCILAB*. Retrieved from https://www.scilab.org.

Wazed Miah, M. A. (1982). *Fundamentals of Electromagnetics*. Noida: New Delhi: Tata McGraw-Hill.

Index